The Electrochemistry of Novel Materials

Frontiers of Electrochemistry

Series Editors

Jacek Lipkowski
Department of Chemistry and
  Biochemistry
University of Guelph
Guelph, Ontario N1G 2W1
Canada

Philip N. Ross
1 Cyclotron Road
Lawrence Berkeley Laboratory
University of California
Berkeley, CA 94720, USA

Advisory Board

D. M. Kolb, Ulm, Germany
M. Van Hove, Berkeley, CA
W. Schmickler, Ulm, Germany
R. Guidelli, Florence, Italy
A. Wieckowski, Urbana, IL

W. O'Grady, Washington, DC
M. J. Weaver, Lafayette, IN
W. R. Fawcett, Davis, CA
R. Parsons, Southampton,
  United Kingdom
S. Trasatti, Milan, Italy

Adsorption of Molecules at Metal Electrodes, Eds. Jacek Lipkowski
  and Philip N. Ross
Structure of Electrified Interfaces, Eds. Jacek Lipkowski and Philip
  N. Ross (Forthcoming)

*Solid Polymer Electrolytes: Fundamentals and Technological
  Applications,* F. M. Gray
*Electrode Kinetics for Chemists, Chemical Engineers, and
  Materials Scientists,* E. Gileadi

The Electrochemistry of Novel Materials

EDITORS

Jacek Lipkowski and Philip N. Ross

New York • Chichester • Weinheim • Brisbane • Singapore • Toronto

Jacek Lipkowski
Department of Chemistry and Biochemistry
University of Guelph
Guelph, Ontario N1G 2W1
CANADA

Philip N. Ross
1 Cyclotron Road
Lawrence Berkeley Laboratory
University of California
Berkeley, CA 94720, USA

This book is printed on acid-free paper. ∞

Library of Congress Cataloging-in-Publication Data

The Electrochemistry of novel materials / editors Jacek Lipkowski and
 Philip N. Ross.
  p. cm. — (Frontiers of electrochemistry)
 Includes bibliographical references and index.
 ISBN 0-89573-788-4 (alk. paper)
  1. Electric batteries—Electrodes—Materials. 2. Electrodes—
 Materials. 3. Electrochemistry. I. Lipkowski, Jacek. II. Ross,
 Philip N. III. Series.
 TK2901.E435   1994
 660'.29724—dc20   93-40037
                                                                  CIP

A NOTE TO THE READER
This book has been electronically reproduced from
digital information stored at John Wiley & Sons, Inc. We
are pleased that the use of this new technology will
enable us to keep works of enduring scholarly value in
print as long as there is reasonable demand for them. The
content of this book is identical to previous printings.

Copyright © 1994 by John Wiley & Sons, Inc. All rights reserved.

Originally published as ISBN 0-89573-788-4

No part of this publication may be reproduced, stored in a retrieval system,
or transmitted in any form or by any means, electronic, mechanical,
photocopying, recording, scanning or otherwise, except as permitted under
Sections 107 and 108 of the 1976 United States Copyright Act, without
either the prior written permission of the Publisher, or authorization
through payment of the appropriate per-copy fee to the Copyright
Clearance Center, 222 Rosewood Drive, Danvers, MA 01923, (978) 750-
8400, fax (978) 750-4744. Requests to the Publisher for permission should
be addressed to the Permissions Department, John Wiley & Sons, Inc.,
605 Third Avenue, New York, NY 10158-0012. (212) 850-6011, fax (212)
850-6008, E-mail PERMREQ@WILEY.COM.

For ordering and customer service, call 1-800-CALL-WILEY.

Printed in the United States of America.
10 9 8 7 6 5 4 3

Preface

The objective of this volume is to review the recent progress in the study of the electrochemical properties of novel materials. Major technological developments often follow from the discovery of a new class of materials or from the discovery of new properties when known materials are modified and/or examined in a new way. In this volume the electrochemical properties of materials that have emerged in recent years in conjunction with other (i.e., nonelectrochemical) technologies are reviewed (e.g., the electrochemical behavior of new materials for lithium batteries and new electronically conducting polymers). In addition, reviews of progress in the development of new materials for electrochemical technologies are presented (e.g., new materials for photoelectrodes, new ionic and polymeric electrolytes, new materials for lithium battery cathodes, zeolite and clay electrodes, and electrochemistry of nuclear fuels).

This volume provides a critical summary of recent progress in the very rapidly moving field of materials chemistry and is addressed to a wide audience of scientists interested in the chemistry and physics of electrified phases. Each chapter provides sufficient background material so that it can be read and appreciated by specialists and nonspecialists alike.

Guelph, Ontario, Canada
November 1993

Jacek Lipkowski

Berkeley, CA, USA
November 1993

Philip N. Ross

Contents

1. Electrode Materials and Strategies for Photoelectrochemistry 1
by N. Alonso-Vante and H. Tributsch

  1.1 Introduction and Scope 1
  1.2 Mechanism of Interfacial Reactions 2
  1.3 Parameters Affecting Electrode Behavior 6
  1.4 Electrocatalytic Processes in Semiconducting Materials 22
  1.5 Photoelectrocatalytic Process (Fuel Generation) 36
  1.6 Photoconversion Process (Current Generation) 42
  1.7 Challenges for Research and Application 49
  1.8 Summary and Outlook 54
  Acknowledgments 55
  References 55

2. Polymeric Materials for Lithium Batteries 65
by M. Armand, J. Y. Sanchez, M. Gauthier, and Y. Choquette

  2.1 Introduction 65
  2.2 Principles and Requirements for Electrochemical Energy Storage 66
  2.3 Polymers as Solid-State Ionizing Solvents 69
  2.4 Polymers as Electrode Materials for Lithium Batteries 85

2.5 Cell Geometry and Power Densities   90
  2.6 EMFs and Stability Windows   93
  2.7 Realizations and Prototypes   96
  2.8 Safety   102
  2.9 Conclusions   105
      References   106

3. Insertion Compounds for Lithium Rocking Chair Batteries   111
by B. Scrosati

  3.1 Introduction   111
  3.2 Lithium Batteries   112
  3.3 Criteria for the Selection of Insertion Electrodes for Rocking Chair Batteries   116
  3.4 Carbon Insertion Materials   117
  3.5 Layered Lithium Metal Oxides   125
  3.6 Manganese Oxides   132
  3.7 Other Types of Rocking Chair Configurations   135
  3.8 Conclusions   136
      References   137

4. Thin Polymer Films on Electrodes: A Physicochemical Approach   141
by K. Doblhofer

  4.1 Introduction   141
  4.2 The Permeability of Nonionic Polymers   148
  4.3 Ionic Polymers on Electrodes   165
  4.4 Electronically Conducting Polymer Films   189
      Acknowledgments   200
      References   201

5. Transition Metal Oxides: Versatile Materials for Electrocatalysis   207
by S. Trasatti

  5.1 Introduction   207
  5.2 Properties of Oxides for Electrodes   210
  5.3 Interfacial Properties   219
  5.4 Electrocatalytic Properties   238
  5.5 Factors of Electrocatalysis   259
  5.6 Problems of Electrode Stability   262
  5.7 Conclusions and Prospects   271
      References   275

6. Electrochemistry of UO_2 Nuclear Fuel   297
by D. W. Shoesmith, S. Sunder, and W. H. Hocking

   6.1 Introduction   297
   6.2 Fuel Composition   299
   6.3 Structural Properties   299
   6.4 Electrical Properties   301
   6.5 Electrochemical Properties   303
   6.6 Thermodynamic Properties   305
   6.7 Surface Composition Under Electrochemical Conditions   307
   6.8 Anodic Dissolution   313
   6.9 Redox Reactions on UO_2 Surfaces   321
  6.10 Electrochemical Reactivity   331
       Acknowledgments   332
       References   332

7. Electrochemistry of Clays with Zeolites   339
by M. D. Baker and C. Senaratne

   7.1 Introduction   339
   7.2 Zeolites and Clays: Structure and Properties Pertaining to Electrode Modification   340
   7.3 Fabrication of Electrodes   344
   7.4 Historical Perspective   347
   7.5 Mechanism of Electrochemistry Occurring at Clay- and Zeolite-Modified Electrodes   348
   7.6 Analytical Applications   352
   7.7 Electrocatalysis   359
   7.8 Ion Exchange in Clays and Zeolites   367
   7.9 Molecular Wires   371
  7.10 Layered Double Hydroxides (Hydrotalcite Clays)   374
  7.11 Studies of Diffusion   376
  7.12 Conclusion   376
       References   376

Index   381

Contributors

N. Alonso-Vante. Hahn-Meitner-Institut, Abteilung Solare Energetik, Glienicker Strasse 100, 1000 Berlin 39, Germany.

M. Armand. Laboratoire d'Ionique et d'Electrochimie du Solide, ENSEE/Grenoble INPG, B.P. 75 3402 Saint-Martin-d'Hères, France.

Mark D. Baker. Department of Chemistry and Biochemistry, University of Guelph, Guelph, Ontario, Canada N1G 2W1.

Y. Choquette. Institut de Recherche d'Hydro-Québec, 2000 Montée Sainte-Julie Varennes, Québec, Canada.

Karl Doblhofer. Fritz-Haber-Institut der Max-Planck-Gesellschaft, Faradayweg 4-6, 1000 Berlin 33, Germany.

M. Gauthier. Institut de Recherche d'Hydro-Québec, 2000 Montée Sainte-Julie Varennes, Québec, Canada.

W. H. Hocking. AECL Research, Whiteshell Laboratories, Pinawa, Manitoba, R0E 1L0, Canada.

J. Y. Sanchez. Laboratoire d'Ionique et d'Electrochimie du Solide, ENSEE/Grenoble INPG, B.P. 75 3402 Saint-Martin-d'Hères, France.

Bruno Scrosati. Dipartimento di Chimica, Universitá di Roma "La Sapienza," 00185 Rome, Italy.

Chandana Senaratne. Department of Chemistry and Biochemistry, University of Guelph, Guelph, Ontario, Canada N1G 2W1.

D. W. Shoesmith. AECL Research, Whiteshell Laboratories, Pinawa, Manitoba, R0E 1L0, Canada.

S. Sunder. AECL Research, Whiteshell Laboratories, Pinawa, Manitoba, R0E 1L0, Canada.

Sergio Trasatti. Department of Physical Chemistry and Electrochemistry, University of Milan, Via Venezian 21, 20133 Milan, Italy.

Helmuth Tributsch. Hahn-Meitner-Institut, Abteilung Solare Energetik, Glienicker Strasse 100, 1000 Berlin 39, Germany.

CHAPTER 1

Electrode Materials and Strategies for Photoelectrochemistry

N. Alonso-Vante and H. Tributsch

1.1 Introduction and Scope

Electrodes in electrochemical cells are devices where thermodynamic forces (electrical potentials, light intensities, temperature gradients, chemical potentials) are located to generate fluxes (electrical currents, chemical mass transport), or where fluxes are generated to produce thermodynamic forces. Such processes require heterogeneous structures that fulfill multiple functions: They provide interfaces between different phases (typically solid and liquid), they provide pathways for the conduction of electrical charges and mediate the transition between electrical conduction in solid and liquid phases. In addition, they provide interfaces that are characterized by chemical turnover, mass transport, and convection.

Electrode electrochemistry is a fully grown scientific discipline with a long tradition and deep roots in science and technology. Therefore no review can claim to provide a comprehensive treatise on electrode materials in electrochemistry, since too many scientific and empirical details would have to be evaluated. This is also true for the much younger and much more interdisciplinary field of photoelectrochemistry. Therefore our contribution can be only an outline of some relevant experiences, promising concepts and unsolved problems. However, it will attempt to pinpoint some of the most relevant elements we consider to be key factors determining scientific and technological progress in the field of photoelectrochemical energy conversion and fuel production.

Light incident on electrodes that undergo photoelectrochemical reactions generates thermodynamic forces. Thus the electrodes must have electronic and interfacial properties that facilitate harvesting photon energy for the purpose of generating chemicals and producing electrical energy. A key precondition is the existence of any energy gap that delays energy release via phonons or photons so long that charge transfer becomes probable for the buildup of photopotentials, equivalent to shifts in the Fermi levels of minority carriers. Only semiconductors and doped insulators can provide such conditions, whereas metals facilitate a much too rapid charge recombination for significant photoelectrochemical responses. The path of photogenerated charge carriers through the interface is the second critical step. It determines whether photoelectrodes are corroding, reacting with the electrolyte, or remaining reasonably stable during the photoreaction. Most classical semiconductors facilitate interfacial minority carrier reactions that are accompanied by photocorrosion (e.g., Si, CdS, GaAs). They are, of course, of limited value of photoelectrodes for energy conversion and catalysis. This review concentrates mostly on photosensitive electrode materials that contain transition metals and provide energy bands derived from transition metal d-states. Even though many properties of these transition metal compounds for photoelectrochemistry and (photo)catalysis are still not evaluated, they provide an attractive strategy toward product-oriented interfacial mechanisms. They also permit the generation, by photon processes, of interfacial coordination chemical mechanisms that can provide the molecular basis for molecular interaction desirable for mechanisms of photoelectrochemistry and photocatalysis.

This review will be an attempt to guide the reader toward those materials and interfacial electrode concepts that provide some special promise for the future and deserve particular scientific attention.

1.2 Mechanism of Interfacial Reactions

Understanding the basic mechanism of electron transfer at the solid–electrolyte interface is of paramount importance as a precondition for gaining some insights into more complex processes, such as photosynthesis. A fundamental difference between a semiconductor and a metal consists in the presence of an energy gap for the semiconductor and the presence of a continuous distribution of electron energies for the metal. Independent of whether the solid is a semiconductor or a metal, the rate of electron transfer can be determined for both solids from the current flows. The extensive property of the electrical current makes no clear distinction between a noncatalytic and a catalytic process. As defined in classical electrochemistry [1], the most important role of an electrocatalyst is to bring the standard electrode potentials of individual electron transfer steps as close as possible to the redox potential of the overall reaction. Reactions of fundamental and technical interest are water oxidation, oxygen reduction, carbon dioxide reduction, and nitrogen fixation. In this respect, photoelectrochemistry, based on concepts derived from electrochemistry, may be considered a special area of energy conversion science, since its main goal is to minimize the dissipation of excitation energy into heat while recovering a maximum of electrical or chemical energy. The resulting electron exchange between electron donors, or acceptors based on weak or strong interactions, depends on the nature of the (photo)electrode material. This latter aspect, which involves chemisorption and catalysis, where kinetic and energetic factors play a major role, is so complex that no specific concept or model has been developed until now, although some efforts in this direction have been reported [2,3]. Nevertheless, this field has a profound influence and represents a challenge for the development of new materials that tailor efficient (photo)electrocatalysis (see later).

Before discussing electron transfer at semiconductor/electrolyte interfaces, we can assume that weak and strong interactions may be due mainly to (1) physical factors (i.e., where the electronic system of the semiconductor and electrolyte are very little disturbed) and (2) chemical factors, or creation of new species that correspond to formation of new energy states, respectively. Among the physical factors we may mention (1) relative energetic position of semiconductor electron energies with respect to the redox system in solution at the surface, (2) thermal activation energy, and (3) barrier for quantum mechanical tunneling. Among the chemical factors we may specify (1) chemical bond formation, (2) nature of activated complex vis-à-vis the nature of the electrode surface, (3) steric and dynamic factors, and (4) activation energy balance for adsorption and desorption. Depending on whether there is weak or strong interaction between the electrode surface and redox species, the chemical nature of electrode materials and the reaction mechanisms are very different.

1.2.1 Interfacial Photoreaction Based on Weak Interactions

A negligible interaction between the redox system and the semiconductor surface may be termed a weak interaction; that is, (photo)electrocatalysis is absent. The description of electron transfer of this type has been the object of a significant theoretical research effort in the 1960s [4–10] and 1970s [11,12]. As very recently reviewed by Koval and Howard [13], the understanding and questions pertaining to this type of electron transfer have not been completely answered. This may be due to the fact that there is not a defined frontier between the so-called weak and strong interaction. This might be the reason that most experiments are still interpreted within the frame developed by Gerischer [4,5]. For reasons of clarity we wish to outline briefly Gerischer's approach.

The following considerations for heterogeneous electron transfer are made: (1) One deals with a nondegenerate semiconductor (i.e., absence of surface state); (2) electronic equilibrium in the semiconductors is rapidly attained; (3) the applied voltage drops only in the semiconductor side V_b (no consideration of Helmholtz potential drop in the electrolyte side is made); (4) donors and acceptors of redox species are distributed randomly in the electrolyte solution; (5) rapid mass transport occurs in the solution electrolyte; and (6) the process is considered bimolecular [i.e., first order with respect to concentration of redox species and first order with respect to the electron (or hole) concentration in the solid]. Energy conservation demands that electron transfer be carried out between an occupied and an empty state. This electron exchange proceeds (fluctuation energy) within an energy of about kT. According to the Frank-Condon principle, the transition is assumed to be

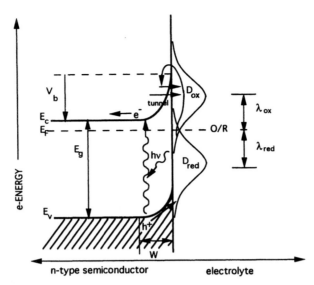

Figure 1.1. Energy bands and energy levels for a n-type semiconductor interface in contact with an (O/R) electrolyte solution with equimolar concentration. Several interfacial electron transfer pathways are described.

fast compared to the relaxation time of the surrounding solvent molecules. For the sake of clarity, let us focus our attention on a classical n-type semiconductor/electrolyte interface, Fig. 1.1, in which the main points outlined earlier are indicated. The two bell-shaped distributions of reduced and oxidized states around the thermodynamic potential of the redox systems separated by 2λ are given by

$$D_{\text{red}}(E) = \exp[-(E - E_{\text{redox}} - \lambda)^2/4kT\lambda] \qquad (1.1)$$

$$D_{\text{ox}}(E) = \exp[-(E - E_{\text{redox}} + \lambda)^2/4kT\lambda] \qquad (1.2)$$

where λ characterizes the reorganization energy. The anodic current density j_{an}, obtained under the condition that a degree of overlapping between the distribution of unoccupied states in the electrode, which follows the Fermi function $\rho(E) = [1 - f(E)]$, and the distribution of occupied states in the redox system, D_{red}, exists, is expressed by

$$j_{\text{an}} = Z_c(\pi kT\lambda)^{-1/2} \int_{-\infty}^{+\infty} \kappa(E)\rho(E)[1 - f(E)]D_{\text{red}}(E)\,dE \qquad (1.3)$$

where Z_c denotes the number of redox molecules reaching the electrode surface, $\kappa(E)$ is the transmission coefficient for the electron, and $(\pi kT\lambda)^{-1/2}$ defines a normalization factor. The cathodic current density j_{cat} is described in an analogous way, but an overlapping between the occupied states $\rho(E)f(E)$ in the electrode and the unoccupied states in the redox system D_{ox} must be considered. Recently, Gerischer himself [14] has refined the kinetics of electron transfer at the semiconductor/electrolyte interface by taking into account the statistics of forming a reaction pair (electron at the surface of the semiconductor with an electron acceptor in the electrolyte at an interacting distance δ) similar to the treatment made by Marcus [15] for a donor–acceptor electron transfer process at the interface between two immiscible liquids. However, the matching of experimental data with the theory depends on the knowledge of the reorganization energies; thus it allows the calculation of exchange current densities of one-electron reactions at the semiconductor electrodes [16–21] and metals [22,23]. Their magnitudes for various aquo and complex ion systems have been deduced from homogeneous kinetics studies of one-electron outer-sphere reactions [17]. As inferred from Eq. (1.3), the driving force that can control the rate constant from the solution side is not taken into account. In realistic situations, however, surface states, or better-named interfacial states, are always present at the semiconductor surface because of its different chemical state (formed or induced) in comparison to the bulk. They may change completely the interfacial behavior per se, and probably the magnitude of the reorganization energies of the species interacting with the semiconductor surface, which can give rise to deviation from the weak interaction case.

1.2.2 Interfacial Photoreaction Based on Strong Interactions

The first attempt to explain the origin of a strong interaction is also due to Gerischer [3]. He pointed out that this kind of interaction is common to oxidation or reduction of a semiconductor crystal surface. Furthermore, with regard to catalysis, the strong interaction is due to surface states with unpaired electrons (see Fig. 1.2). This bond formation accompanied by a shift of the energy levels (bonding and antibonding orbitals) relates to the so-called coordination chemistry. Electron transfer catalysis via surface states was explained recently as due to an increased electronic coupling with the redox species [14]. These theoretical approaches to semiconductor/

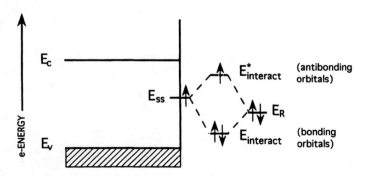

Figure 1.2. Electronic levels shift because of a strong interaction between a reactant with energy level E_R and an interface state with energy level E_{SS}. (Adapted from Ref. 3.)

electrolyte interfaces remain, however, within the frame of the classical model (Schottky). These results challenge the necessity to develop research in this direction, which is unfortunately not free of mathematical complications. Some experimental evidence supports the idea that coordination chemistry (i.e., strong interaction) can be accompanied by a profound change of the interfacial state during charge transfer. One might expect that the bell-shaped distribution of the redox species could change from the characteristic broad distribution of energy levels (unperturbed redox system) to a narrow asymmetrical δ-function of surface states (Fig. 1.3). The consequence can be the formation or modification of interfacial states which could be responsible for a redistribution of the applied electrode potential at the interface. A crucial factor will be its ability to mediate or store charges. Experimental evidence in this respect has been obtained from photoelectrochemical studies on d-band transition metal semiconductors. (See Section 1.6.)

1.3 Parameters Affecting Electrode Behavior

In electrode electrochemistry, and especially in semiconductor electrochemistry, besides the problem of controlling bulk properties of new electrode materials, one

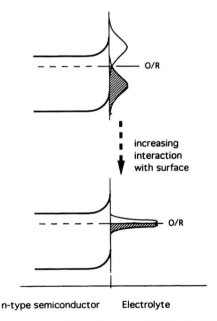

Figure 1.3. The evolution of occupied and unoccupied electronic levels of a redox couple during an enhanced interaction with the electrode surface (cf. Fig. 1.2). The narrowing is defined by the degree of interaction with the interface.

ELECTRODE MATERIALS AND STRATEGIES FOR PHOTOELECTROCHEMISTRY

must deal with heterogeneous electrode reactions, which can limit the effective reaction rate. For a desired reaction the application of a moderate overpotential can enhance the reaction rate, but considerably higher overpotentials would lead to side reactions, making the process nonselective and inefficient. The mass-transport-limited current plateau is limited at the low-overpotential side by the charge-transfer-controlled region, and on the high-overpotential side by the transition to a new reaction, usually the hydrogen evolution reaction (HER) on the cathodic side or the oxygen evolution reaction (OER) on the anodic side. Our intention in this section is to recall some interesting aspects related to the parameters that affect heterogeneous electron transfer at electrode systems.

1.3.1 Surface Morphology

The surface morphology of an electrode controls numerous physical and chemical functions in (photo)electrochemical processes. Such functions, which take place in pores or on adsorption sites of electrode surfaces, can be (1) adsorption or coordination, (2) absorption of light, (3) energy transfer, (4) electron-hole pair generation, (5) reaction of adsorbed or coordinated species, and (6) desorption of products. As already known from practical electrochemistry, the creation of a maximum possible area of interface between electrode and electrolyte gave rise to the development of porous electrodes. This aspect is responsible for the geometric-to-real-surface ratio (high real surface area) in comparison to the planar electrode. The advantage of porous electrodes lies in the high rates of reactions; consequently, the reaction is usually arranged to be under limiting conditions. Further, another consequence of the internal surface area is the ability to perform three-phase electrochemistry. (The reader interested in this latter aspect is referred to Ref. 24.)

Catalytic reactions are, on the other hand, structure-dependent reactions (where *structure* refers to crystal planes [25] porosity [24], particle size [26]). It is becoming clear that such phenomena also play a role in photoelectrochemistry (see Sections 1.5 and 1.7). It has recently been shown that realistically high solar energy conversion efficiencies can be obtained via the sensitization process of dyes on large-gap oxides, provided colloidal or highly structured oxide substrates are used. There is no doubt that little-understood surface morphology of these oxide layers as well as the nature of electrochemical mechanism proceeding within the pores of these highly dispersed materials will play a key role in controlling stability and efficiency of these systems.

A modern approach to analyze the electrode surface morphology is that of applying the concept of fractal geometry. The concept of fractal geometry [27] has been introduced in electrochemistry for impedance analysis on porous electrodes [28–30], and extended to blocking electrodes [31,32], diffusion processes of electroactive substances [33–35], image analysis on particular systems [36,37], and heterogeneous chemistry [38–40], to mention only a few examples. This concept is based on the self-similarity geometry (or scaling) of an object—that is, to the fact that after various magnifications (or resolutions) of the object, its geometrical properties may look similar or indistinguishable. This results in a simple power-law relation between the magnification power M and a measurable geometric feature such as length L:

$$L \propto M^D \tag{1.4}$$

The fractal dimension D carries information on the degree of geometric irregularity of the object. Taking as an example a line, whose topological dimension is 1, two extreme cases can be visualized: (1) For a straight, smooth line, its magnification by, say, a factor of 2 will double simply its resolution when $D = 1$; (2) for an irregular line filling the plane, D tends to be 2. The measure of the irregularity of a line is clearly seen when the value of D ranges between 1 and 2. This effect is illustrated in Fig. 1.4. This simple reasoning represents an intuitive link between the fractional dimension allocated to the curve and its apparent structure in a two-dimensional space. In perfectly regular fractal objects, such as Koch or Peano curves [27], their dimension D can be easily calculated. A real material characterized by a bulk, a surface, or an interface does not necessarily follow the same fractal rules. However, its characteristics can have a range of scaling ratio that is self-similar in a statistical sense, that is, the magnification of one part looks like another part somewhere else. Aside from self-similar fractals, one may also consider self-affine fractals, in which the scaling ratio factor is nonuniform. This means that a change of scale is accompanied by a distortion, as found in the case of the Brownian motion. The mathematical classification of fractals based on the nature of their dilational symmetry operations and their statistical properties was recently reviewed by Voss [41].

The translation of equation (1.1) to various problems related to heterogeneous chemistry demonstrates the applicability of the fractal geometry. Defining M as a set

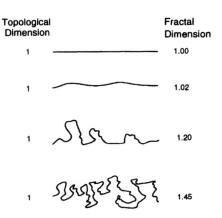

Figure 1.4. The classical topological dimension versus the fractal dimension of a line. (Reproduced with permission from Ref. 37.)

of yardsticks with varying size (e.g., radii of particles, diffusional distances, frequencies, etc.), one can observe how L, the measurable property (e.g., molecule–surface interaction, catalytic turnover, faradaic current, admittance, etc.,) changes. In this connection the approach given by Pajkossy and Nyikos for blocking electrodes [31,32] and diffusion process [33–35] in electrochemical reactions and by Avnir et al [39–41] for catalysis in heterogeneous chemistry deserves some attention and is briefly mentioned here. In the absence of a faradaic process the simple picture of a homogeneous double-layer capacity describing the interfacial impedance failed using solid electrodes because of the so-called constant phase element (CPE) behavior. This phenomenon was absent in a perfectly smooth electrode like mercury. This peculiar CPE property of solid electrodes causes a frequency-independent phase shift between the ac applied potential and its current response. With the pionering work of Le Méauté et al. [28–30] it was recognized that this CPE behavior was due to a surface roughness that presents self-similar scaling characteristics in a sufficiently wide range of length. According to Pajkossy and Nyikos [31], who considered the problem in a more specific way, the admittance of such interfaces can be well described by $Y = 1/Z = \sigma(i\omega)^\beta$, where $\beta = 1/(D - 1)$. For a perfectly smooth surface, $D = 2$ at all scales, and $\beta = 1$ (i.e., pure capacitive behavior is recovered). Their results were verified by model experiments done by themselves [32] and theoretically confirmed by Sluyters et al. [43], who obtained the same result in a different way. Using this approach of Pajkossy and Nyikos [31], Grätzel et al. [44] arrived at the conclusion that their porous TiO_2 electrode employed in the sensitization process (see Section 1.6.3) could be characterized in terms of a fractal-type dimension. Diffusion-controlled charge transfer reactions to fractal interfaces were also analyzed [33–35] and represent another example in which temporal phenomena are coupled to the geometrical properties of the interface. The resulting generalized Cottrel expression (or Warburg impedance) describing this dependence is of the form $j(t) = \sigma t^{-\beta}$, with $\beta = (D - 1)/2$. Interestingly, this latter result holds for rough electrodes with completely active ($D > 2$) or partially active surfaces ($D < 2$). This latter can be found in systems with dispersed (photo)catalysts supported on SiO_2 or TiO_2 substrates [40]. In this connection the combined electrochemical/structure-sensitive investigations in an electrolyte environment by means of the STM technique can be considered as a highly valuable tool, since it can deliver a local monitoring of the substrate structure and morphology [45] information that is necessary to assess the previous discussion in a dynamic way. Exploration of fractal surface properties of solids at molecular scales have also been reported [39–41]. The catalytic activity a and the particle size R are also describable in terms of power laws, such as $a \propto R^{D_r}$. The reaction dimension, D_r, is a characteristic parameter of the catalytic reaction, since it provided a way of evaluating comparatively the degree of catalyst structure sensitivity as found on various systems, including Pt, Pd, Ir, Rh, Fe, Ni, and bimetallic catalysts dispersed on SiO_2, Al_2O_3, TiO_2, MgO, and charcoals. As roughly surveyed in this section, a fractal generality is present in many systems, and therefore it deserves a potential application and understanding specially in the field of (photo)electrochemistry of new materials.

1.3.2 Electronic Structure and Interfacial Reactivity

For transition metals, where the d band is the major electron donor and acceptor, the effective density of states term may be as much as an order of magnutide higher than in metals in which the s band predominates [46]. In this connection, the property of transition metals vis-à-vis many electrode reaction rates can be a function of the nature of the electrode surface for electrocatalysis. The same principle might also be valid for photoelectrocatalytic systems in which transition metal-containing semiconductors are employed (see Section 1.5.). The density of states (DOS) which is a kind of weighting factor for each electron level, appearing in the integrated equation similar to eq. (1.3), but in which the energy interval is of ca. $\pm kT$, around the Fermi level, is a factor in the overall reaction rate in metals [47]. For semiconductors the classical approach (tunneling process) predicts the reaction rate, which can be calculated from Eq. (1.3). However, the electronic properties of transition metal chalcogenides may determine (photo)electrocatalytic processes due to their content of transition elements able to coordinate. Therefore their behavior may not follow the classical prediction, as was briefly discussed in Section 1.2.1 (see also Sections 1.5 and 1.6). The bulk properties of transition metal chalcogenides are affected by the valence electrons, electrons forming electron bands, and localized electronic states. In a first approximation, the corrected atomic ionization potentials (IP) determine the degree of ionicity or covalency of the formed bonds. The values of ionization potentials have been used in a qualitative manner in order to compare the energy values of the atomic orbitals involved in a chemical bonding [48,49]. In this way the interaction of the transition metal valence d-, p-, and s-electrons and the chalcogenide valence p-electrons (Fig. 1.5) must be considered for the formation of valence and conduction bands in the solid state. The contribution of valence s-electrons of chalcogenides localized at too low binding energies is not important. For the chalcogenide bonding several factors are considered: (1) The increasing mixing of metal d-states with chalcogenide p levels going from O to Te, which leads to a higher covalency of the bonds. Further, the tendency of transition elements to form chalcogenides containing S^{2-} ions and S_2^{2-} (group VIII) has been explained due to a decreasing energy difference of the atomic eigenvalues that stabilizes the formation of anions pairs. (2) The second factor is attributed to the type and degree of electron interaction: intra-atomic and interatomic. The former is encountered on oxide and rare earth compounds [50], whereas the latter describes the majority of the chalcogenide compounds and determines the width of the bands and their energetic position. This is also affected by the internuclear distances and symmetries that define the electronic overlap. Unlike oxides, chalcogenide p-levels hybridize with metal d-states. Such hybridization increases the metal-metal interaction and, concomitantly, the bandwidth. The degree of filling of the metal d-based band determines the electric properties of most chalcogenides. In this way semiconductors are obtained for d-subbands that are completely filled by the number of d-electrons remaining after subtraction from the formal metal charge, and metallic conductors are obtained for uncompletely filled bands. Charge carriers mobilities are reasonably high and the magnetism of these compounds is determined by cooperative interactions.

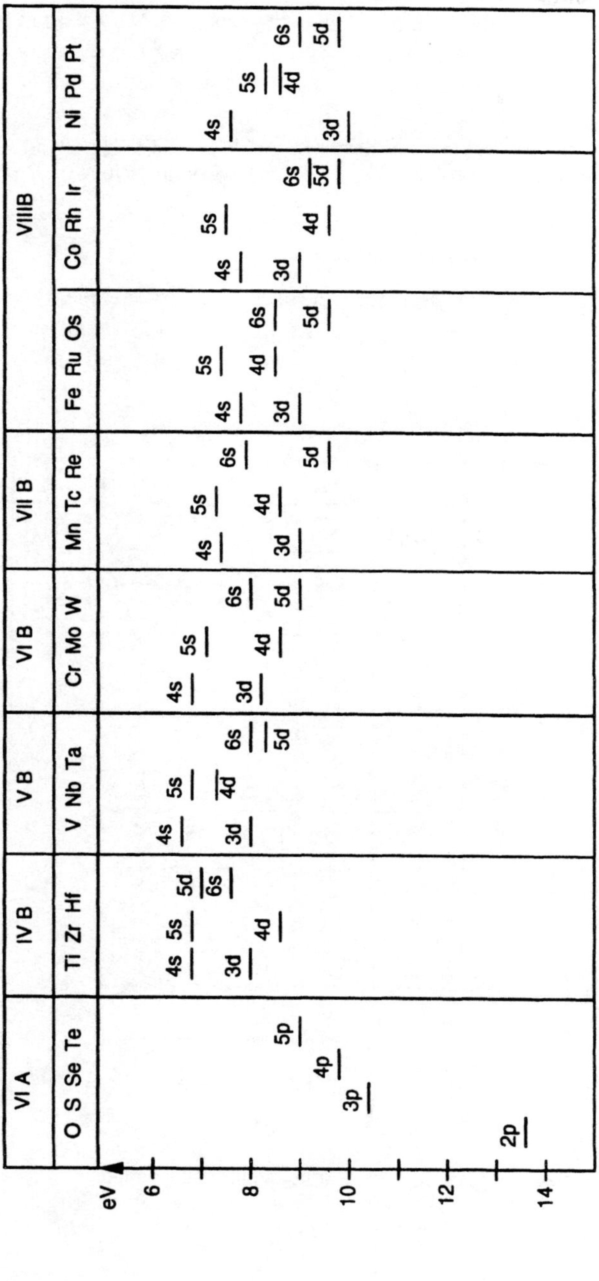

Figure 1.5. Atomic ionization potentials for the chalcogens and transition metals. (From Refs. 48 and 49.)

Transition metal dichalcogenide MX_2 (X = S, Se, Te) phases crystallize in either two-dimensional or three-dimensional structures. With respect to the metal location in the periodic table, these phases can be separated into two groups [51]. Excepting manganese and platinum, the first group, situated on the left-hand side (see Fig. 1.6), contains the IVb to VIIb elements, which constitutes layered phases with the formula $M^{4+}(X^{2-})_2$ (e.g., ZrS_2, MoS_2). The elements on the right side exhibit the three-dimensional structure, such as a pyrite-type structure with the formula $M^{2+}(X_2)^{2-}$. As previously pointed out [6], these two groups represent the frontier of destabilization of the higher oxidation states of the cation because of the redox competition between the metal and anion d- and sp band levels, respectively.

1.3.2.1 Layered Compounds

Some differences also exist within the crystal symmetry (octahedral and trigonal prismatic), depending upon the position of the transition metal in the periodic system. Figure 1.7 shows a simplified electronic structure of layered transition metal chalcogenides belonging to five different groups. An energy band charac-

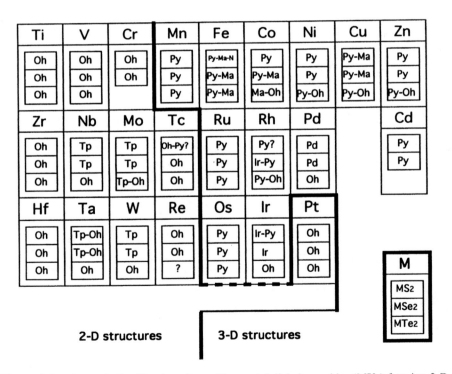

Figure 1.6. A rough classification of transition metal dichalcogenides (MX_2) forming 2-D structure layered compounds (left) and pyrite 3-D structure compounds (right). The thick line shows this separation. The coordination symmetry is also shown [e.g., octahedral (Oh); trigonal prismatic (Tp), pyrite (Py), etc.]. (Adapted from Ref. 51 with permission.)

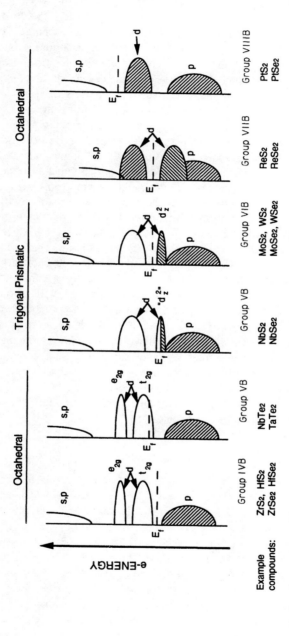

Figure 1.7. Comparison of band schemes of layered transition metal dichalcogenides belonging to different groups of the periodic table (cf. Fig. 1.6). (Adapted from Refs. 52–55.)

terized as t_{2g} derived from the transition metal d-states placed above a lower energy band made up of p-states derived from chalcogen are the main features for the octahedral coordination symmetry. In the trigonal prismatic coordination a small structure deviation leads to a splitting of the t_{2g} energy band into a lower d_z^2 energy band and a higher d-band derived from d_{xy} and $d_{x^2-y^2}^2$ states separated by a hybridization gap. This schematic picture corresponds to a quantitative band structure calculation [56,57]. Zirconium and hafnium dichalcogenides from group V (octahedral coordination) are semiconductors with valence bands derived from sulfur p-states in which the conduction band is made of d-states, with d_z^2 states forming the lower band edge. The compounds of group V presenting the two coordination symmetry (octahedral: $NbTe_2$, $TaTe_2$) (trigonal prismatic: NbS_2, $NbSe_2$) are metallic. Molybdenum and tungsten dichalcogenides from group VI (trigonal prismatic coordination) are semiconductors. Their upper region of the valence band is basically made of transiton metal d_z^2 states and the conduction band of transition metal d_{xy} and $d_{x^2-y^2}^2$ states. The rhenium dichalcogenides from group VII are also semiconductors. They have a triclinic symmetry and can be viewed as a distorted cadmium iodide structure. This distortion is probably due to the metal–metal interaction, which leads to a Re_4 cluster with a bond distance of 2.65 Å. The electronic consequence of metal–metal bonding is a restructurization of energy bands leading to semiconducting properties. The partly filled threefold degenerate t_{2g} levels are split leading to a filled subband with d-character. PtS_2 is also a semiconducting layer type compound from group VIII. Its forbidden energy gap is formed between energy bands made up by d-states. In contrast to the layer compounds of group VI, the d_z^2 band is now situated above the d_{xy} and $d_{x^2-y^2}^2$ bands due to trigonal distortion. All layer-type semiconducting compounds have energy gaps ranging between 1 and 2 eV. As seen in Fig. 1.7, group IV layer semiconducting compounds do not possess a valence band derived from transition metal d-states. This difference in the electronic structure has been the basis to model the photoelectrochemical interfacial reactivity and long-term stability of layer-type semiconducting surfaces, the bands of which are mostly derived from d-states [53,54].

1.3.2.2 Pyrite-Type Dichalcogenides

The pyrite structure differs from many other structural forms of transition metal dichalcogenides by the presence of paired S sites separated by a distance close to the bond length of an S_2 molecule. A considerable number of theoretical and experimental studies with respect to the electronic structure on pyrite sulfides series MS_2 (M ≡ Mn, Fe, Co, Ni, Cu, Zn) [58–62] and (M ≡ Ru) [63–65] has been reported. An interesting approach of the band structure calculation of the sulfide series, using a partial self-consistent scheme, was provided by Bullet [60]. Lauer et al. [61] have improved this band structure calculation using a semiempirical and self-consistent LCAO tight binding method. These authors concluded that the use of the S_2-anion pairs is extremely important for the MO-cluster calculation in the pyrite system. The presence of these S_2-anion pairs gives pyrite far from a close-packed structure. However, inserting empty spheres into the unit cell for the band structure calculation, using the so-called muffin-tin approximation, which is adequate for describing packed structure, provided good agreement with experimental data, as reported by Folkerts et al. [62]. The electronic structure of pyrite materials is, in principle, very similar: chalcogen p-band widths of roughly 5 eV, narrow crystal-field split transition metal d bands, and weak hybridization between these bands. However, differences in the electronic, magnetic, and optical properties are obtained in the sulfide series materials going from Mn to Zn (i.e., the transition-metal ions take up formal valence configurations increasing from d^6 to d^{10}, respectively). In this way, FeS_2 is a semiconductor (diamagnetic, with Van Vleck paramagnetism), CoS_2 is metallic (ferromagnetic) because of a partly filled e_g band, NiS_2 is characterized as a Mott semiconductor [66], whereas CuS_2 exhibits metallic character. Finally, ZnS_2 is a semiconductor with a completely filled d band. In this latter, the sulfur p band forms the top of the valence band and the sulfur $p\sigma*$ antibonding band forms the bottom of the conduction band.

The subject matter in this section is mainly focused on FeS_2 and RuS_2 semiconductors. RuS_2 also has an indirect intraband transition. The reason is that although both materials have a strong structural similarity, their electrochemical interfacial behavior is rather different and understanding it is a challenge, as shall be described. Iron as well as ruthenium is in a nearly octahedral environment, surrounded by six nearest-neighbors S atoms. The S_2 atoms are located in nearly tetrahedral sites and surrounded by three nearest-neighbors Fe or Ru atoms and one nearest-neighbor S atom. In an octahedral field, fivefold degenerate d atomic orbitals are split into threefold degenerate t_{2g} orbitals and twofold degenerate e_g orbitals. The t_{2g} orbitals are essentially nonbonding. From theoretical works [61,62,64], the schematic band structure of FeS_2 and RuS_2 is represented in Fig. 1.8. In the case of FeS_2 a very broad band mainly of S 3p character is superposed on a narrow band derived from the metal t_{2g} levels. The empty conduction band has mainly iron 3d e_g character with some admixture of sulfur $3p\sigma*$. The calculated energy gap of ~ 0.95 eV for pyrite and indirect intraband transition [60–62] appears consistent as reported from different experimental techniques: optical absorption [58,59], electrical resistivity measurements [67], and photoelectrochemical action spectra [68]. A similar picture is observed for RuS_2. The Ru 4d t_{2g} band, which is below the Fermi level, and an important admixture of the bonding S3p-Ru4d e_g band have a width of ~ 5.4 eV. The unoccupied antibonding S $3p\sigma*$-Ru 4d e_g^* has a width of ~ 4.5 eV. To account for the total bandwidths the theoretical DOS are also shown in Figure 1.8. The agreement between XPS experiment spectra [64,69] and calculation [64] is very good. A discrepancy exists, however, between the calculated (0.84 eV [64]) and the experimentally obtained indirect band gap of RuS_2. According to recent experimental data obtained by optical absorption [70], and by photocurrent action spectra [65,71,72] measurements, this figure turns out to be ~ 1.3 eV.

1.3.2.3 Cluster Compounds (Chevrel Phase)

The present knowledge of the electronic properties of the Chevrel phase has been gained through band structure calculations [73–78]. One semiquantitative analysis

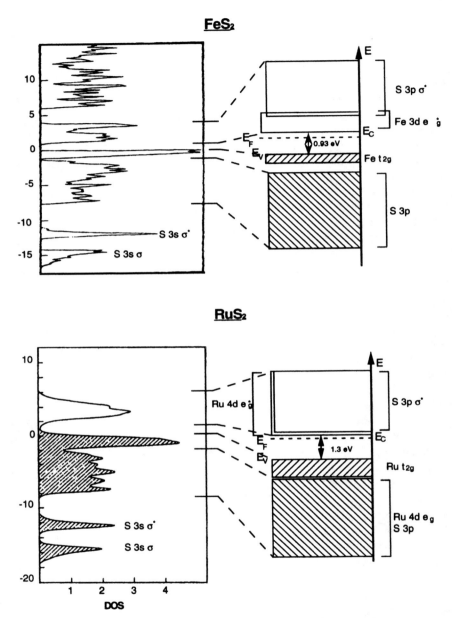

Figure 1.8. Theoretical density of states (DOS) of FeS_2 (top) (adapted from Ref. 62) and of RuS_2 (bottom) (adapted from Ref. 61). Their corresponding energy level scheme and their experimental obtained energy gaps are shown, respectively, at the right.

of the molybdenum chalcogenide Mo_6S_8 was developed by Hugbanks and Hoffmann [78]. A more rigorous analysis of the energy orbitals and gap energy determination was, however, done by Mattheiss and Fong [73]. These authors employed the ab initio augmented plane wave calculation: APW–LCAO. The former method, based on ligand field symmetry, has the advantage of being more accessible to the chemist. This approach is briefly illustrated later.

Figure 1.9 (left) shows the relative position of bonding and antibonding molecular orbitals (MO) for Mo_6S_8. These levels reflect primarily the d-character of molybdenum. In general, the d–d interaction is weaker than p–d or p–p interactions, so that the resulting molecular orbitals from these d–d interactions are less split. The indicated occupation of the orbitals corresponds to $[Mo_6S_8]^{4-}$ species. This means that 12 Mo d levels are filled, which is consistent with the assignment of 24 e^- cluster Mo orbitals based on an oxidation state of (2–) for the sulfur. The chalcogen block begins just below the lowest a_{1g} level shown. Further, the intercluster interaction of $(S^{2-})_6$ with $[Mo_6S_8]^{4-}$, shown in Fig. 1.9 (center), can supply a hint about the interaction of the cluster with the species present during an electrochemical reaction.

In the basic structure (Mo_6X_8) 20 d-electrons are available, and the Fermi level is found in the bonding states leading to metallic conduction (see Fig. 1.10). The transfer of electrons to the empty bonding states resulting from intercalated metal atoms (into channels of the structure; cf. Fig. 1.14) or substitution of Mo by Ru or Re (e.g., in $Mo_4Ru_2X_8$, and $Mo_2Re_4X_8$) shift the Fermi level toward the gap. Semiconducting properties (24 cluster electrons) are expected as experimentally verified [79,80]. The density of state (DOS) distribution for the Mo_6S_8 crystal (Figure 1.10) also indicates that the bands near the Fermi level possess more than 90% d character. Photoemisson spectra lead to similar schemes [80,81], as shown in Fig. 1.9 (right) for $Mo_4Ru_2Se_8$.

1.3.3 Electrochemical Electrode Reactivity and Stability

A transition metal chalcogenide semiconducting material (layered, pyrite, or Chevrel cluster structure) as electrode in an electrochemical cell, (photo)reacts anodically with water to produce mainly M_xO_y, XO_y^{n-} species, for instance, as indicated by reactions on WSe_2 (1.5), FeS_2 (1.6), and on $Mo_4Ru_2Se_8$ (1.7):

$$WSe_2 + 9H_2O + 14h^+ \rightarrow WO_3 + 2SeO_3^{2-} + 18\ H^+ \tag{1.5}$$

$$FeS_2 + 8H_2O + 15h^+ \rightarrow Fe^{3+} + 2SO_4^{2-} + 16\ H^+ \tag{1.6}$$

$$Mo_4Ru_2Se_8 \rightarrow Mo_{4-x}Ru_2Se_8 + xMo^{3+} + x\ e^- \tag{1.7a}$$

$$xMo^{3+} + yH_2O \leftrightarrow Mo_xO_y + 3x\ e^- + 2y\ H^+ \tag{1.7b}$$

For the oxidation process it is relevant to understand the surface chemistry, which can strongly influence the electronic properties (e.g., interfacial states formation) at the semiconductor surface. Several factors have been considered to influence photocorrosion on chalcogenide materials: (1) electronic structure [69,82–84], (2) an-

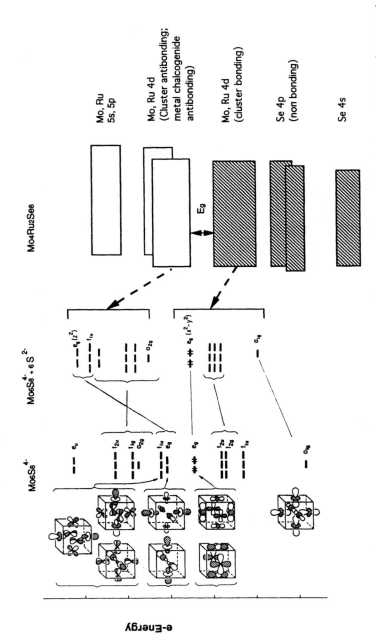

Figure 1.9. The molecular orbital representation of Chevrel phases (left) and the corresponding energy level scheme (right). (Left part reprinted with permission from Ref. 78. Copyright 1983, American Chemical Society.)

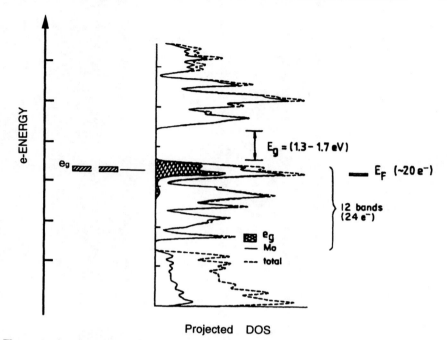

Figure 1.10. The projected density of states (DOS) of the Mo_6S_8 unit. The Fermi level (20 e^-) was ajusted to one-sixth of the surface belonging to $24e^-$. (Adapted from Ref. 78. Reprinted with permission from the American Chemical Society.)

isotropy [85] in the case of layered compounds, and (3) the inability of the transition metal to reach higher oxidation states. In the first case, from photoemission measurements, a correlation between the increasing degree of contribution of the chalcogenide p level to the top of the valence band and (photo)corrosion products was found that, in electrochemical terms, can be understood as a mixed potential development because of the easy oxidation of X when going from S, Se, and Te. The second point stressed that photocorrosion products, as shown in Eq. (1.5), increase in proportion to the area of the parallel c axis face exposed to the electrolyte or to dislocations and defects at the electrode surface leading to interfacial states formation. This effect results in a distortion of the potential between the space charge region and the Helmholtz layer [85] (see Fig. 1.11). Finally, the third point emerges when comparing FeS_2 and its structure homologue RuS_2 (cf. Fig. 1.8). The highest oxidation state reached by ruthenium, for exmaple, is (8+) in RuO_4, whereas iron is limited to (3+) in Fe_2O_3 [eq. (1.6)]. Higher oxidation states of iron are of practically no importance. We can conclude that there is not a unified picture about the photocorrosion process on chalcogenide materials. The oxidation process on these semiconducting surfaces is probably determined by a primary chemical adsorption of water molecules or OH^-. The intermediate steps following this process may be controlled by the nature of the elements present in the material surface,

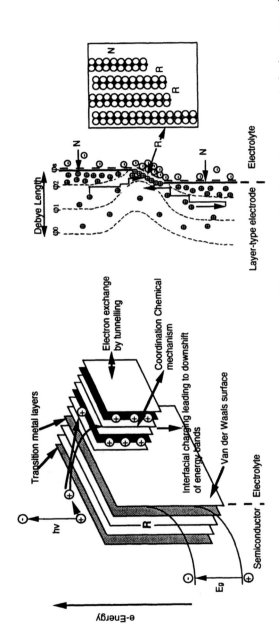

Figure 1.11. A simplified picture of a layer-type semiconductor with van der Waals and reactive (R) surfaces, respectively, in contact with electrolyte (left); the effect of the reactive site on the space charge layer is shown. (Adapted from Refs. 53 and 85.)

which are found at energies determined by the crystal structure. From a microscopic point of view, this point remains up to now a challenging problem to be solved. Nevertheless, more insight has been gained by means of ex situ surface spectroscopic analysis of such electrode surfaces after electrochemical treatment in acid medium. Table 1.1 summarizes the electrochemical oxidation products on some chalcogenides investigated in our laboratory (cf. also Table 1.3). It is worthwhile to mention that oxide formation (about one or two monolayers) on RuS_2 and OsS_2 was independent of the charge passed during water oxidation in acid medium. Their behavior is comparable to that of the layered compound, PtS_2. This was not the case for FeS_2. On the other hand, with Chevrel phase compounds, the Ru–cluster was the most stable when exposed to air. In oxygen-saturated solution, all these materials oxidize when the applied potential bias was more positive than the open-circuit potential. However, they were stable for the oxygen reduction process (see Section 1.4.2).

From a thermodynamic viewpoint, the semiconductor–electrolyte interface is characterized by a decomposition potential E_D [87,88]. This decomposition potential value should be more positive than the valence band edge in order to avoid photocorrosion, which has as yet not been identified in a real system, with the exception of some large gap oxides. The concept of kinetic stability has been a good approach on II–VI compounds [89], in which holes arriving at the surface can be transferred rapidly to the electrolyte before destabilizing the crystalline lattice. Further, against corrosion a variety of approaches have been applied, such as surface derivatization, chemisorption of metal ions, inorganic or organic thin-film

Table 1.1 XPS Analysis of Electrochemical Oxidation Products of Some Chalcogenide Compounds Electrode Surfaces.

Compound	Products	References
MoS_2	MoO_3; S_x^{2-}; SO_4^{2-}	82
$MoSe_2$	MoO_3; SeO_3^{2-}	
$MoTe_2$	MoO_3; TeO_3	
WSe_2	WO_3; SeO_3^{2-}	82
WTe_2	WO_3; TeO_3	86
$ZrSe_2$	ZrO_2; Se	86
PtS_2	$Pt(OH)_4$; SO_4^{2-}	82
RuS_2	RuO_2; SO_4^{2-}	83
$RuSe_2$	RuO_2; SeO_2	
$RuTe_2$	RuO_3; RuO_4; TeO_3	
FeS_2	Fe_2O_3; SO_4^{2-}	68
OsS_2	OsO_x; SO_4^{2-}	84
$OsSe_2$	OsO_x; SeO_2	
$OsTe_2$	TeO_3	
Mo_6Se_8	MoO_3; SeO_2	80, 81
$Mo_4Ru_2Se_8$	MoO_3; RuO_3; SeO_2	
$Mo_2Re_4Se_8$	MoO_3; Re_2O_7; SeO_2	

deposition, and so on (see Sections 1.4.3, 1.5, and 1.6). Negatively charged electron donors such as I^-, Br^- are easily oxidized and stabilize transition metal compounds efficiently. On the other hand, positively charged electron donors such as Fe^{2+} and Mn^{2+} or complexes like $Fe(phen)_3^{2+}$ are less efficient to stabilize the transition metal chalcogenides [90,91]. Such electron transfer cannot be explained by a tunnel process but in terms of interfacial coordination electrochemistry (see Section 1.6).

1.4 Electrocatalytic Processes in Semiconducting Materials

For any interfacial electrochemical mechanism reaction barriers have to be surmounted that manifest themselves in overpotentials. A catalytic process is any mechanism that aids in decreasing these activation barriers by opening up alternative electrochemical mechanisms. Catalysis is specially useful and desirable when several electronic charge carriers have to be transferred to obtain a desired product. In this case catalytic mechanism typically proceeds via intermediate complexes that can accommodate several electrons in different oxidation or reduction states.

1.4.1 Oxide Compounds

During the last three decades semiconducting oxides have been investigated mainly with respect to their catalytic properties, especially for oxidation reactions. Among such oxides were TiO_2 [92,93], ZnO [94], WO_3 [95,96], $SrTiO_3$ [97,98], Co_3O_4 [99–101], MnO_2 [102], Fe_2O_3 [92], PbO [103], NiO [104], Cu_2O [105]. A wide range of properties, from corrosive behavior to specific catalytic activity and photoreactivity with water, has been reported [106,107] (see also Table 1.2). Since photogenerated holes have to be transported to the surface of oxides via the valence band, the electronic structure and energetic position of these bands seem to be of special importance for the catalytic reactivity of oxides. In the case of TiO_2 the energetic position of the valence band is so deep below the thermodynamic oxidation potential of water that formation of an adsorbed $OH^.$ radical is possible. This is the origin of a photoelectrochemically induced radical chemistry leading to the formation of molecular oxygen and to the generation of a wide variety of oxidation products of organic substances. For this electrode the very positive potential of photogenerated holes is a basis for catalysis. The fact that they are generated from oxygen p-states is not a major disadvantage because $OH^.$ radicals are generated via one–hole transfer reactions (see Fig. 1.12). This is a simple catalytic process. In the case of Fe_2O_3, the energy of the valence band is more negatively shifted and radical formation from water encounters difficulties. Oxygen evolution only becomes possible because the oxidation involves an increase of the oxidation state of iron in the semiconductor surface. Oxygen from water binds to iron together with oxygen from ligand sites. Thus the oxygen species may be combined to form molecular oxygen, which is found to be released from anodically polarized iron electrodes. The imag-

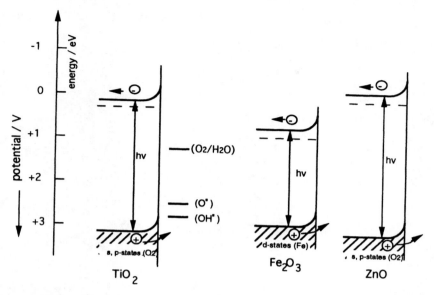

Figure 1.12. Comparison of energy band position of TiO_2, Fe_2O_3, and ZnO. The energetic position of water oxidation (via radical pathway) is also contrasted at the TiO_2 interface. For comparison, the flatband positions of these semiconductors were adjusted to pH = 0.

inable reaction sequence is illustrated in Fig. 1.13. Fe(III) in Fe_2O_3 surface reacts with OH^- and a hole to form Fe(III) hydroxide. The proton from the OH group is released when a photogenerated hole is captured. Now a head-on peroxo complex with iron (IV) is formed that captures another hole and complexes another OH^- group to form iron (IV) and free molecular oxygen. This unstable iron (IV) complex releases a proton to react back to the original iron (III) surface complex. It should be pointed out that an oxygen vacancy in the iron oxide interface is playing a crucial role in the suggested mechanism. (If sulfur were a ligand to iron, a mechanism of this type could not proceed.) Such a photooxidation mechanism of water is only possible when the transition metal d-states significantly contribute to the valence band, as is the case for Fe_2O_3. This is the main difference as compared with TiO_2, where the valence band is made of oxygen p-states. Oxides of transition metals that allow higher oxidation states are even more favorable for oxygen evolution because they facilitate the formation of peroxo groups. This may explain why oxides of ruthenium, iridium, or rhodium show a higher catalytic activity. Unfortunately, these materials have metallic properties and cannot be used as photoelectrodes. These considerations may be completed by considering the photoelectrocatalytic properties of ZnO [108]. This material has an energy band position similar to that of TiO_2 and, equally, a valence band derived from oxygen p-states. It is, however, thermodynamically unstable, in contrast to TiO_2. The consequence of this is that photogenerated holes weaken chemical bonds in the ZnO surface and lead to a photodecomposition of the material [109]. Interesting oxidation reactions can still be observed because of the radical nature of oxygen released from the ZnO lattice.

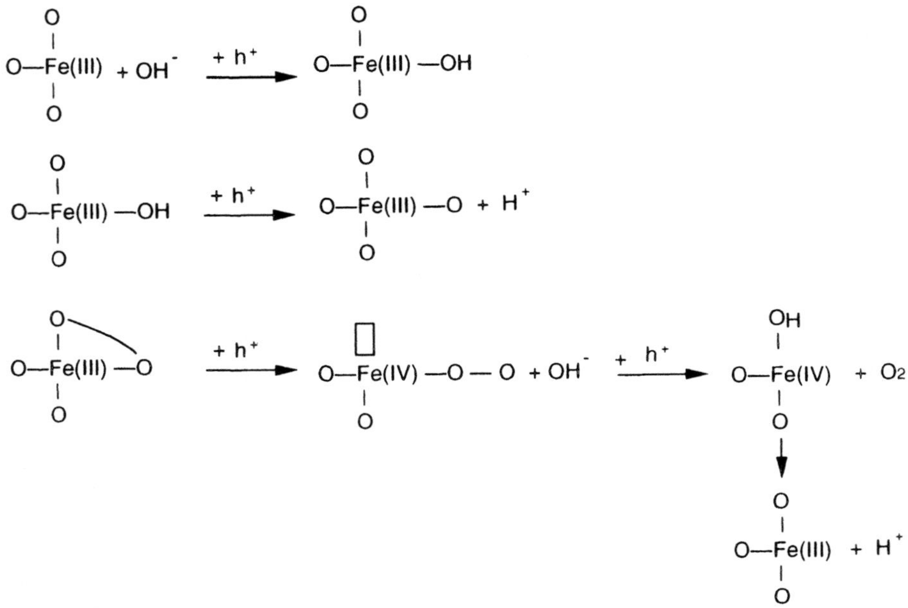

$$\begin{array}{c}\text{O}\\|\\\text{O}\text{—Fe(III)}\ +\ \text{OH}^-\\|\\\text{O}\end{array}\quad\xrightarrow{+\ h^+}\quad\begin{array}{c}\text{O}\\|\\\text{O}\text{—Fe(III)}\text{—OH}\\|\\\text{O}\end{array}$$

$$\begin{array}{c}\text{O}\\|\\\text{O}\text{—Fe(III)}\text{—OH}\\|\\\text{O}\end{array}\quad\xrightarrow{+\ h^+}\quad\begin{array}{c}\text{O}\\|\\\text{O}\text{—Fe(III)}\text{—O}\ +\ \text{H}^+\\|\\\text{O}\end{array}$$

$$\begin{array}{c}\text{O}\\|\\\text{O}\text{—Fe(III)}\text{—O}\\|\\\text{O}\end{array}\xrightarrow{+\ h^+}\begin{array}{c}\square\\\text{O}\text{—Fe(IV)}\text{—O—O}\ +\ \text{OH}^-\\|\\\text{O}\end{array}\xrightarrow{+\ h^+}\begin{array}{c}\text{OH}\\|\\\text{O}\text{—Fe(IV)}\ +\ \text{O}_2\\|\\\text{O}\end{array}$$

$$\downarrow$$

$$\begin{array}{c}\text{O}\\|\\\text{O}\text{—Fe(III)}\ +\ \text{H}^+\\|\\\text{O}\end{array}$$

Figure 1.13. Possible reaction pathway for oxygen evolution on iron oxide sites.

When negative potentials are applied to oxide electrodes, reduction processes occur, with the exception of thermodynamically stable oxides like TiO_2, at which a hydrogen insertion reaction may occur [110]. During electrochemical reaction, metal aggregates or films may form on the electrode surface, which explains why many transition metal oxides are catalytic for hydrogen evolution.

Recently, mixtures of transition metal oxides (RuO_2, TiO_2, MoO_2, Co_2O_4, Rh_2O_3) have been found to be good catalysts for carbon dioxide reduction to methanol in acid medium, however, only in the presence of low current densities [111]. The principal component (45% to 65%) in all these experiments was the n-type semiconductor TiO_2. It is known to be a poor catalyst for hydrogen evolution, which is a precondition for efficient carbon dioxide reduction. RuO_2 is apparently the catalytic agent [112].

Since only a few p-type oxides have been developed up to now, photoreduction processes have been little explored. They are typically investigated with classical semiconductors modified by deposition of noble metals or catalytically active chemical species (see Section 1.4.3). A much more specific survey on electrocatalytic oxide electrodes is given by Trasatti in this publication.

1.4.2 Chevrel-Type Cluster Structure Compounds

For efficient electrocatalysis, it appears to be necessary for electrical charge carriers to react interfacially via transition metal d-states. Such a mechanism automatically

ELECTRODE MATERIALS AND STRATEGIES FOR PHOTOELECTROCHEMISTRY

leads to a coordination chemical bonding. This is a precondition for efficient chemisorption and electrocatalysis. In most oxides the transition metals are separated by oxygen. This implies that chemical adsorption and electrocatalytic reaction sites are too distant from each other for cooperation. On the other hand, it would seem to be favorable to provide adjacent sites for adsorption and desorption of reactant species. The presence of two complementary functioning metal atoms in close neighborship (bimetallic catalysts) would seem to be a very favorable condition for complicated catalytic mechanisms. Only materials that contain transition metal clusters within their structure appear to be able to provide such conditions. But to supply charge carriers in neighboring metals they also have to provide electronic properties that guarantee electron conduction via clusters.

The Chevrel phase—Mo_6X_8, where X can be sulfur, selenium, or tellurium—represents an interesting family of cluster materials that has been tested for oxygen reduction in acid medium [80,113–116]. Its crystal structure is shown in Fig. 1.14. As observed, this material is characterized by an octahedron of molybdenum surrounded by a cube of chalcogen atoms. Up to about 40 compounds with the stoichiometry $M'Mo_6X_8$ reported in the literature [117,118] are formed by insertion of M' atoms (from alkaline, transition, and rare earth series) occupying the three-

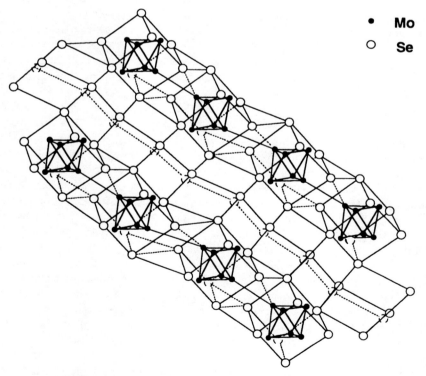

Figure 1.14. Crystal structure of Chevrel phase (Mo_6Se_8) unit.

dimensional channels of the basic structure. These compounds, most of them metallic, have been named ternary cluster materials. The metal–metal (Mo–Mo) distance depends on the number of metal d-electrons within the cluster unit. For example, when the electron number is increased from 20 (Mo_6X_8) to 24 ($Cu_4Mo_6X_8$), the bonding distance is reduced by about 15% [119]. The compounds of formula ($Mo_{6-x}M_x)X_8$ are named pseudo-ternary clusters, because one or more Mo atoms in the binary Mo_6X_8 are replaced by other metals, such as Ru and Re (e.g., $Mo_4Ru_2X_8$ and $Mo_2Re_4X_8$). In this case the channels remain unoccupied. These latter compounds and some ternary compounds, such as $Cu_4Mo_6X_8$, reach 24 e^- counting per cluster unit. A semiconductor with an energy gap ranging between 0.11 and 0.13 Ry (1.49–1.76 eV) [73] is obtained by filling up the valence band with 24 e^-. The energy gap of $Mo_4Ru_2Se_8$ was determined to be 1.3 eV [80]. Because of the possibility of delocalization of electrons in the clusters, the relaxation of electronic states due to electron transfer would be attenuated, and therefore these materials are interesting models for multielectron charge transfer reactions [e.g., oxygen reduction to water (4 e^- charge transfer)]. Their poor p-type semiconducting properties (highly degenerate, Fermi level close to the edge of the valence band) allowed only the electrochemical catalysis of oxygen reduction in acid medium in darkness. For this reaction platinum is by far the best catalyst [120] not being, up to now, easily replaceable in acid fuel cells. Investigation of oxygen reduction performed with $Mo_4Ru_2Se_8$ electrodes has shown that this material is catalytically comparable with platinum. It reduces oxygen directly to water in a four-electron transfer reaction with less than 4% of hydrogen peroxide formation [80]. Rotating disc electrode measurements for oxygen reduction in acid medium on sintered cluster samples is depicted in Fig. 1.15. As observed, a significant improvement is obtained in the direction from the Mo metal to Mo_6Se_8, $Mo_2Re_4Se_8$, and $Mo_4Ru_2Se_8$. The remarkable oxygen reduction properties of $Mo_4Ru_2Se_8$, which have previously been discussed [114] cannot be approached when replacing Ru with Re. This mixed semiconducting cluster compound, however, is still significantly more catalytic than the pure Mo-containing metallic cluster compound. Because of the difficulty of compacting materials, the electrocatalytic investigation of other ternary Chevrel cluster compounds with metallic character (20–23 e^- per cluster unit) was performed using carbon paste as substrate [115]. The main conclusion was that electronic charge carriers were channeled into bimetallic interfacial clusters or cluster–metal associations. The key problem with these materials in acid medium remains the corrosion stability of the catalyst interface, as, for example, the Cu–(Mo-cluster) complex, which loses Cu through anodic extraction. The advantage of pseudo-ternary clusters might come from the presence of two neighboring different transition metals, and in part from the Fermi-level displacement toward the edge of the valence band, even though a corrosion (mixed potential) in the case of $Mo_4Ru_2Se_8$ is not totally absent. The electrochemistry in nonaqueous electrolytes demonstrated that similar kinetic behavior is obtained with this cluster in comparison to platinum in acetonitrile [116]. A remarkable result was that the selectivity of the cluster material surface with respect to the oxygen reduction in methanol-containing electrolyte or pure methanol solution was not altered [115,116]. This specificity could be an interesting

ELECTRODE MATERIALS AND STRATEGIES FOR PHOTOELECTROCHEMISTRY

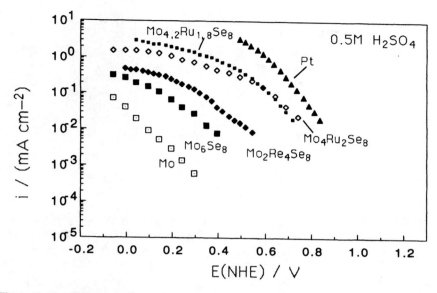

Figure 1.15. Tafel plot for molecular oxygen reduction at sintered materials in 0.5 M H_2SO_4. Pt and Mo metals are also presented. (From Ref. 115.)

advantage for a possible technical realization of a methanol fuel cell, since a physical separation of methanol and oxygen through a separator (membrane) would not be necessary. The catalytic differences observed between $Mo_4Ru_2Se_8$ and $Mo_2Re_4Se_8$ have clearly shown that the chemistry of transition metal clusters is an important factor. For catalytic purposes, it is interesting to produce this type of compound in thin layers. One strategy was to produce electrocatalytic layers fom the thermal decomposition of some neutral carbonyl metals [121]. These layers were obtained by reacting molybdenum hexacarbonyl, triruthenium dodecacarbonyl mixed with selenium powder in xylene under argon atmosphere at about 140°C for 20 hours. The layers with a thickness of less than 1 μm were deposited on different substrates (ITO, porous carbon, glassy carbon, etc.). This relatively easy procedure leads to the creation of powder or deposition of a layer with an amorphous structure according to the x-ray measurements. However, the analysis with a high-resolution transmission electron microscope revealed nanocrystalline regions of ~40–100 Å. Statistical EDAX of these crystalline aggregates as well as RBS analysis on these layers, deposited on glassy carbon, indicated a composition of $(Ru_{1-x}Mo_x)_ySeO_z$ where $0.02 < x < 0.04$; $1 < y < 3$ and $z \approx 2y$ [122], which is different from $Mo_4Ru_2Se_8$. The power spectrum analysis of the aggregates revealed, however, a distance plane of 2.9 Å typical of a metal–metal interaction. Thus it was concluded that in $(Ru_{1-x}Mo_x)_ySeO_z$, with an unknown structure, a bimetallic transition metal compound exists. This latter is the precondition, as discussed earlier, for a favorable oxygen reduction catalysis. These layers also showed semiconducting behavior according to the temperature-dependent conductivity measurements. To gain an understanding of the kinetics of oxygen reduction on these layers, rotating-ring disk electrode (RRDE) measurements were performed using a disk of glassy carbon as substrate to evaluate the rate constants k_1, k_2, and k_3 [122], following the treatment of Srinivasan et al. [123], on the basis of the following model:

$$O_{2,b} \rightarrow \overset{\overset{\displaystyle k_1}{\overbrace{}}}{O_{2,ad} \overset{k_2}{\rightarrow} H_2O_{2,ad} \overset{k_3}{\rightarrow} H_2O} \\ \downarrow \\ H_2O_{2,b}$$

As observed in Figure 1.16, at 0V/SCE the ratio k_1 to k_2 is about 2.5, and it is potential dependent. Since k_1 is larger than k_2, at higher overpotentials, O_2 is mainly reduced to H_2O via the direct four-electron transfer reaction path, as also observed in Figure 1.17. This research direction seems to be interesting because of the possibility of identifying new transition metal combinations in order to tailor new (photo)electrocatalysts.

The development of photoactive cluster material in single crystals is still in its infancy. However, some work has been undertaken in this direction to identify new cluster materials for (photo)electrocatalytic purposes. This led to the consideration of rhenium compounds, since from the photochemical literature $Re(CO)_3(bipy)Cl$

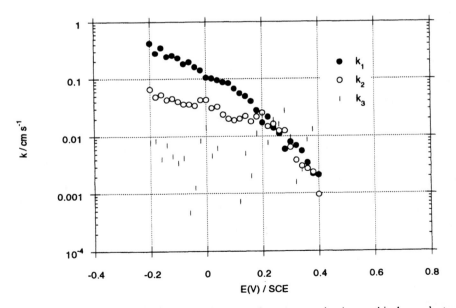

Figure 1.16. Rate constants of intermediate steps for oxygen reduction on thin-layer cluster materials $(MoRu)_2Se$, in 0.5 M H_2SO_4. (Calculated according to the model derived from Ref. 123.)

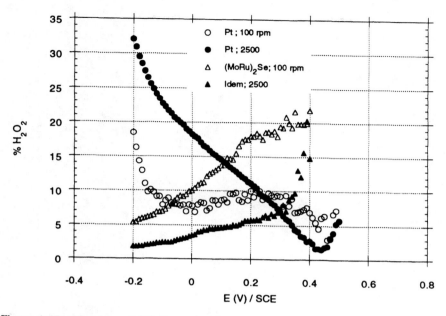

Figure 1.17. Percentage of H_2O_2 produced on cluster thin layers, Mo-Ru-Se compared to platinum electrode in the same condition as measured by RRDE.

has been found to be a selective catalyst for the electroreduction of CO_2 [124], in a 2 e^- process (as most other CO_2 reduction catalysts) in a low-water-containing dimethylformamide (DMF) electrolyte. The first material synthesis in this series was reported in early 1980 by Bronger et al. [125] [e.g., $A_2Re_3S_6$ (A = Na, K), $M_2Re_6S_{11}$ (M = Ba, Sr)]. Five years later, Perrin et al. reported the synthesis of $Re_6Se_8Cl_2$, $Re_6Se_8Br_2$ [126], which were identified as semiconductors, consisting of Re_6Se_8 building blocks with a Chevrel-phase-like structure (cf. Fig. 1.14). The first photoelectrochemical investigation of $Re_6Se_8Cl_2$ was reported by Le Nagard et al. [127]. This cluster material showed n-type conduction. Recently, research in this direction in our group led to the synthesis of $Re_6Se_8Cl_2$ and $Re_6Se_8Br_2$, as well as the identification of novel compounds—$Re_6S_8Br_2$ [128], and $Re_6S_8Cl_2$ [129]. Surprisingly, $Re_6S_8Cl_2$ is not a lamellar compound in comparison to $Re_6Se_8Cl_2$ [126]. On the other hand, $Re_6S_8Br_2$, $Re_6S_8Cl_2$, and $Re_6Se_8Br_2$ are isomorphous to each other (see Fig. 1.18). The preliminary photoelectrochemical activity of the best novel $Re_6S_8Br_2$ samples vis-à-vis the hydrogen evolution process is depicted in Fig. 1.19a, and compared in the same condition with the photoresponse of $Re_6Se_8Br_2$ (Fig. 1.19b). The positive sign of the photopotential, ~0.2 V in 0.5 M H_2SO_4, was typically observed. This effect, together with the shape of the photocurrent–voltage characteristic, is in agreement with the photoresponse expected for a p-type semiconductor. These materials are unstable in the region of the photoanodic response; hence at an electrode potential more positive than 0.6 V/NHE the corrosion process was enhanced under illumination. This induced a variation of the shape of the

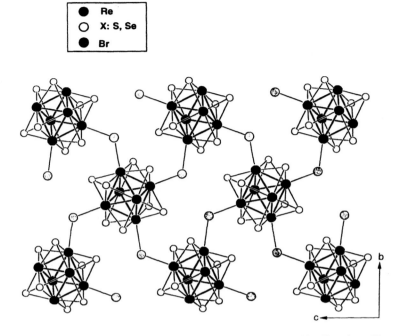

Figure 1.18. Crystal structure of $Re_6X_8Hal_2$ projected in the [100] direction. (From Ref. 128.)

photocurrent–voltage characteristic. Evidence of Re_2O_7 traces was revealed by ESCA analysis on a comparable rhenium cluster chalcogenide ($Mo_2Re_4Se_8$) [81]. The photocurrent action spectrum of the cluster compound $Re_6S_8Br_2$ is depicted in Fig. 1.20. The determined band gap energy for this material based on a probable indirect optical transition ($n = \frac{1}{2}$) was 1.7 eV. This energy gap confirms the fact that in this type of semiconducting cluster material, there is an agreement predicted from the band structure calculations for molybdenum Chevrel phases with 24 e^- per cluster unit $[Mo_6S_8]^{4-}$. However, much effort should be devoted to understanding the type of optical transition (bulk properties) of these cluster materials as well as their interfacial behavior. Nevertheless, it should also be mentioned that $Re_6Se_8Cl_2$, with a p-type behavior, in photoelectrochemical experiments is able to photoelectrocatalyze the reduction of oxygen in acid medium under visible light [130]. How this process proceeds at this interface remains to be investigated. The ensemble of the experimental evidence opens new directions in the understanding and development of new (photo)electrocatalytic material based on the concept of multielectron charge transfer via cluster centers. Furthermore, we believe that more sophisticated spectroscopic in situ techniques, such as EXAFS, may shed some light on the dynamics of the electrocatalytic interfaces. In fact, research in this direction has, at least qualitatively, confirmed that during electrocatalysis of the

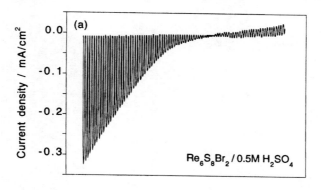

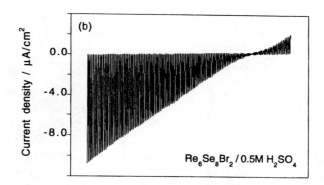

Figure 1.19. Current-potential curves under white intermittent illumination (~800 mW cm^{-2}) of (a) Re$_6$S$_8$Br$_2$ and (b) Re$_6$Se$_8$Br$_2$ clusters in 0.5 M H$_2$SO$_4$. (From Ref. 128.)

oxygen reduction with (MoRu)$_2$Se layers a dynamic reversible structure change during multielectron transfer [131] is involved.

1.4.3 Catalytically Modified Electrodes

The relatively easy availability of classical n- and p-type high-quality semiconductors in single-crystal forms—such as Si (E_g = 1.1 eV), InP (E_g = 1.35 eV), and GaAs (E_g = 1.42eV)—is related to electronic industry developments. These semiconductors have been analyzed with respect to their electrochemical stability [132–134] as well as to their photoelectrocatalytic properties [135–136]. As confirmed from the literature, these compounds are unstable for photoanodic processes and

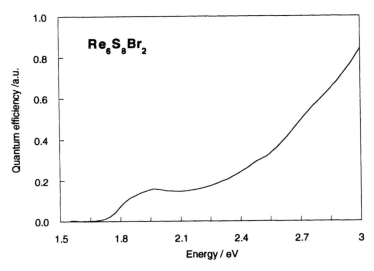

Figure 1.20. Photocurrent action spectra of $Re_6S_8Br_2$ recorded in 0.5 M H_2SO_4. (From Ref. 128.)

poor catalysts for photocathodic processes. Because of the better matching to the solar spectrum by these semiconductors, several strategies were developed to circumvent these problems (e.g., to enhance the surface stability and increase the catalytic activity via immobilization of redox systems [138–140], chemical derivatization, [141–144] *adsorbed redox* systems [145,146], and chemical–electrochemical deposits of metals [132–137,147,148] (see Fig. 1.21). For catalytic purposes, chemical derivatization and electrodeposits of catalytic material, such as Pt on

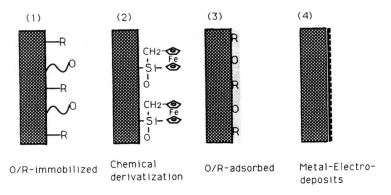

Figure 1.21. Some strategies for preparing modified photoelectrode—for example, (1) inmobilization of redox systems, (2) chemical derivatization, (3) adsorbed redox systems, (4) electrodeposits.

semiconductor surfaces, were successfully applied. The other strategies were mainly employed to stabilize the semiconductor surface against photocorrosion.

The favorable combination of surface-attached particles on semiconductors, in the specific case of silicon [148] or indium phosphide [136], has improved the electron transfer properties needed for photoelectrocatalysis of hydrogen evolution. However, the picture of the model of the interface presented in these works is still related to the Schottky model; that is, the rate-determining step is the potential dependent transport of carriers to the semiconductor surface, and the interfacial region between semiconductor and solution (Helmholtz potential drop) plays a minimal role in this process. This statement implies that the Fermi level is unpinned and that at the flat band position there would be no current. This is reasonable, as long as the interfacial state concentration is very small. In this connection, the pioneering work of Nakato and Tsubomura [132], who investigated the metal islands effects for hydrogen production and photovoltaic conversion, was interpreted as a structure change of the space charge region induced by the metal islands (Schottky barrier) (see Fig. 1.22). By changing the Fermi level within the metal, the reaction rate would be changed, and if the potential drop in the semiconductor ($\Delta\phi_{sc}$) was increased, the reaction would be enhanced as experimentally supported. This "nonelectrocatalytic" viewpoint was also followed by Heller et al. [136]. They also added another concept about the variation of the work function of metals [149] as a function of hydrogenation in order to explain their experimental results. However, Kautek et al. [150] concluded that the effect of electrodeposited metal is purely electrocatalytic. This statement implies that the rate-determining step of the reaction is dependent not on the semiconductor bulk properties but on the surface, which is the first hinting at the importance of the Helmholtz potential drop at modified semiconductor electrolyte interfaces. More work in this direction was undertaken by Bockris et al. [137]. Figures 1.23a and 1.23b contrast their main results obtained by measuring p–Si electrodes modified with various metals. In comparison to the naked Si electrode, Pt, Au, and Ni enhance the photoevolution rate of hydrogen, whereas Cd and Pb decrease it. Furthermore, the shape of the photocharacteristics was affected according to the nature of the electrodeposited metal islands. These authors [137] have found that a correlation between the log of exchange current densities of the corresponding massive metals and the potential shift (ΔV) at a photocurrent density of 1 mA cm^{-2} induced by the same submonolayer of electrodeposited metals on Si for the hydrogen evolution exists. This fact is again strong evidence for a surface catalytic phenomenon (i.e., in favor of a Helmholtz-like model). However, a direct correlation of the interfacial states concentration on the capacity of the Helmholtz layer has not been made with such systems in association, for example, with different semiconductor dopant concentrations. One method that allows such an approach is the electrochemical impedance spectroscopy (EIS). In this direction, some works have recently been reported on p-InP decorated with platinum islands by Kühne and Schefold [151]. These authors confirm earlier statements [137,150] that the interfacial kinetics is determined by the potential drop in the Helmholtz layer acting as the driving force in photocatalytic systems. The Schottky barrier maintained in their system is apparently a limiting case of a kinetics

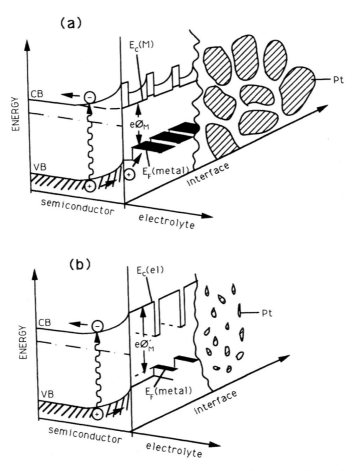

Figure 1.22. Effect of modification of the space charge region induced by platinum islands. (a) Large platinum islands produce a smaller barrier height independent of the electrolyte, in comparison to (b) tiny particles that produce a higher barrier dependent on the redox species. (Adapted from Ref. 132.)

interfacial behavior probably determined by the quality of the semiconducting material. A nondefined frontier between the "nonelectrocatalytic" view with the "electrocatalytic" one still exists. More insight can be gained when varying not only potential or illumination levels but also the temperature of the photoelectrocatalytic system [152] (see Section 1.5.1). Although this part alludes to what will be discussed in the next section, the topic of modified semiconductor surface constitutes a separate chapter. Surfaces of large band gap semiconductors have also been modified [e.g., TiO_2 (E_g = 3.0 eV)]. As emphasized in Section 1.4.1, this semiconductor realizes photoelectrocatalytic reactions based on the generation of OH^- radicals produced upon UV illumination. That is why many well-known reactions from

ELECTRODE MATERIALS AND STRATEGIES FOR PHOTOELECTROCHEMISTRY

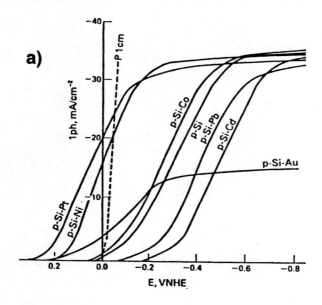

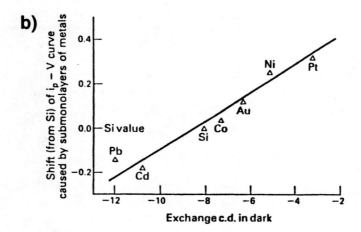

Figure 1.23. Current-potential photocharacteristics on metal modified p-Si electrodes recorded in 0.5 M H_2SO_4 (a); and the potential shift ΔV (at 1 mA cm-2) variation in 1 M H_2SO_4 on log exchange current density for hydrogen evolution of the corresponding massive metals (b). (From Ref. 137.)

classical radiation chemistry can be obtained with this material. Surface-adsorbed radical pairs (OH_{ad}, H_{ad}) are easily generated in this manner; they recombine with themselves, producing H_2O_2 and H_2, respectively, liberating thermal energy, or reacting, when diffusing from the surface, with chemical quenchers or traps present in the region near the bulk solution. In modifying microscopic TiO_2 particle surfaces with platinum particles, the reductive addition of hydrogen (hydrogenation) can be mediated. Electron transfer reactions from the edges of valence and conduction bands from this system are in this way possible [93,153,154]. (See Fig. 1.24, where radical photoelectrochemical pathways are shown.)

The macroscopic and the microscopic approaches describing a modified semiconductor surface for photoelectrocatalytic purposes differ, in that the former can be contacted electrically. Therefore known potential differences can be applied across the interface. For the latter a conservation of charge is required to balance oxidation and reduction on each particle. These systems have always been analyzed in the frame of a classical model (Schottky); thus a physical relation (electron transfer, relaxation of electron–hole pairs, etc.) between both systems is plausible. A key problem, however, is to determine the most fundamental features of each interface system in a quantitative manner.

1.5 Photoelectrocatalytic Process (Fuel Generation)

The main goal of photoelectrocatalysis is to provide chemical products via low-energy activation pathways of charge transfer at the semiconductor–electrolyte interface. The process related to (photo)electrocatalysis is linked to a multistep electron transfer reaction in which intermediates are formed. The outstanding role of a (photo)electrocatalyst is to bring the standard redox potential of individual electron transfer steps as close as possible to the redox potential of the overall reaction. Several systems demonstrating this principle on some important reactions, such as the photoelectrolysis of water, carbon dioxide reduction, and nitrogen fixation, have already been reported (see Table 1.2).

1.5.1 Photoelectrolysis of Water (HER/OER)

The hydrogen evolution reaction (HER) on photocathodes in photoelectrochemical cells is an important strategy in the cleavage of water into hydrogen, at the semiconductor, and oxygen, at the metal counterelectrode:

$$\text{(at p-sc)} \quad H_2O + 2e^- \rightarrow H_2 + 2OH^- \quad\quad -0.42 \text{ V/NHE} \quad\quad (1.8a)$$

$$\text{(at met)} \quad 2OH^- + 2h^+ \rightarrow H_2O + 1/2 O_2 \quad\quad 0.81 \quad\quad (1.8b)$$

Most p-type semiconductors—such as Si, GaP, and InP, WS_2—are effective photocatalysts for the hydrogen evolution reaction (HER) when their surfaces are properly modified [149,155–160] (see Table 1.2). The main requirements in the case of metal modified p–Si electrodes for efficient solar to hydrogen conversion have recently been defined [148]. On the other hand, the oxygen evolution reaction (OER) has mostly been accomplished with large band gap oxides (e.g., TiO_2,

ELECTRODE MATERIALS AND STRATEGIES FOR PHOTOELECTROCHEMISTRY

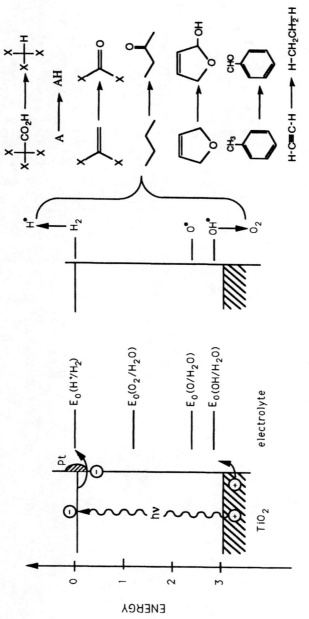

Figure 1.24. Some photoelectrochemical radical pathways on TiO_2. (From Refs. 93, 153, 154.)

Table 1.2 Different Materials Used for (Photo)electrocatalysis

Substrate	Chemical condition	Catalytic reaction	Refs.
n-Fe_2O_3	—	OER(l)	188
	Pt disp. (electrolyte)		189
	Pt, Rh deposited		163
$Fe_2O_3(H_2O)_n$ +	Colloidal gel	NFR	185
$TiO_2(H_2O)_n$			
$Sm_2O_3(H_2O)_n$ +			186
$Eu_2O_3(H_2O)_n$			
$Fe_xTi_{1-x}O_y$	Amorphous sc	OER(?)	190
	$0.1 < x < 0.9$	HER(?)	
Fe-xRhO, PdO	High-energ. med. agents	OER(?)	165
n-$SrTiO_3$	In NaOH	OER (l)	8,191
NiO-$SrTiO_3$	Powder	OER	192
		HER	
n-$BaTiO_3$	Bare electrode	OER(l)	193
	sc susp. in $AgNO_3$	OER(l)	194
n-TiO_2	Bare electrode	OER(l)	188,195,196
	RuO_2 deposited	OER(l)	162
	Pt, Rh deposited	OER(l)	163
	(Ru-doped) susp.	NFR(l)	112
	Sol–gel–film	OER(d)	197
	sc susp. in $AgNO_3$ soln	OER(l)	194
	Fe_2O_3-doped	NFR	183
(TiO_2,$SrTiO_3$,CdS	Mixed cat.	NFR(l)	182
or GaP)			
n-SnO_2	Bare electrode	OER(l)	198
ZnO	Sc susp. in $AgNO_3$ soln.	OER(l)	194
WO_3			
p-$Co_{1.8}O_3$	React. sputtering	OER(d)	199
(La,Ba)CoO_3	Sintered powder	OER(d)	200
(La,Sr)CoO_3	semiconductors		
La_2O_3-M_xO_y(M =	Oxide paste on Ti or W	OER(l)	164
Cr,Ru,Pt,Au,Co)	discs		
High-Tc oxides	High. conc. alk. soln	OER(d)	201
p-type sc.			
n-CdS	RuO_2 + cond poly	OER(l)	202
n-CdSe	-(bithiophene),		
	-(pyrrole)		
	-(vinyl chloride)		
Pt-CdS-RuO_2	Powder in aq. susp.	CDR(l)	203
p-InP	Metal(M)-coated	HER(l)	149,158,159
	M = e.g. Pt,Au		
	Dehydrogenaze	CDR	173
	enzyme		
n-Si	Granular metal films	OER(l)	134
	coating		
p-Si	Bare	CDR(l)	202
	Metal coated	HER(l)	155
	HPA coated	HER(l)	156
	Tetra-azamacroc.	CDR(l)	170

(Continued)

Table 1.2 *(Continued)*

Substrate	Chemical condition	Catalytic reaction	Refs.
pGaP	Re(CO)$_3$(v-bipy)Cl	CDR(l)	174
	Bare electrode	CDR(l)	175
	Al^{3+} complex	NFR(l)	187
	Metal coated	HER(l)	155
	Nafion+cluster mat.	HER(l)	157
	Ni (Cyclam)$^{2+}$	CDR(l)	171
p-GaAs	Ni (Cyclam)$^{2+}$	CDR(l)	205
	V(II)-V(III)chloride	CDR(l)	172
p-CdTe	Metal complexes in non-aqueous soln.	CDR(l)	176,178
n-RuS$_2$	Sintered powder	OER(l)	166
	Single crsytals		72
OsS$_2$	Pressed powder	OER(d)	84
n-PtS$_2$	Single crystals	OER(l)	167
p-WSe$_2$	Metal coated	HER(l)	150
	Re(CO)$_3$(v-bipy)Cl	CDR	174
p-WS$_2$	Pt coating	HER(l)	160

HER: hydrogen evolution reaction; OER: oxygen evolution reaction (this term includes water oxidation and oxygen evolution process key words; CDR: carbon dioxide reduction; NFR: nitrogen fixation reaction. (v-bipy) ≡ (4 vinyl-4′-methyl-2,2′-bipy); (cyclam) ≡ (1,4,8,11-tetraazacyclotetradecane).
(l) and (d): light and dark.

SrTiO$_3$, SnO$_2$, etc.), (see Table 1.2 and Section 1.4.1). One should recall that the condition for the fuel-forming photoelectrolysis reaction with these semiconductors to take place is that a sufficient illumination level be provided to get the needed free energy. The reaction is thus driven in the nonspontaneous direction ($\Delta G > 0$) which is characteristic of the so-called photoelectrosynthetic cell [161]. Photo-electrocatalysis is then fulfilled when the reaction is driven in the spontaneous direction ($\Delta G < 0$) with the light energy used to overcome the energy of activation of the process. This can be the case of oxide-modified semiconductor photoanodes [162–165] and of metal sulfide semiconductors [72,84,166,167] (see Table 1.2). A combination of electrosynthesis and electrocatalysis is, of course, possible.

At the anodic side, because of problems of photocorrosion (see Section 1.3.3) there are few photoanodes that can sustain the oxygen evolution reaction (OER) under visible light illumination without deterioration. However, n-RuS$_2$ has been identified as one of the most active, efficient, and stable photoelectrocatalysts [168]. This process was reinvestigated very recently in darkness and under illumination in a wide range of temperature (300–180 K) and with different levels of doping [169]. One typical example of a measurement is shown in a semilogarithmic plot in Fig. 1.25 for a photoactive sample. As observed, the process in darkness as well as under illumination is activated, indicating that holes are probably conveyed to the reaction center via surface states. With the help of electrochemical impedance spectroscopy [152], it was found that the essential driving force for OER is the presence of a significant Helmholtz layer potential drop at the interface.

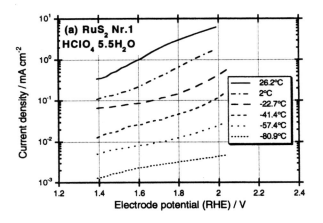

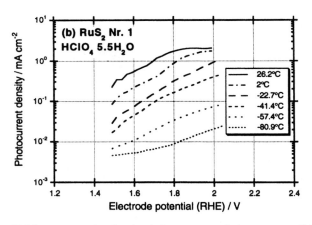

Figure 1.25. Tafel representation for the dark process and photoprocess (15 mW cm^{-2}) of oxygen evolution reaction on RuS$_2$ at different temperatures in HClO$_4$ 5.5H$_2$O, with melting point at $-45°C$.

1.5.2 Carbon Dioxide Reduction (CDR)

The reduction pathways between CO$_2$ and H$_2$O (solution with pH 7) to formic acid, formaldehyde, methanol, and methane require formally less energy with an increasing number of electrons transferred as follows:

$$CO_2(g) + 2H^+ + 2e^- \rightarrow HCOOH(aq) \qquad -0.61 \text{ V/NHE} \qquad (1.9)$$

$$CO_2(g) + 2H^+ + 2e^- \rightarrow CO(g) + H_2O \qquad -0.52 \qquad (1.10)$$

$$CO_2(g) + 4H^+ + 4e^- \rightarrow HCHO(aq) + H_2O \qquad -0.48 \qquad (1.11)$$

$$CO_2(g) + 6H^+ + 6e^- \rightarrow CH_3OH(aq) + H_2O \qquad -0.38 \qquad (1.12)$$

$$CO_2(g) + 8H^+ + 8e^- \rightarrow CH_4(g) + 2H_2O \qquad -0.24 \qquad (1.13)$$

In nonaqueous solvents, $CO_2(g)$ reduces to carbon monoxide and carbonate ions:

$$CO_2(g) + 2e^- \rightarrow CO(g) + CO_3^{2-} \qquad -1.07 \text{ V/NHE} \qquad (1.14)$$

As observed thermodynamically, but not kinetically, CO_2 reduction is comparable in difficulty to hydrogen evolution [eq. (1.8a)]. For this reason some attempts to use p-type semiconductors as multielectron donors have been reported. These are again summarized in Table 1.2. High selectivity at reasonably low overpotentials has been reported when a transition metal complex such as $Ni(cyclam)^{2+}$ is present in the electrolyte solution [170–172] and dehydrogenase enzyme [173] or $[Re(CO_3)(v\text{-}bipy)Cl]$ is fixed or incorporated onto the electrode surface [174]. However, CO [170–172] and HCOOH [173] were the predominant products obtained. On naked photoelectrodes the CDR process is no longer favorable energetically [175,176]. However, an ample variety of products, such as CH_3OH, HCOOH, and HCHO, has been identified. The very low concentration determined for these products ($1 \cdot 10^{-2}$ to $1 \cdot 10^{-4}$ M) was suggested to be the result of the photolysis of cell materials. CO_2 reduction has also been performed on metal electrodes, semiconductor suspensions, iron sulfur clusters [177], and metal complexes [178]. The photoactivity of semiconductor materials alone as reported so far for CDR does not seem to represent an advantage. A challenge should be seen in combining the coordinating properties of the transition metal complexes with photoactivity in one material. Furthermore, additional insight about the factors that govern competitive reaction paths for photogenerated electrons may be important to elucidate the CDR mechanism. This can be obtained by doing in situ measurements (product(s) detection as a function of applied electrode potential under illumination) on semiconducting membranes by means of differential electrochemical mass spectroscopy (DEMS), which has been successfully applied on metal film electrodes [179a]. First on-line measurements of photoproducts (e.g., O_2, Cl_2) have been recently reported using TiO_2 (anatase) films as a model system [179b]. The key problem is, up to now, the preparation of other semiconducting membranes (presently under investigation in our laboratory).

1.5.3 Nitrogen Fixation Reaction (NFR)

The photocatalytic fixation of dinitrogen to ammonia, a process of vital practical importance, remains up to now one of the most challenging problems. Apart from the heterogeneous catalytic synthesis of ammonia (at high temperatures and pressures) and the enzymatic nitrogen fixation, the pioneering work by Shilov and co-workers [180] on chemical nitrogen fixation to N_2H_4 and NH_3 in protic media, using strong reducing agents such as $Ti(OH)_3$, and recently, the electrosynthesis of ammonia using W- or Mo-containing mediators by Pickett and Talarmin [181], show the feasibility of such reaction in mild conditions (ambient temperature and pressure).

As with CO_2 reduction, the required energy input decreases with the increasing number of electrons:

$$N_2 + H^+ + 1e^- \rightarrow N_2H \qquad -3.2 \text{ V/NHE} \qquad (1.15)$$

$$N_2 + 2H^+ + 2e^- \rightarrow N_2H_2 \qquad -1.09 \qquad (1.16)$$

$$N_2 + 4H^+ + 4e^- \rightarrow N_2H_4 \qquad -0.3 \qquad (1.17)$$

$$N_2 + 6H^+ + 6e^- \rightarrow 2NH_3 \qquad 0.092 \qquad (1.18)$$

According to these reaction pathways, the preconditon for an efficient semiconductor catalyst is to have a sufficiently negative flat band potential and favorable chemisorption of nitrogen. Particulate systems based on titanium dioxide (mixed with $SrTiO_3$, CdS or GaP) [182], iron doped [183] or ruthenium doped [112] under UV irradiation, yield traces of ammonia and hydrazine (see Table 1.2). A systematic investigation of irradiated TiO_2 loaded with different transition metal catalysts, has confirmed this process [184]. Other approaches employing composites of Fe(III) and Ti(IV) [185] or Sm(III) and Eu(III) hydrous oxides [186] were found to photocatalyze reduction of molecular nitrogen to ammonia with visible light. For the former system, the authors report that nitrogen is preferentially adsorbed on ferric sites and that the optimum yield was attained at a pH of 10, which favors the negative shift of the flat band potential position and thus the nitrogen reduction. A similar conclusion was drawn for Sm(II) hydrous oxide. However, because of the complex nature of these composite systems, the mechanism of electron-hole separation is far from being understood. This problem can be circumvented with the use of p-type catalytic crystalline semiconductor materials. Experiments performed 16 years ago with p-GaP [187] fulfilled part of the requirements, since aluminum involved in the reaction was used as a sacrificial donor. We expect that cluster materials could have favorable properties for the catalysis of such multielectron transfer reactions. Experiments have shown that hydrogen evolution is catalyzed by individual catalytic atoms, whereas multielectron transfer reactions (e.g., oxygen reduction) required the cooperation of adjacent metal sites (bimetallic sites, metal clusters). Such basic requirements may also have to be provided at modified and nonmodified semiconductor surfaces. This may explain why only a few semiconductor catalysts were found to be very active for multielectron transfer catalysis [162,185,200]. A better understanding of multielectron transfer processes may help to obtain a more selective choice of semiconducting catalysts in the future.

1.6 Photoconversion Process (Current Generation)

1.6.1 Pyrite-Type Semiconductor Compounds

Among the pyrite structure types, FeS_2 and RuS_2 are semiconducting compounds materials of interest. However, a theoretical and experimental understanding of their solid state as well as of their interfacial behavior is still far from being completed.

ELECTRODE MATERIALS AND STRATEGIES FOR PHOTOELECTROCHEMISTRY

When operating in a regenerative way (current generation mode), both materials have similar photoelectrochemical characteristics; that is,

1. the dark current is significant and depends on the nature and concentration of the redox species;
2. there is a significant shift of the photocurrent onset (~0.5 V for FeS_2 and >1 V for RuS_2 (see, e.g., Fig. 1.26), due to the pinning of the Fermi level—hence the photopotential obtained is largely independent of the redox species used; and
3. experimental data provide evidence that interfacial states are localized at midgap.

Some differences exist, however, between both compounds regarding the quantum yield and the resistance against photocorrosion, as discussed in Section 1.3.3. The quantum yield, measured under the same conditions, turned out to be at least a factor 2 bigger for n-FeS_2 than for n-RuS_2 in the presence of I^- ions [206]. This fact supports the idea that the strength of the atomic interaction could be responsible for the difference of the photocurrent. RuS_2 is, on the other hand, a stable photoanode for oxygen evolution at moderate electrode potential bias (see Section 1.5.1), whereas FeS_2 photocorrodes in absence of suitable redox systems. If one compares the photoelectrochemical activity with that of layered semiconductor materials, one arrives at the conclusion that the same trend in the photoelectrochemical behavior is also present (see Section 1.6.2). This might be due to the role of interfacial states formed (on pyrite materials) and to the exposed transition metal centers (on steps in the case of layered compounds; cf. Figure 1.11).

The electrochemical instability of FeS_2 electrode in acid electrolyte (Section 1.3.3) can be diminished or even completely reduced when an adequate electron donor is added to the electrolyte. This effect was demonstrated when large concen-

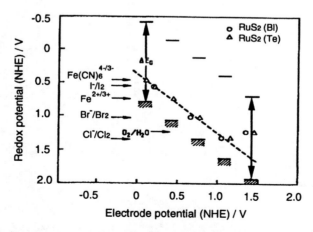

Figure 1.26. Positive energy band shift of RuS_2 induced by the presence of different redox potentials of electron donors species. Data obtained at a comparable photocurrent density (10 mA cm^{-2}). (From Ref. 53.)

trations of negatively charged ions (e.g., iodide, bromide) were present in the electrolyte using rotating-ring disk technique (RRDE) [207]. No evidence of photocorrosion was observed after a passage of ~623,000 C cm^{-2} in a long-term experiment using I/I_3^- [68]. A solar energy efficiency output of 2.8% at AM0 illumination was obtained [208]. Stability can also be obtained by mediation with positive iron complexes through a bridge formed with Cl^- ions [209]. Further, other species in relatively high concentration were also able to stabilize this interface in a purely kinetic sense. As Fig. 1.27 shows, this was not the case for positive charged ions. In the case of iron species, stability higher than 80% is obtained at a low illumination level (60 mW/cm^2). At higher illumination levels (up to 8-10 suns) the stability effect breaks down for iron ions in a manner similar to the other positive species, being maintained, on the other hand, by the negative charged ions. The selectivity vis-à-vis the negatively charged species points to the fact that the charge transfer and stabilization are both effectively performed through coordination chemistry. Additional support indicating complex formation for photoelectrochemical electron transfer has been obtained from the reaction entropy through photopotential measurements as a function of temperature [210]. Furthermore, the I^- species was found to interact differently on FeS_2 single-crystal (100) and (111) surfaces. More work in this direction is at present being performed in our laboratory. Up to now it is difficult to speculate about the channeling of the electronic charge carriers through transition metal surfaces on pyrite. A thorough account on the physicochemical aspect of pyrite material can be found in Ref. 211.

1.6.2 Layer-Type Semiconductor Compounds

Layer-type compounds, in which transition metals (M = Mo, W, Ta, Zr, Ti, Nb, Pt) are sandwiched between layers of chalcogens (X = sulfur, selenium, or tellurium) to give stacks of MTM that are separated by van der Waals gaps, have been

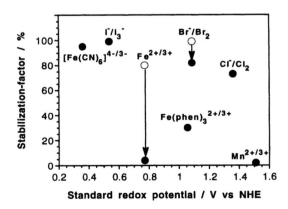

Figure 1.27. Stabilization effect of various electron donors on FeS_2/electrolyte interface as determined by RRDE measurements. (From Ref. 211.)

investigated quite intensively [54,212,213]. Among them semiconductors with reasonable energy gaps for solar energy conversion can be found (e.g., WSe_2, $MoSe_2$) for which energy conversion efficiencies of up to 15% have been demonstrated. They owe their favorable properties mainly to the fact that their energy bands are derived from nonbonding d-states, which explains their high stability in the presence of suitable redox agents (e.g., I^-/I_3^-) and in the presence of a van der Waals surface, which is characterized by a low concentration of surface states. Samples that have been prepared by careful cleaving have a low concentration of step sites with surface states, to which majority carriers may drift to contribute to the dark current. The smaller the concentration of step sites, the smaller the dark current, which explains why cleaved crystals exhibit such favorable properties. The larger the contribution of van der Waals surfaces, the smaller the dark current and consequently the better the energy conversion efficiency. Step sites, on the other hand, are also the sites where transition metal states are exposed for reaction. Here coordination-type interaction with redox species occurs as well as formation of oxides, which may lead to a certain degree of passivation. More specific, complexing agents (e.g., EDTA) as well as other suitable molecular species may be used to partially neutralize these step sites and to improve the power output characteristic of such electrodes.

Basically, the photoelectrochemical behavior of layer compounds with d-energy bands (MoS_2, WS_2) is very similar to the behavior of d-band semiconductors with pyrite structure (FeS_2, RuS_2). The crucial (photo)electrochemical mechanisms are controlled by interfacial transition metal states, which catalyze the interaction with water. Those transition metal chalcogenides, which derive their energy gap from chalcogen p-states, on the other hand, are not able to react with water and corrode to liberate sulfur and selenium, respectively (e.g., ZrS_2, $ZrSe_2$, see Table 1.3).

There is one important difference between layer type and pyrite structured transition metal chalcogenides with comparable electronic structure: the van der Waals

Table 1.3 Oxidation Products and Photopotentials in Contact with Redox Systems*

Compound	Energy Gap/eV	Photopotential/V	Oxidation Prod.
ZrS_2	1.68–1.80	0.4	So,Zr(IV)
$ZrSe_2$	1.05–1.22		Se,Zr(IV)
MoS_2	1.3	0.6–0.7	MoO_4^{2-},SO_4^{2-}
$MoSe_2$	1.1	0.5–0.6	MoO_4^{2-},SeO_4^{2-}
$MoTe_2$	1.0	0.4	MoO_4^{2-},TeO_4^{2-}
WS_2	1.35	0.6–0.7	WO_3,SO_4^{2-}
p-WS_2	1.35	0.7–0.8	—
WSe_2	1.2	0.6–0.8	WO_3,SeO_4^{2-}
ReS_2	1.33	0.4	Re_2O_7,O_2,SO_4^{2-}
$ReSe_2$	1.2	0.4	Re_2O_7,O_2,SeO_4^{2-}
PtS_2	0.95		O_2

* For layer-type structure transition metal dichalcogenides with energy gaps between 1 and 2 eV according to different sources [54,160,230,231].

surface of layer compounds suppresses the dark current, because of the absence of suitable reactive surface states that are only available at step sites. This leads to a significant improvement of the photocurrent characteristics and to the formation of high photopotentials. This explains why practically all favorable energy conversion efficiencies for layer compounds have been reported for cleaved crystals. Thin layers of these interesting semiconductor materials produced by MOCVD, by sputtering or by melt growth [214,215], always tend to show microcrystalling growth away from the interface and rosette-type morphologies, which add large concentrations of step sites. This causes the photopotential and the charge carrier collection efficiency to decrease. Efforts to control the concentration of step sites in such layers have up to now not been sufficiently successful.

Layer-type materials have therefore only a chance of becoming of practical interest when large-scale surfaces with a high contribution of van der Waals surfaces can be generated. This is presently attempted with epitaxial growth techniques on suitable substrates [216], but only the future may show whether a technologically practical way can be found.

1.6.3 Inorganic–Organic Sensitization

Spectral sensitization of semiconducting substrates by organic dyes is a process that has long been applied for making photographic emulsions sensitive for low-energy visible and infrared light. Beginning in 1967, large-gap oxide electrodes (e.g., ZnO) and organic molecular crystals were spectrally sensitized with organic dyes in photoelectrochemical cells [217]. Spectral sensitization, where electrons (holes) from excited molecules are injected into energy bands of suitable electrodes, bears a resemblance to the energy conversion process of photosynthetic membranes, where excited chlorophyll injects electrons into the electron transfer path through specialized proteins along electron transfer chains. It has, in fact, been possible to demonstrate that chlorophyll molecules can act as sensitizing electron pumps between molecular electron donors and large-gap oxide electrodes [218]. Twenty-two years ago the principle and first working prototypes of sensitization solar cells were described [219]. They consisted of a large-gap oxide electrode (e.g., ZnO) to which broadly light-absorbing sensitizing molecules [chlorophyll derivate, (Na_3)-mesochlorine] were adsorbed in contact with an electrolyte containing a suitable redox couple (hydroquinone–quinone, iodide–ioidine, etc.). A metal electrode is used as a counter electrode. In principle, generation of photoelectrical energy from a dye-sensitized photochemical reaction can be accomplished by separation of electronic reaction steps at two electrodes of an electrochemical cell, analogous to the way in which batteries can be built by separating electron transfer of a redox reaction and collecting charge carriers via electrodes. One advantage of the sensitization solar cell over conventional solar cells consists in a separation of charge collection and electronic carrier transport. A minority carrier is rapidly transformed into a majority carrier that cannot be easily lost due to recombination.

Even though, two decades ago, quantum efficiencies for sensitization of the order of 10% and in special cases of up to 100% have been observed [217,220,233,235,

243,244], and even though dyes adsorbed on porous sintered oxide electrodes yielded energy conversion efficiencies of the order of 1–2% [221–223], efforts to develop this type of solar cell further faded because of the observed photochemical instability of the sensitizing molecules. Regardless of the observation that a sensitizating dye could mediate the photoinduced pumping of electrons from a suitable molecular electron donor into the conduction band of an oxide semiconductor more than 10^5 times without side reaction, this appeared to be insufficient to guarantee adequate stability for solar cell operation. For such a purpose stability exceeding a period of a year would be a minimum requirement. However, side reactions and oxidation indicated degradation of energy output occur within days and weeks.

The last three years have seen a revival of interest in sensitization solar cells, mainly because of the work of Graetzel and collaborators [224]. The basis for their progress was (1) the development of highly structured TiO_2-anatase layers, (2) the use of ruthenium-based sensitizer complexes, and (3) the chemical attachment of these complexes to the TiO_2 surface via carboxyl groups. Quantum efficiencies of the order of 80% and solar energy conversion efficiencies of 7% in direct sunlight and 12% in diffuse sunlight were claimed [225]. Sufficient long-term stability was also claimed to justify development of technical solar cells [226].

Remarkable features of these recently developed TiO_2-based sensitization cells are not only the highly structured oxide electrode, which allows the absorption of a large fraction of the incident solar energy by surface-attached ruthenium complexes, but also the relatively complex composition of the redox electrolyte used. It contains not only the redox couple (I^-/I_2), but also organic components (propylene carbonate, acetonitrile). It is not yet clear why this sensitization mechanism should be significantly more stable than previously investigated ones. It also has to be established why electrochemical mass transport necessary for the regeneration of the sensitizer is still so active in microscopic and submicroscopic pores. In addition, the fact that the light-absorbing electrolyte is photochemically stable (the UV component of light has been filtered out) is puzzling. Obviously, careful studies, including long-term investigations on stability, will be required to answer these questions.

These described progress reports with sensitization solar cells have led to a revival of interest and research on sensitization solar cells. This is justified because of the unique property of these cells of mediating extremely rapidly the transfer of excited electrons into electrodes, where they form majority carriers and can safely be collected. A wide variety of large-gap materials is accessible for spectral sensitization (Table 1.4). With many of these, high quantum efficiencies of 10% to nearly 100% per absorbed photon have been reported, and total energy conversion efficiencies of 1–2.5% with ZnO and up to 10% with highly structured TiO_2 have been found [225].

Since long-term photoelectrochemical stability of dyes and metal–organic complexes appeared to be the most critical factor in the development of sensitization solar cells, attempts were made in our laboratory to replace the sensitizing species by thin semiconducting layers. First, experiments in which ZnO was sensitized by depositing a very thin film of the p-type polymer polyfurandivinylene were successful [227], but it remained clear that only highly absorbing semiconductor materials

Table 1.4 Large Band Gap Material for Which Sensitization with Dyes and Complexes Has Been Demonstrated

Substrate	Sample	Sensitizer	References
ZnO	Single crystal	Rose bengal	232,233
		Rhodamine B	237
		Chlorophyl	234
	Ceramic	Ru complex	221
		Rose bengal	235
TiO_2	Colloid	Ru complex	225
		Chlorophylin	236
SnO_2	Thin films	Rhodamine B	238
		Xanthene	239
$SrTiO_3$	Microcrystalline	Pynacianol	240
$BaTiO_3$		Rose bengal	
CdS	Single crystal	Rhodamine B	241
SnS_2	Single crystal	Various dyes	242
WS_2, $MoSe_2$, WSe_2	Single crystals	Various dyes	243
GaP	Single crystal	Methylene blue, rhodamine B	244

could be of any practical value. This was the main reason that much effort was put into development of FeS_2 (pyrite) for application in solar cells. The high absorption coefficient of 6×10^5 cm^{-1}, together with the possibility of light concentration through scattering processes on the basis of nonimaging optics, makes it possible to conceive of a sensitization solar cell in which a deposited pyrite film only 10 nm (100 Å) thick sensitizes electron injection into TiO_2. This type of solar cell could indeed be demonstrated and shows the typically favorable current voltage characteristics known for sensitization processes [228]. Much work, however, will be needed to develop both the ultrathin semiconductor layer and the interfaces through which excited electrons have to be passed. The main advantage of sensitization solar cells in which ultrathin semiconductors act as sensitizers of large-gap oxides is the easy and rapid production of the films by MOCVD or sputtering techniques, as well as the possibility of controlling electron regeneration via a solid/liquid or a solid/solid contact. Because of the extremely high absorption coefficient no highly structured oxide substrate is required, which avoids problems with mass transport in pores.

Recently, in our institute, semiconductor colloids of CdS and PbS have been used as sensitizers for highly structured TiO_2 [229]. An interesting promise is the possibility of tailoring the band gap of sensitizers by selecting appropriate colloid dimensions. A big problem is the chemical and photochemical instability of colloids, which tend to coagulate and appear to have the same photocorrosion problems as macroscopic CdS or related materials. A way to improve this would be to develop materials that involve nonbonding d-states during photoelectrochemical reactions. However, it is not easy to imagine why colloids could become more stable than molecules where chemical bonding guarantees stable configurations and where changes in dimension and composition easily allow one to introduce changes in spectral sensitivity.

1.7 Challenges for Research and Application

1.7.1 Nanocrystalline Electrode Materials

When the dimensions of solid-state materials are made smaller and smaller and molecular dimensions are reached, significant changes in the energetic distribution of orbitals are observed. These quantization phenomena [245,246] observed with colloids of semiconductors and metals have attracted significant interest with respect to new material properties and photochemical behavior. With semiconducting compounds, for example, a systematic widening of the energy gap is possible with decreasing size of the colloids. Cadmium phosphide, for example, is black as a macroscopic semiconductor material with an energy gap of 0.6 eV. With decreasing dimensions of colloids in the range of 50–10 Å the material changes from brown to red and pale yellow, with its energy gap increasing to 3 eV.

Colloids may be stabilized with appropriate chemical ligands and may be dispersed in solids or fabricated into solids. It is intended that the peculiar optoelectronic properties of colloids is thereby maintained to obtain solids with tailored properties. For electrodes the large surface area of colloid structures may also be attractive. New approaches have been attempted in photoelectrochemistry using electrodes prepared from colloids. Titanium dioxide electrodes prepared by a sol-gel technique were used as large surface electrodes for sensitization with Ru complexes [44,225,247] or with CdS or PbS colloids [229] (see Fig. 1.28).

Key problems that have to be handled with colloid electrodes are stability against coagulation of quantum size particles, stability against interfacial corrosion, and charge carrier transport between colloidal particles within a solid-state electrode. To some extent colloids combine the properties of molecules with the intricate properties of semiconductor surfaces. When pressed into solids, charge transfer is typically controlled by hopping processes with high activation energy, as in samples prepared from organic molecules. On the one hand, they basically consist of surfaces, which in the case of CdS or PbS can, under illumination, not easily remain stable in contact with an electrolyte. On the other hand, very efficient charge transfer in the picosecond and subpicosecond range has been observed between colloids of different composition, formed into a sandwich. This is an interesting perspective for energy conversion. The large surface area as well as tailored electronic states also make colloids interesting for catalysis. Since individual particles are so small that individual electronic charge carriers may significantly shift the energetic position of electronic states, nonlinear far-from-equilibrium phenomena are also to be expected. This adds an interesting aspect to catalytic properties, since autocatalytic mechanisms may be possible, which may induce improved electron transfer processes [248]. Summarizing, colloid electrodes offer many interesting

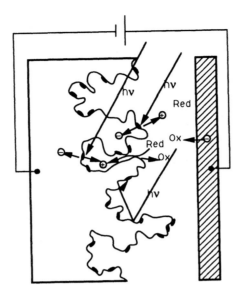

Figure 1.28. Dye (or colloid) sensitization cell utilizing a very large surface area for efficient light collection.

challenges, but significant problems have to be overcome before they may attain practical interest.

1.7.2 Electrode Microstructure, Light Collection, and Energy Conversion Efficiency

Light absorption and light collection by an electrode material are not only controlled by its bulk absorption coefficient. In addition, the morphology and composition of its surface and of its bulk are critical. The phenomena produced range from antireflective surface effects to improved coupling of quantum energy into electrodes and light concentration via scattering phenomena. On the one hand, it is important to know such phenomena, since they may occur unintentionally during photoelectrochemical treatment of electrodes. On the other hand, systematic development of such mechanisms and incorporating them into material developments may provide us with better and cheaper materials for solar energy conversion and photoelectrochemistry. When tailoring materials, therefore, consideration and improvement of relevant surface and bulk morphological structures may bring significant advantages.

1.7.2.1 Selective Absorber Surfaces

A most efficient way to reduce light reflection from a semiconductor interface is accessible by generating a smooth gradient of refractive index from one phase to the

ELECTRODE MATERIALS AND STRATEGIES FOR PHOTOELECTROCHEMISTRY   51

other. Such an interface can sometimes be created by etching, when etching pyramides, or nipples are produced within dimensions and distances smaller than the wavelength of light (Fig. 1.29). When the generated pyramidal structure has a larger dimension, ranging from 2 to 50 μm, multiple reflection processes can, as computer simulation has confirmed [249], improve the capture of light. Dentritic surfaces with such properties can be produced by carefully chosen deposition or etching

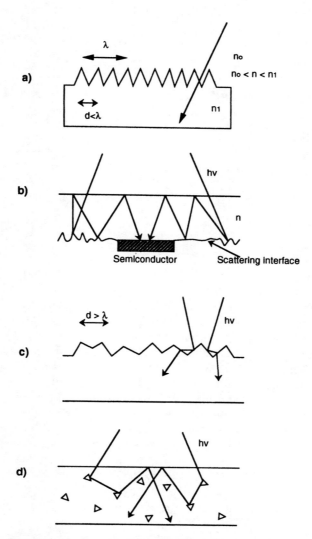

Figure 1.29. Material structures for improved light collection: (a) dendrites providing a gradual transition of the refractive index, (b) light coupled into oxide layer by scattering interfaces, (c) light collection through textured interfaces, and (d) light concentration by scattered light collection.

techniques. Many useful semiconductors have high refractive indexes, which gives high reflection losses. Thin films grown or deposited on such semiconductors, with appropriate optical properties, may efficiently reduce reflection. Metal-dielectric composite coatings, which consist of very fine metal particles in a dielectric host, offer a high degree of flexibility, since selectivity may be optimized with regard to the constituents, coating thickness, particle concentration, and size, shape, and orientation of particles. Such layers can be produced electrochemically. An example is Ni-pigmented anodic Al_2O_3 coating on Al, which is commercially produced by dc anodization of Al sheet in dilute phosphoric acid followed by ac electrolysis in a bath containing $NiSO_4$ [250]. A porous surface structure partially filled with Ni particles is obtained.

1.7.2.2 Scattered Light Collection Through Nonimaging Optics

For efficient light collection it is not necessary to use geometrical optics, which has been designed to transform images without distortion. From basic thermodynamics it can be derived that the maximum concentration C for a concentrator with its exit aperture Θ_2 immersed in a medium of refractive index n and with a maximum incident angle for light of Θ_1 is

$$C \leq \frac{n^2 \sin^2 \Theta_2}{\sin^2 \Theta_1} \tag{1.19}$$

Most conventional imaging systems, such as Fresnel lenses and parabolic reflectors, fall short of equation (1.19) by a factor of 4 or more. However, cone- or trough- (in two dimensions) shaped compound parabolic concentrators (Winston collector) can approach this limit [251]. If such a collector with $\Theta_1 = 42°$ and $\Theta_2 = 90°$ is just filled with air, light concentration at the bottom is 2.2, and if it is placed in a fixed orientation, it may collect approximately six hours during the day. If the concentrator is filled with a material of refractive index $n = 1.5$, the concentration is still approximately $n^2 = 2.2$, but the system will now accept light over 2π solid angle and acts as a collector for diffuse light.

Nonimaging optics has significant consequences for the design of light-absorbing materials and solar energy-converting electrodes, especially since many photoactive materials have high refractive indices ($n^2 = 10-25$). Some of these are depicted in Fig. 1.29. First, a refractive layer (e.g., oxide layer) above a light-sensitive material can capture light and provide locally more light intensity than that provided in the incident beam (Fig. 1.29a). When a material is deposited on a rough surface with many scattering centers, concentration of light within this layer is subject to relation (1.19) (Fig. 1.29b). The same is true with a semiconducting layer providing a structured surface and scattering internal surfaces (Fig. 1.29c). They can efficiently confine radiation and significantly reduce the layer thickness needed for light absorption, as calculated from the Lambert-Beer law. An important practical example is crystallized silicon, which has to be provided as a layer approximately 300 μm thick for efficient light absorption. If sufficient light-scattering centers could be incorporated into the silicon material (see Fig. 1.29), or into its surface, a concentration of light within silicon proportional to $n^2 = 12$ should be possible. This means that efficient solar cells with silicon layers as thin as 50 μm or less are possible, which has stimulated significant research with the aim of reducing material costs.

The design of photoactive materials for solar energy conversion should consider the advantages to be drawn from incorporating mechanisms of nonimaging optics. Highly absorbing materials such as pyrite, with an absorption coefficient of $\alpha = 5 \times 10^5$ cm^{-1}, could, in combination with light collection through scattering processes, serve for solar light collection with a layer thickness of 5–20 nm. Sensitization solar cells thus become possible in principle in which the photoactive semiconductor acts as light absorber and electronic charge carriers are injected into contact materials, which generate the photopotentials. In such a system charge generation and charge transport are separated, which may be a significant advantage in efforts to develop economic solar energy converting devices.

1.7.3 High-Current-Density Electrodes and Irreversible Thermodynamics

A significant challenge for applied research is the development of electrode materials that can sustain large current densities. Such electrodes are needed for processes like electrolysis, electrowinning, fuel cell reactions, or (photo)catalysis of fuel production. A necessary precondition is the presence of large reaction surfaces. But additional requirements are efficient transport of reactants and products, including fast interfacial reaction steps. The practical problems encountered are manifold. Conventional electrolysis cells show, for example, a high degree of electrode instability and corrosion when powered by, daily for short time periods, fluctuating photovoltaic electricity. The conclusion was reached that the fluctuating bias was responsible for the observed problems [252]. A permanently applied bias voltage indeed helped to counteract this problem.

High-current-density electrodes are systems that operate quite far from equilibrium with high thermodynamic fluxes. It may therefore be inappropriate to rely completely on electrochemical theoretical formalisms that are basically derived from reversible thermodynamics. It may be interesting to explore the applicability of concepts and mechanisms of irreversible thermodynamics to (photo)electrochemical systems, which has already been attempted with electron transfer involving autocatalysis [248,253]. In these publications electron transfer mechanisms were calculated that involve autocatalysis. This means that intermediates are formed that have a reverse (autocatalytic) effect on charge transfer parameters of preceding reactants. Such a mechanism can also be imagined on submolecular scale. Such autocatalytic steps push the system far from equilibrium, where conventional electron transfer theories are not applicable (nonlinear range of irreversible thermodynamics). It is found that in this range new mechanisms become possible (e.g., stimulated and cooperative mechanisms of electron transfer). Such cooperative mechanisms may better allow the control of entropy fluxes, which determine stability and reactivity of interfaces. Autocatalytic processes allow, in principle, the export of entropy and the maintenance or buildup of order at the expense of the overall energy-dissipating mechanism [254]. Such a strategy of an operating mechanism far from equilibrium, which has been applied widely in biological structures, which maintain their order by dissipating energy, may also be explored in interfacial electrochemistry to obtain improved stability of electrodes. However, such goals cannot realistically be achieved in the foreseeable future. Therefore the search for new material combinations with high interfacial reactivity and large surface areas will mostly remain empirical.

1.8 Summary and Outlook

Selection and development of semiconducting materials for photoelectrochemistry and photocatalysis have been influenced by historical circumstances. At the beginning of photoelectrochemical research in the 1960s and 1970s semiconductors from electronics clearly dominated. The high quality of available semiconductor materials, such as silicon, gallium arsenide and cadmium sulfide, was the basis for establishing the foundations of classical photoelectrochemistry. The observed photoinduced interfacial mechanism, of course, was determined by properties typical for these materials, such as pronounced corrosion due to photoinduced bond breaking and moderate catalytic activity due to the chemical nature of the reacting elements. Subsequently, the stability of electrode materials was improved by optimizing redox electrolytes and catalytic properties enhanced by surface modifications of electrodes and attachment of catalysts. Although significant progress was achieved, the incomplete stability of photoelectrodes remained a limiting factor for the technical application of photoelectrochemical systems. In the case of water photoelectrolysis, research was dominated by the favorable and interesting properties of TiO_2, which could not be surpassed by any other oxide investigated. It biased photoelectrochemical efforts toward a radical photoelectrochemistry, which turned out to be very useful for photooxidation of organic chemicals but is much too energy intensive for water photoelectrolysis. No radicals should be formed during such a mechanism, but water splitting should occur near the thermodynamic potential of 1.23 V. Many photoactive materials have been studied during the last three decades, and many valuable results on interfacial mechanisms have been obtained, but little systematic work has been done. A main reason may be the difficulty encountered with material preparation, which requires a long-term commitment.

An attempt at a systematic approach—development of electrode materials capable of inducing photogenerated interfacial coordination chemical mechanism—has been outlined in this review. Both remarkable stable photoelectrodes for current generation and highly catalytic semiconductor electrodes have been obtained in this way. This experience shows that tailoring the electronic and chemical properties of photoelectrodes is a helpful strategy that should be pursued and expanded. It is, however, a relatively work-intensive strategy, because many new compounds have to be synthesized and tested. Failures are programmed, because elements often cannot be steered into desired crystal structures. This has been observed, for example, with semiconducting cluster compounds of the compositon: $(Mo_{6-x}M'_x) X_8$, where Mo could not be replaced by W and M' = Ru, Re not by element combinations including Mn or Fe.

Photoactive electrodes that permit light-induced interfacial coordination chemical mechanisms allow not only improved electron transfer mechanisms via chemical bonds formed between transition metals in the illuminated surface and suitable electron donors, but also the formation of interfacial complexes that can temporarily store minority carriers and chemically bind reactants (e.g., OH^-) during multielectron transfer catalysis. Such metal–centered photoelectrochemistry, which channels minority carriers via changes in oxidation states, thus suppresing side reactions, may also explain the reported improved stability of sensitization cells based on TiO_2 and Ru complexes. Sensitization solar cells should be reexamined with special attention paid to the surface morphology of oxides, which controls collection efficiency, and to the attachment and nature of the sensitization complexes as well as reducing agents used. Electrode materials for sensitization solar cells must combine optimized electric conductivity with low defect state concentrations in the forbidden energy regions, large surface area, and favorable conditions for light scattering and light collection. This outlines the challenge for material science only for the special application of photoelectrodes.

Research on photoactive electrode materials has a long way to go. The target is the use of typically thermodynamically unstable materials for the generation of high photocurrent fluxes, which are generating well-defined products. Kinetic stabilization of such electrodes is required, since high entropy fluxes have to be maintained without increasing the entropy of the system. Nature has managed to solve this problem by making use of the mechanisms of nonlinear irreversible thermodynamics, which are complicated and involve autocatalytic processes. Development of materials and interfaces for photoreactions may, in the long term, follow a similar pathway.

Acknowledgments

The authors would like to thank Prof. D. Haneman for critically reading the manuscript. The research performed in our laboratory has been generously supported by the BMFT, which is gratefully acknowledged.

References

1. Vetter, K. J. "Electrochemical Kinetics." Academic, New York, 1967, pp. 537–540, 615–641.
2. Wolkenstein, T. "The Electronic Theory of Catalysis on Semiconductors." Pergamon, Oxford, 1963.
3. Gerischer, H. *Surf. Sci.* **1969**, *18*, 97.

4. Gerischer, H. *Z. Phys. Chem., N.F.* **1960**, *26*, 223, 325.
5. Gerischer, H. In "Physical Chemistry: An Advanced Treatise," Eyring, H.; Henderson, D.; Jost, W., eds.; Academic, New York, 1970; Vol. 9, Part A, p. 463.
6. Hush, N. S. *Trans. Faraday Soc.* **1961**, *57*, 557.
7. Marcus, R. A. *J. Chem. Phys.* **1963**. *67*, 853.
8. Marcus, R. A. *J. Chem. Phys.* **1965**, *43*, 679.
9. Levich, V. G. In "Advances in Electrochemistry and Electrochemical Engineering," Delahay, P., ed. Interscience, New York, 1966, Vol. 14, p. 249.
10. Dogonadze, R. R.; Chizmadzhev, Y. A. *Dokl. Akad. Nauk USSR* **1962**, *144*, 463; **1963**, *145*, 563.
11. Memming, R. In "Electroanalytical Chemistry," Bard, A. J., ed. Marcel Dekker, New York, 1979, Vol. 11, p. 1.
12. Gerischer, H. In Topics App. Phys. **1979**, *31*, 115.
13. Koval, C. A.; Howard, J. N. *Chem. Rev.* **1992**, *92*, 411.
14. Gerischer, H. *J. Phys. Chem.* **1991**, *95*, 1356.
15. Marcus, R. A. *J. Phys. Chem.* **1990**, *94*, 1050.
16. Morrison, S. R. "Electrochemistry at Semiconductor and Oxidized Metal Electrodes," Plenum, New York, 1980.
17. Frese, K. W. *J. Phys. Chem.* **1981**, *85*, 3916.
18. Frese, K. W. *J. Electrochem. Soc.* **1983**, *130*, 28.
19. Nakabayashi, S.; Fujishima, A.; Honda, K. *J. Phys. Chem.* **1983**, *87*, 3487.
20. Lorenz, W. *J. Phys. Chem.* **1991**, *95*, 10566.
21. Lewis, N. S. *Annu. Rev. Phys. Chem.* **1991**, *42*, 543.
22. Weaver, M. J. *J. Phys. Chem.* **1980**, *84*, 568.
23. Frese K. W. *J. Electroanal. Chem.* **1988**, *249*, 15.
24. Hampson, N. A.; McNeil, A. J. S. In "Electrochemistry: A Specialist Periodical Report." The Royal Society of Chemistry, 1984, Vol. 9, p. 1.
25. Somorjai, G. A. In "Photocatalysis: Fundamentals and Applications." Serpone, N.; Pelizzetti, E., eds. John Wiley, New York, 1989, pp. 251–310.
26. Boudart, M. *Adv. Catal.* **1969**, *20*, 153.
27. Mandelbrot, B. B. "The Fractal Geometry of Nature." Freeman, San Francisco, 1982.
28. Le Méaute, A.; Crepy, G. *Compt. Rend Acad. Sci. Paris* **1982**, *294*, 685.
29. Le Méaute, A.; Crepy, G. *Solid State Ionics* **1983**, *9–10*, 17.
30. Le Méaute, A.; Dugast, A. *J. Power Sources* **1983**, *9*, 359.
31. Nyikos, L.; Pajkossy, T. *Electrochim. Acta* **1985**, *30*, 1533.
32. Pajkossy, T.; Nyikos, L. *J. Electrochem. Soc.* **1986**, *133*, 2061.
33. Nyikos, L.; Pajkossy, T. *Electrochim. Acta* **1986**, *31*, 1347.
34. Pajkossy, T.; Nyikos, L. *Electrochim. Acta* **1989**, *34*, 171.
35. Pajkossy, T.; Nyikos, L. *Electrochim. Acta* **1989**, *34*, 181.

36. Kaye, B. H. *Powder Technol.* **1985**, *46*, 245.
37. Kaye, B. H. "A Random Walk Through Fractal Dimension." VCH Publishers, Weinheim, 1989.
38. Van Damme, H. In Ref. 25, pp. 175–215.
39. Seri-Levy, A.; Samuel, J.; Farin, D.; Avnir, D. In "Photochemistry on Solid Surface," Anpo, M.; Matsura, T., eds. Elsevier, Amsterdam, 1989, pp. 353–374.
40. Farin, D.; Avnir, D. *J. Am. Chem. Soc.* **1988**, *110*, 2039.
41. Seri-Levy, A.; Avnir, D. *Surf. Sci.* **1991**, *248*, 258.
42. Voss, R. F. In "The Science of Fractal Images," Peitgen, H.-O.; Saupe, D., eds. Springer-Verlag, New York, 1988, p. 21.
43. Mulder, W. H.; Sluyters, J. H. *Electrochim. Acta* **1988**, *33*, 303.
44. Vlachopoulos, N.; Liska, P.; Augustynsky, J.; Grätzel, M. *J. Am. Chem. Soc.* **1988**, *110*, 1216.
45. Siegenthaler, H.; Christoph, R. In "Scanning Tunneling Microscopy and Related Methods," Behm, R. J.; Garcia, N.1; Rohrer, H., eds. Kluwer, Dordrecht, 1990, p. 315.
46. Kittel, C. "Introduction to Solid State Physics," 6th ed. John Wiley, New York, 1986.
47. Gerischer, H. *Electrochim. Acta* **1968**, *13*, 1467.
48. Lotz, W. *J. Opt. Soc. Am.* **1970**, *60*, 206.
49. Huheey, J. E. "Inorganic Chemistry: Principles of Structure and Reactivity," 3rd ed. Harper International, Cambridge, 1983, p. 42.
50. Goodenough, J. B. "Magnetism and the Chemical Bond," Interscience, New York, 1963.
51. Jobic, S.; Brec, R.; Rouxel, J. *J. Alloys Comp.* **1992**, *178*, 253.
52. Brec, R.; Rouxel, J. In "Intercalation in Layered Materials," Dresselhaus, M. S., ed. Plenum, New York, 1986, p. 75.
53. Tributsch, H. In "Photoelectrochemistry and Photovoltaics of Layered Semiconductors," Aruchamy, A., ed. Kluwer, Dordrecht, 1992, Vol. 14, p. 83.
54. Tributsch, H. *Structure and Bonding: Solar Energy Materials* **1982**, *49*, 127.
55. Liang, W. Y. In Ref. 52, p. 31.
56. Coehoorn, R.; Maas, C.; Dijkstra, J.; Flipse, C. J. F.; de Groot, R. A.; Wold, A. *Phys. Rev. B*, **1987**, *35*, 6195, 6203.
57. Grasso, V.; Mondio, G. In "Electronic Structure and Electronic Transitions in Layered Materials," Grasso, V., ed. Reidel, Dordrecht, 1986.
58. Schlegel, A.; Wachter, P. *J. Phys. C: Solid State Phys.* **1976**, *9*, 3363.
59. Kou, W. W.; Seehra, M. S. *Phys. Rev. B* **1978**, *18*, 7062.
60. Bullet, D. W. *J. Phys. C: Solid State Phys.* **1982**, *15*, 6163.
61. Lauer, S.; Trautwein, A. X.; Harris, F. E. *Phys. Rev. B* **1984**, *29*, 6774.
62. Folkerts, W.; Sawatzky, G. A.; Haas, C.; de Groot, R. A.; Hillebrecht, F. U. *J. Phys. C: Solid State Phys.* **1987**, *20*, 4135.
63. Herm, D.; Wetzel, H.; Tributsch, H. *Surf. Sci.* **1985**, *163*, 13.
64. Holzwarth, N. A. W.; Harris, S.; Liang, K. S. *Phys. Rev. B* **1985**, *32*, 3745.
65. Huang, Y.-S.; Chen, Y.-F. *Phys. Rev. B* **1988**, *38*, 7997.

66. Endo, S.; Mitsui, T.; Miyadai, T. *Phys. Lett.* **1973**, *46A*, 29.
67. Karguppikar, A. M.; Vedeshwar, A. G. *Phys. Stat. Sol.* **1988**, *109*, 549.
68. Ennaoui, A.; Fiechter, S.; Jaegermann, W.; Tributsch, H. *J. Electrochem. Soc.* **1986**, *133*, 97.
69. Kühne, H.-M.; Jaegermann, W.; Tributsch, H. *Chem. Phys. Lett.* **1984**, *112*, 160.
70. Bichsel, R.; Levy, F.; Berger, H. *J. Phys. C* **1984**, *17*, L19.
71. Piazza, S.; Kühne, H.-M.; Tributsch, H. *J. Electroanal. Chem.* **1985**, *196*, 53.
72. Kühne, H.-M.; Tributsch, H. *J. Electroanal. Chem.* **1986**, *201*, 263.
73. Mattheiss, L. F.; Fong, C. Y. *Phys. Rev. B: Solid State* **1977**, *15*, 1760.
74. Andersen, O. K.; Klose, W.; Nohl, H. *Phys. Rev. B: Solid State* **1978**, *17*, 1209.
75. Nohl, H.; Klose, W.; Andersen, O. K. In "Superconductivity in Ternary Compounds"; Fischer, Ø.; Maple, M. B., eds. Springer-Verlag Berlin Heidelberg New York 1981; pp. 165–220.
76. Nohl, H.; Andersen, O. K. In "Superconductivity of d- f-Band Met.," Proc. Conf. Fourth, Buckel, W., Weber, W., eds. KFK: Karlsruhe, Germany 1982; p. 165.
77. Burdett, J. K.; Lin, J.-H. *Inor. Chem.* **1982**, *21*, 5.
78. Hughbanks, T.; Hoffmann, R. *J. Am. Chem. Soc.* **1983**, *105*, 1150.
79. Perrin, A.; Chevrel, R.; Sergent, M.; Fischer, Ø. *J. Solid State Chem.* **1980**, *33*, 43.
80. Alonso-Vante, N.; Jaegermann, W.; Tributsch, H.; Hönle, W.; Yvon, K. *J. Am. Chem. Soc.* **1987**, *109*, 3251.
81. Jaegermann, W.; Pettenkofer, Ch.; Alonso-Vante, N.; Schwarzlose, Th.; Tributsch, H. *Ber. Bunsenges. Phys. Chem.* **1990**, *94*, 513.
82. Jaegermann, W.; Schmeisser, D. *Surf. Sci.* **1986**, *165*, 143.
83. Jaegermann, W.; Kühne, H.-M. *Appl. Surf. Sci.* **1986**, *26*, 1.
84. Schwarzlose, Th.; Fiechter, S.; Jaegermann, W. *Ber. Bunsenges. Phys. Chem.* **1992**, *96*, 887.
85. Kautek, W.; Gerischer, H.; Tributsch, H. *Ber. Bunsenges. Phys. Chem.* **1979**, *83*, 1000.
86. Jaegermann, W. Habilitation thesis, Free University, Berlin, 1992.
87. Gerischer, H. *J. Vac. Sci. Technol.* **1978**, *15*, 1422.
88. Bard, A.; Wrighton, M. S. *J. Electrochem. Soc.* **1977**, *124*, 1706.
89. Hodes, G.; Miller, B. *J. Electrochem. Soc.* **1986**, *133*, 2177.
90. Li, X.-P.; Alonso-Vante, N.; Tributsch, H. *J. Electroanal. Chem.* **1986**, *242*, 255.
91. Ennaoui, A.; Tributsch, H. *J. Electroanal. Chem.* **1986**, *204*, 185.
92. Salama, S.; Kennedy, J. H. *J. Electrochem. Soc.* **1989**, *136*, 2906.
93. Fox, M.-A. In Ref. 25, pp. 421–455.
94. Esser, P.; Feierabend, R.; Gopel, W. *Ber. Bunsenges. Phys. Chem.* **1981**, *85*, 447.
95. Butler, M. A.; Nasby, R. D.; Quinn, R. K. *Solid State Comm.* **1976**, *19*, 1011.
96. Thampi, R. R.; Reddy, T. V.; Ramakrishnan, V.; Kuriacose, J. C. *J. Electroanal. Chem.* **1983**, *157*, 381.
97. Kumar, A.; Santangelo, P. G.; Lewis, N. S. *J. Phys. Chem.* **1992**, *96*, 834.
98. Matsumura, M.; Hiramoto, M.; Tsubomura, H. *J. Electrochem. Soc.* **1983**, *130*, 326.

99. Singh, R.; Hamdani, M.; Koenig, J. F.; Poillerat, G.; Gautier, J.-L.; Chartier, P. *J. Appl. Electrochem.* **1990,** *20,* 442.
100. Trasatti, S. *Croat. Chem. Acta* **1990,** *63,* 313.
101. Conway, B. E.; Liu, T. C. *Mater. Chem. Phys.* **1989,** *22,* 163.
102. Grzegorzewski, A.; Heusler, K. E. *J. Electroanal. Chem.* **1987,** *228,* 455.
103. Hardee, K. L.; Bard, A. J. *J. Electrochem. Soc.* **1977,** *124,* 215.
104. Tongchang, L.; Conway, B. E. *J. Appl. Electrochem.* **1987,** *17,* 983.
105. Williams, R. *J. Vac. Sci. Technol.* **1976,** *13,* 12.
106. Maruska, H. P.; Ghosh, A. K. *Solar Energy* **1978,** *20,* 443.
107. Schultze, J. W. *Dechema-Monographien* **1985,** *98,* 417.
108. Wolkenstein, T. "Physico-Chimie de la Surface des Semi-conducteurs." MIR, Moscow, 1977, p. 215.
109. Inoue, T.; Fujishira, A.; Honda, K. *Bull. Chem. Soc. Jpn.* **1979,** *52,* 3217.
110. Betz, G.; Tributsch, H.; Marchand, R. *Ber. Bunsenges. Phys. Chem.* **1984,** *14,* 315.
111. Bandi, A. *J. Electrochem. Soc.* **1990,** *137,* 1990.
112. Sakata, T.; Hashimoto, K.; Kawai, T. *J. Phys. Chem.* **1984,** *88,* 5214.
113. Alonso-Vante, N.; Tributsch, H. *Nature (London)* **1986,** *323,* 431.
114. Alonso-Vante, N.; Schubert, B.; Tributsch, H.; Perrin, A. *J. Catal.* **1988,** *112,* 384.
115. Alonso-Vante, N.; Schubert, B.; Tributsch, H. *Mater. Chem. Phys.* **1989,** *22,* 281.
116. Bungs, M.; Alonso-Vante, N.; Tributsch, H. *Ber. Bunsenges. Phys. Chem.* **1990,** *94,* 521.
117. Fischer, Ø. *Appl. Phys.* **1978,** *16,* 1.
118. Chevrel, R.; Sergent, M. In *Topics in Current Physics;* Fischer, Ø.; Maple, M. B., eds.; Springer, New York, 1982; Vol. 32, p. 25.
119. Yvon, K. In Ref. 118, p. 87.
120. Appelby, A. J.; Foulkes, F. R. "Fuel Cell Handbook." Van Nostrand Reinhold, New York, 1990.
121. Alonso-Vante, N.; Tributsch, H. *J. Electrochem. Soc.* **1991,** *138,* 639.
122. Solorza-Feria, O.; Ellmer, K.; Giersig, M.; Alonso-Vante, N. *Electrochim. Acta.* in press.
123. Hsueh, K.-L.; Chin, K. L.; Srinivasan, S. *J. Electroanal. Chem.* **1983,** *153,* 79.
124. Hawecker, J.; Lehn, J.-M.; Ziessel, R.; *J. Chem. Soc., Chem. Comm.* **1984,** 728.
125. Bronger, W.; Spangenberg, M. *J. Less-Common Met.* **1980,** *76,* 73; Bronger, W.; Miessen, H.-J. *J. Less-Common Met.* **1982,** *83,* 29.
126. Pilet, J. C.; Le Traon, F.; Perrin, C.; Perrin, A.; Leduc, L.; Sergent, M. *Surf. Sci.* **1985,** *156,* 359.
127. Le Nagard, N.; Perrin, A.; Sergent, M.; Lévy-Clement, C. *Mater. Res. Bull.* **1985,** *20,* 835.
128. Fischer, C.; Alonso-Vante, N.; Fiechter, S.; Tributsch, H.; Reck, G.; Schultz, W. *J. Alloys Comp.* **1992,** *178,* 305.
129. Fischer, C.; Fiechter, S.; Tributsch, H.; Reck, G.; Schultz, W. *Ber. Bunsenges. Phys. Chem.* **1992,** *96,* 1652.
130. Fischer, C. et al. in preparation.

131. Alonso-Vante, N.; Tributsch, H.; Gruzdkov, Yu. A.; Nikitenko, S. G.; Kochubey, D. I.; Parmon, V. N. in preparation.
132. Nakato, Y.; Ohnishi, T.; Tsubomura, H. *Chem. Lett.* **1975**, 883.
133. Nakato, Y.; Tsubomura, H. *Ber. Bunsenges. Phys. Chem.* **1976**, 80, 1002.
134. Morisaki, H.; Ono, H.; Yasawa, K. *Solar Energy Mater.* **1986**, *14*, 13.
135. Nakato, Y.; Tonomura, S.; Tsubomura, H. *Ber. Bunsenges. Phys. Chem.* **1976**, *80*, 1286.
136. Heller, A.; Vadimsky, R. G. *Phys. Rev. Lett.* **1982**, *46*, 1153.
137. Bockris, J. O'M. *Int. J. Hydrogen Energy* **1988**, *13*, 489.
138. Murray, R. W. In "Electroanalytical Chemistry," Bard, A. J., ed. Marcel Dekker, New York, 1983; Vol. 13, p. 191.
139. Gningue, D.; Horowitz, G.; Garnier, F. *J. Electrochem. Soc.* **1988**, *135*, 1695.
140. Rubin, H. D.; Arendt, D. J.; Humphrey, B. D.; Bocarsly, A. B. *J. Electrochem. Soc.* **1987**, *134*, 93.
141. Fox, M. A.; Nobs, F. J.; Voynick, T. A. *J. Am. Chem. Soc.* **1980**, *102*, 4029.
142. Hohman, J. R.; Fox, M. A. *J. Am. Chem. Soc.* **1982**, *104*, 401.
143. Boltz, J. M.; Wrighton, M. S. *J. Am. Chem. Soc.* **1978**, *100*, 5257.
144. Wrighton, M. S. In "Chemistry Modified Surfaces in Catalysis and Electrocatalysis," Miller, J. S., ed. American Chemical Society, Washington, D.C., 1982, Vol. 192, p. 99.
145. Vrachnou, E.; Grätzel, M.; McEvoy, A. *J. Electroanal. Chem.* **1989**, *258*, 193.
146. Ryan, M. A.; Spitler, M. T. *Langmuir* **1988**, *4*, 861.
147. Tsubomura, H.; Nakato, Y. *Nouv. J. Chim.* **1987**, *11*, 167.
148. Nakato, Y.; Tsubomura, H. *Electrochim. Acta* **1992**, *37*, 897.
149. Heller, A.; Aharon-Shalom, E.; Bonner, W. A.; Miller, B. *J. Am. Chem. Soc.* **1982**, *104*, 6942.
150. Kautek, W.; Gobrecht, J.; Gerischer, H. *Ber. Bunsenges. Phys. Chem.* **1980**, *84*, 1034.
151. Schefold, J.; Kühne, H.-M. *J. Electroanal. Chem.* **1991**, *300*, 211.
152. Colell, H.; Alonso-Vante, N. *Electrochim. Acta*, **1993**, *38*, 1929.
153. Fox, M. A. *Nouv. J. Chim.* **1987**, *11*, 129.
154. Sakata, T. In ref. 25, p. 311.
155. Nakato, Y.; Tsubomura, H. *J. Photochem.* **1985**, *29*, 257.
156. Keita, B.; Nadjo, L. *J. Electroanal. Chem.* **1986**, *199*, 229.
157. Alonso-Vante, N.; Tributsch, H. *J. Electroanal. Chem.* **1987**, *229*, 223.
158. Kühne, H.-M.; Schefold, J. *J. Electrochem. Soc.* **1990**, *137*, 568.
159. Szklarczyk, M.; Bockris, J. O'M. *J. Phys. Chem.* **1984**, *88*, 5241.
160. Baglio, J.; Calabrese, G. S.; Harrison, D. J.; Kamieniecki, E.; Ricco, A. J.; Wrighton, M. S.; Zoski, G. D. *J. Am. Chem. Soc.* **1983**, *105*, 2246.
161. Bard, A. J.; Faulkner, L. R. "Electrochemical Methods: Fundamentals and Applications," John Wiley, New York, 1980.
162. Ileperuma, O. A.; Weerasinghe, F. N. S.; Bandara, T. S. L. *Solar Energy Mater.* **1989**, *19*, 409.

163. Richardson, P.; Ang, P.; Sammells, A. *Adv. Hydrogen Energy* **1982**, *2*, 805.

164. Guruswamy, V.; Murphy, O. J.; Young, V.; Hildreth, G.; Bockris, J. O'M. *Solar Energy Mater.* **1981**, *6*, 59.

165. Rauh, R. D.; Alkaitis, S. A.; Buzby, J. M., Schiff, *Sci. Tech. Aerosp. Rep.* (NASA-CR-163586) **1980**, *18*,

166. Heindl, R.; Parsons, R.; Redon, A. M.; Tributsch, H.; Vigneron, J. *Surf. Sci.* **1982**, *115*, 91.

167. Tributsch, H.; Gorochov, U. *Electrochim. Acta* **1982**, *27*, 215.

168. Kühne, H.-M.; Tributsch, H. *J. Electroanal. Chem.* **1986**, *201*, 263.

169. Alonso-Vante, N.; Colell, H.; Tributsch, H. *J. Phys. Chem.* **1993**, *97*, 8261.

170. Bradley, M. G.; Tysak, T.; Graves, D. J.; Vlachopoulos, N. A. *J. Chem. Soc., Chem. Comm.* **1983**, 349.

171. Petit, J.-P.; Chartier, P.; Beley, M.; Sauvage, J.-P. *New J. Chem.* **1987**, *11*, 751.

172. Zafrir, M.; Ulman, M.; Zuckerman, Y.; Halmann, M. *J. Electroanal. Chem.* **1983**, *159*, 373.

173. Parkinson, B. A.; Weaver, P. F. *Nature (London)* **1984**, *309*, 148.

174. Cabrera, C. R.; Abruña, H. D. *J. Electroanal. Chem.* **1986**, *209*, 101.

175. Halmann, M. *Nature (London)* **1978**, *275*, 115.

176. Bockris, J. O'M.; Wass, J. C. *J. Electrochem. Soc.* **1989**, *136*, 2521.

177. Taniguchi, I. In "Modern Aspects of Electrochemistry," Bockris, J. O'M; White, R. E.; Conway, B., eds. Plenum, New York, 1989, Vol. 20, p. 327.

178. Bockris, J. O'M.; Wass, J. C. *Mater. Chem. Phys.* **1989**, *22*, 249.

179. (a) Bittins-Cattaneo, B.; Cattaneo, E.; Vielstich, W. In "Electroanalytical Chemistry," Bard, A. J., ed. Marcel Dekker, New York, 1991, Vol 17, p. 181.; (b) Bogdanoff, P.; Alonso-Vante, N. Ber. Bunsenges. Phys. Chem. **1993**, *97*, 940.

180. Shilov, A. E. In "Energy Resources Through Photochemistry and Catalysis," Grätzel, M., ed. Academic, New York, 1983, p. 535.

181. Pickett, C. J.; Talarmin, J. *Nature (London)* **1985**, *317*, 652.

182. Toray Industries, Inc., Jpn. Kokai Tokkyo Koho, **1981**, 4 pp.

183. Schrauzer, G. N.; Guth, T. D. *J. Am. Chem. Soc.* **1977**, *99*, 7189.

184. Endoh, E.; Bard, A. J. *Nouv. J. Chim.* **1987**, *11*, 217.

185. Tennakone, K.; Fernando, C. A. N.; Wickramanaye, S.; Damayanthi, M. W. P.; Silva, L. H. K.; Wijeratne, W.; Ileperuma, O. A.; Punchihewa, *Solar Energy Mater.* **1988**, *17*, 47.

186. Tennakone, K.; Ileperuma, O. A.; Bandara, J. M. S.; Thaminimulla, C. T. K. *Sol. Energy Mater.* **1991**, *22*, 319.

187. Dickson, C. R.; Nozik, A. J. *J. Am. Chem. Soc.* **1978**, *100*, 8007.

188. Handschuh, M.; Lorenz, W.; Aegerter, C.; Katterle, T. *J. Electroanal. Chem.* **1983**, *144*, 99.

189. Sammells, A. F.; Ang, P. G. P. *J. Electrochem. Soc.* **1979**, *126*, 1831.

190. Danzfuss, B.; Stimming, U. *J. Electroanal. Chem.* **1984**, *164*, 89.

191. Wrighton, M. S.; Ellis, A. B.; Wolczanski, P. T.; Morse, D. L.; Abrahamson, H. B.; Ginley, D. S. *J. Am. Chem. Soc.* **1976**, *98*, 2774.

192. Domen, K.; Kudo, A.; Onishi, T.; Kosugi, N.; Kuroda, H. *J. Phys. Chem.* **1986**, *90*, 292.
193. Schleich, D. M.; Derrington, C.; Godek, W.; Weisberg, D.; Wold, A. *Mater. Res. Bull.* **1977**, *12*, 321.
194. Sato, S.; Kadowaki, T. *Denki Kagaku Oyobi Kogyo Butsuri Kagaku* **1989**, *57*, 1151.
195. Fujishima, A.; Honda, K. *Nature (London)* **1972**, *238*, 38.
196. Salvador, P. *J. Electrochem. Soc.* **1981**, *128*, 1895.
197. Yoko, T.; Yuasa, A.; Kamiya, K.; Sakka, S. *J. Electrochem. Soc.* **1991**, *138*, 2279.
198. Wrighton, M. S.; Morse, D. L.; Ellis, A. B.; Ginley, D. S.; Abrahamson, H. A. *J. Am. Chem. Soc.* **1976**, *98*, 44.
199. Schumacher, L. C.; Holzhueter, I. B.; Hill, I. R.; Dignam, M. J. *Electrochim. Acta* **1990**, *35*, 975.
200. Kobussen, A. G. C.; Van Wees, H. J. A. *Mater. Sci. Monogr.* (Energy Ceram.) **1980**, *6*, 1019.
201. Ohkawa, K.; Itoh, K.; Fujishima, A. *Bull. Chem. Soc. Jpn.* **1990**, *63*, 1287.
202. Gningue, D.; Horowitz, G.; Roncali, J.; Garnier, F. *J. Electroanal. Chem.* **1989**, *269*, 337.
203. Taqui, M. M.; Rao, N. N.; Chatterjee, D. *J. Photochem. Photobiol. A: Chem.* **1991**, *60*, 319.
204. Vlachopoulos, N., A. *Diss. Abstr. Int. B* **1985**, *45*, 2248.
205. Beley, M.; Collin, J.-P.; Sauvage, J.-P.; Petit, J.-P.; Chartier, P. *J. Electroanal. Chem.* **1986**, *206*, 333.
206. Büker, K.; Colell, H.; Alonso Vante, N.; Tributsch, H. Eighth International Conference on Photochemical Conversion and Storage of Solar Energy (I.P.S.-8) (Palermo) **1990**, p. 149.
207. Ennaoui, A.; Tributsch, H. *J. Electroanal. Chem.* **1986**, *204*, 185.
208. Ennaoui, A.; Fiechter, S.; Chatzitheodorou, G.; Tributsch, H. Sixth International Conference on Photochemical Conversion and Storage of Solar Energy (I.P.S.-6) (Paris) **1986**, p. D117.
209. Li, X.-P.; Alonso-Vante, N.; Tributsch, H. *J. Electroanal. Chem.* **1988**, *242*, 255.
210. Alonso-Vante, N.; Tributsch, H. *Croat. Chem. Acta* **1990**, *63*, 417.
211. Ennaoui, A.; Fiechter, S.; Pettenkofer, Ch.; Alonso-Vante, N.; Büker, K.; Bronold, M.; Höpfner, Ch.; Tributsch, H. *Solar Energy Mater. Solar Cells*, **1993**, *29*, 289.
212. Wilson, J. A.; Yoffe, A. D. *Adv. Phys.* **1969**, *18*, 193.
213. "Photoelectrochemistry and Photovoltaics of Layered Semiconductors," Aruchamy, A., ed. Kluwer, Dordrecht, 1992.
214. Hofmann, W. K.; Könenkamp, R.; Schwarzlose, Th.; Kunst, M.; Tributsch, H. *Ber. Bunsenges. Phys. Chem.* **1986**, *90*, 824.
215. Chatzitheodorou, G.; Fiechter, S.; Könenkamp, R.; Kunst, M.; Jaegermann, W.; Tributsch, H. *Mater. Res. Bull.* **1986**, *21*, 1481.
216. (a) Koma, A.; Sounouchi, K.; Miyajima, T. *J. Vac. Sci. Technol. B* **1985**, *3*, 724; (b) Jaegermann, W.; in Ref. 213.
217. Tributsch, H. Ph.D. thesis, Technical University, Munich, 1968.
218. Watanabe, T.; Miyasaka, T.; Fujishima, A.; Honda, K. *Chem. Lett. Jpn.* **1978**, 433.
219. Tributsch, H. *Photochem. Photobiol.* **1972**, *16*, 261.
220. Memming, R.; Schröppel, F. *Chem. Phys. Lett.* **1979**, *62*, 207.
221. Alonso-Vante, N.; Chartier, P.; Ern, V. *Rev. Phys. Appl.* **1981**, *16*, 5.

222. Alonso-Vante, N.; Ern, V.; Chartier, P.; Dietrich-Buchecker, C. O.; McMillin, D.; Marnot, P.; Sauvage, J.-P. *Nouv. J. Chim.* **1983**, *7*, 3.
223. Matsumura, M.; Nomura, Y.; Tsubomura, H. *Bull. Chem. Soc. Jpn.* **1977**, *50*, 2533.
224. O'Regan, B.; Moser, J.; Anderson, M.; Grätzel, M. *J. Phys. Chem.* **1990**, *94*, 8720.
225. O'Regan, B.; Grätzel, M. *Nature (London)* **1991**, *353*, 737.
226. Fischer, A. *Bild Wissenschaft*, **1992**, *7*, 30.
227. Li, X.-P.; Tributsch, H. *Photochem. Photobiol.* **1989**, *50*, 531.
228. Ennaoui, A.; Fiechter, S.; Tributsch, H.; Giersig, M.; Vogel, R.; Weller, H. *J. Electrochem. Soc.* **1992**, *139*, 2514.
229. Vogel, R.; Pohl, K.; Weller, H. *Chem. Phys. Lett.* **1990**, *174*, 241.
230. Tributsch, H., *J. Electrochem. Soc.* **1981**, *128*, 1261.
231. Wheeler, B. L.; Leland, J. K.; Bard, A. J. *J. Electrochem. Soc.* **1986**, *133*, 358.
232. Gerischer, H.; Tributsch, H. *Ber. Bunsenges. Phys. Chem.* **1968**, *72*, 437.
233. Nakao, M.; Itoh, K.; Watanabe, T. *Ber. Bunsenges. Phys. Chem.* **1985**, *89*, 134.
234. Tributsch, H.; Calvin, M. *Photochem. Photobiol.* **1971**, *14*, 95.
235. Tsubomura, H.; Matsumura, M.; Nomura, Y.; Amamiya, T. *Nature (London)* **1976**, *261*, 402.
236. Kamat, P. V.; Chauvet, J.-P.; Fessenden, R. W. *J. Phys. Chem.* **1986**, *90*, 1389.
237. Spitler, M. T.; Calvin, M. *J. Chem. Phys.* **1977**, *67*, 5193.
238. Hickman, J.; Wessel, S.; Mackintosh, A.; Colbow, K. *Semicond. Sci. Technol.* **1987**, *2*, 207.
239. Itoh, K.; Nakao, M.; Honda, K. *Chem. Phys. Lett.* **1984**, *111*, 492.
240. Akimov, I. A. *Elektrokhimiya* **1984**, *20*, 1399.
241. Watanabe, T.; Takizawa, T.; Honda, K. *Ber. Bunsenges. Phys. Chem.* **1981**, *85*, 430.
242. Parkinson, B. A. *Langmuir* **1988**, *4*, 967.
243. Spitler, M. T.; Parkinson, B. *Langmuir* **1986**, *2*, 549.
244. Memming, R.; Tributsch, H. *J. Phys. Chem.* **1971**, *75*, 562.
245. Henglein, A. *Chem. Rev.* **1989**, *89*, 1861.
246. Wang, Y.; Herron, N. *J. Phys. Chem.* **1991**, *95*, 525.
247. Enea, O.; Moser, J.; Grätzel, M. *J. Electroanal. Chem.* **1989**, *529*, 59.
248. Tributsch, H. *J. Electroanal. Chem.* **1992**, *331*, 783.
249. Campbell, P.; Green, M. A. *J. Appl. Phys.* **1987**, *62*, 243.
250. Anderson, Å.; Hunderi, O.; Granqvist, C. G. *J. Appl. Phys.* **1980**, *51*, 754.
251. Welfort, W. T.; Winston, R. "High Collection Non-imaging Optics." Academic, New York, 1989.
252. Steeb, H. Lecture presented at the 3rd Seminar "Forschungsverbund Sonnenenrgie," Berlin-Adlershof, Germany, 1992.
253. Tributsch, H.; Pohlmann, L. *Chem. Phys. Lett.* **1992**, *188*, 338.
254. Nicolis, N.; Prigogine, I. "Self-organization in Nonequilibrium Systems." John Wiley, New York, 1977.

CHAPTER

2

Polymeric Materials for Lithium Batteries

M. Armand, J. Y. Sanchez, M. Gauthier, Y. Choquette

Sodium?

Yes, sir. Mixed with mercury, it forms an amalgam which replaces zinc in Bunsen cells. The mercury lasts forever. Only sodium is consumed and the ocean provides it. Furthermore, sodium batteries are the most energetic, and their electromotive force is double that of zinc cells.

Jules Verne, Twenty Thousand Leagues Under the Sea *(Hetzel Publisher, Paris 1870)*

2.1 Introduction

Jules Verne's nineteenth-century state-of-the-art Nautilus roamed under the seas, powered by high-energy-density batteries. It was not until 1871 that the dynamo was invented by Gramme. The fundamental laws of electricity had already been established using batteries as the source of current. More than a century later, in an energy-hungry world, the electrochemical energy of sodium has not been harnessed and developmental sodium–sulfur batteries are still only possible contenders for use in the long-sought electric vehicles. Lithium-based electrochemical cells are more promising in the short term.

To increase the energy density of electrochemical storage devices, the obvious choice is to use couples of light elements having large differences in their electronegativity. However, attempts to utilize halogens or chalcogens with alkali metals have come up against crucial technical difficulties for rechargeable systems due to various undesirable properties (gaseous state, corrosiveness, lack of conductivity, requirement for a catalyst to activate the molecule, etc.) of either the starting materials or their reaction products. The sodium–sulfur batteries may be an exception, taking advantage of the unique properties of β-aluminas as Na^+-conducting electrolytes and relatively low-melting sodium polysulfides. Yet very little flexibility in the design and materials is offered, especially concerning the operating temperature ($\geq 280°C$).

The advent of intercalation electrode materials in the early 1970s [1,2] has markedly changed the orientation and outlook of research in the field of energy storage. Although these compounds were known from the early stages of chemistry (tungsten bronzes, graphite salts) [3,4], their importance in solid-state redox reactions in general and their advantages in electrochemistry had mostly been overlooked. Indeed, one of the most rugged electrodes, the nickel oxide–hydroxide is now recognized as a proton intercalation system. Proton intercalation in MnO_2 is another example of practical importance. Today the vast majority of candidate materials for the positive electrode, at the research or development stage, are intercalation materials or at least function via a transient intercalation intermediate. Another chapter of this book is dedicated to recent considerations in the field of intercalation compounds. The use of macromolecular materials as active components in electrodes is a possibility and will be addressed here.

The concept of an all-solid-state battery is very appealing, in terms of safety and flexibility. However, associating conventional solid electrolytes—which until recently were restricted to either glasses or ceramics—with solid electrode materials raises major technical difficulties. Unlike electronic devices, battery operation implies mass transport and variations of electrode volumes. Keeping contact with solid electrolytes for many cycles presents a technological challenge and no general solution has been offered.

Polymer electrolytes were proposed in 1979 [5] in an attempt to combine the advantages of solid-state electrochemistry with the ease of processing inherent in plastic materials. At the same time, the recently discovered electronically conducting polymers were reported to function as electrode materials, raising high expectations in the scientific community [6]. We address in this chapter the main issues, in terms of science and technology, behind the use of polymeric materials in batteries, with emphasis on lithium systems.

2.2 Principles and Requirements for Electrochemical Energy Storage

Basically, an electrode is the interface between two conductive media, the ionic (electrolyte) and the electronic conductor (current collector), where a chemical potential μ of a given species is imposed. An electrochemical cell is represented by two electrodes joined by a common electrolyte. These can be termed the *source* ($\ominus$) and the *sink* ($\oplus$), in the thermodynamic sense for electropositive species, as schematized in Fig. 2.1.

In such a case the electrolyte serves only as a transducer of ions (M^+) and the passage of current does not change its average composition. Consequently, electrolytes with a transport number of unity for the ions exchanged at the electrode can be used ($t_+ = 1$, $t_- = 0$ where the species exchanged is a cation, as in Fig. 2.1).

On the other hand, if the electrode exchanges with electrolyte species that are chemically different, the latter need to serve as a chemical reservoir as well as an

POLYMERIC MATERIALS FOR LITHIUM BATTERIES

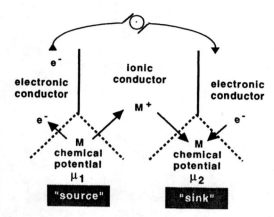

Figure 2.1. Schematic diagram of a battery in which the electrodes serve as source and sink for a common chemical species (M).

ionic transducer. This situation is depicted in Fig. 2.2, as discussed later, where the two electrodes exchange complementary ions, increasing the salt concentration on discharge and depleting it during the charge process (Fig. 2.2).

In many situations the electrolyte is used as a reservoir for ions or is chemically involved in the electrode's reactions balance. Using the electrolyte as a reservoir for either redox species or ions always corresponds to a major weight penalty, as the allowable concentration variations are limited. This is even more restrictive when acceptable conductivity levels have to be kept. The conventional lead–acid battery

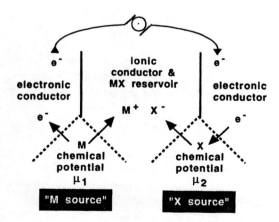

Figure 2.2. Schematic diagram of a battery in which the electrodes exchange complementary ionic species with the electrolyte (M^+, X^-).

illustrates this example as the electrolyte serves as a source of sulfate ions SO_4^{2-} used in the electrode reaction.

Both type of battery operation are possible with intercalation compounds, which possess in bulk the three functions the electrode requires:

1. Electronic conductivity.
2. Ionic conductivity.
3. Chemical potential μ_M (μ_X) imposed at the interface.

Such conditions have to be maintained throughout the electrochemical process and battery operation. This defines a host structure, represented by $\langle H \rangle$, in which vacant ion and electron sites are available. The compound acts as a reservoir for the species M (X, respectively); it exchanges electrons with the current collector and ions with the electrolyte. The finite size of ions requires an open crystallographic structure, whereas electrons can be accommodated in empty atomic orbitals or in band levels, depending on the degree of orbital overlap in the lattice. Thus the possible electrode reactions can be chosen as

$$M^0 \Leftrightarrow M^+ + e^- \quad | \Leftarrow M^+ \Rightarrow | \quad xM^+ + xe^- + \langle H \rangle \Leftrightarrow \langle M_xHX \rangle \tag{2.1}$$

$$\tfrac{1}{2}X_2 + e^- \Leftrightarrow X^- \quad | \Leftarrow X^- \Rightarrow | \quad \langle H \rangle + xX^- \Leftrightarrow \langle HX_x \rangle + xe^- \tag{2.2}$$

If the exchanged species are different at the electrode, the scheme is then

$$M^0 \Leftrightarrow M^+ + e^- \quad | \Leftarrow M^+ X^- \Rightarrow | \quad xX^- + \langle H \rangle \Leftrightarrow \langle HX_x \rangle + xe^- \tag{2.3}$$

The intercalation reaction can in principle be written in a similar form for species able to lose or gain electrons (metal or nonmetals). In practice, as stated earlier, the use of nonmetal represented here as X_2 in the source electrode is unpractical for chemical reasons. On the other hand, the use at the source electrode of the metal M^0 (electrode of the first kind) supposes a chemically and mechanically reversible (100% volume change) dissolution-plating process. With any metal this conditon has proved difficult to meet, because of the tendency for dendrites to form leading to short circuits across the electrolyte. Possible explanations invoke electric field enhancement at local surfaces spurs, leading to catastrophic forward growth; also, the electron injection may actually take place within the electrolyte when the potential corresponding to the $M^+ + e^- \Leftrightarrow M^0$ equilibrium is reached *beneath* the electrolyte, the electron traveling by tunneling (overpotential η at the $\ominus$ electrode, followed by M^0 nucleation).

If a second intercalation compound is used in place of the metal, the problem of dendritic deposition is overcome, again as long as the potential never reaches that of the M^0/M^+ equilibrium. The electrochemical chains, with either M^+ or X^- and as mobile species:

$$\langle M_xH^1\rangle \Leftrightarrow yM^+ + ye^- + \langle M_{x-y}H^1\rangle \;\Big|\Leftarrow M^+ \Rightarrow\Big|\; yM^+ + ye^- + \langle M_zH^2\rangle$$
$$\Leftrightarrow \langle M_{z+y}H^2X\rangle \tag{2.4}$$
$$\langle HX_x\rangle \Leftrightarrow \rangle HX_{x-y}\rangle + xX^- + xe^- \;\Big|\Leftarrow X^- \Rightarrow\Big|\; \langle HX_z\rangle + xX^- + xe^- \Leftrightarrow \langle X_{z+y}H\rangle \tag{2.5}$$

Such systems, termed "rocking-chair" systems, are receiving renewed attention despite two immediate drawbacks: (1) lower energy density due to lower voltage and added equivalent mass; (2) wide operating voltage swings due to the added contributions of sloping voltages versus composition from the two electrodes [7]. These aspects are addressed in detail in another chapter of this book.

The potential of an intercalation compound is written as:

$$e = e° - RT\left[\ln\frac{y_i}{1-y_i} + \ln\frac{y_e}{1-y_e}\right] \tag{2.6}$$

where y_i and y_e represent the fraction of sites occupied by ions and electrons, respectively, in the solid structure. In most cases, $y_i = y_3 = x/x_{max}$; thus

$$e = e° - 2RT\left[\ln\frac{x/x_{max}}{1-x/x_{max}}\right] \tag{2.7}$$

These equations (2.6, 2.7) are the solid-state extensions of Nernst's law.

Both ions (size, coordination) and electrons (bandwidth, Jahn–Teller effect . . .) determine the range of nonstoichiometry $x_{min} < x < x_{max}$ in $\langle M_{x_1}H\rangle$. In this case, the emf versus the composition curves are formed by adjacent nonstoichiometric domains, often overlapping to show voltage plateaus. The utilization of the battery will strongly depend on the titration curve of the electrodes.

2.3 Polymers as Solid-State Ionizing Solvents

2.3.1 General Considerations

Any electrochemical system requires an electrolyte, that is, a medium in which ions can move but that is impervious to electrons within its stability window. Electrolytes are solution of salts either in an ionizing solvent or in the molten state. In solids, ions are usually immobile, with the spectacular exception of specific structures having galleries whose size and connectivity are adapted to ion motion; these include glasses or lamellar sodium aluminate (β-alumina).

Solution chemistry has established that there is a marked difference when ionogenic compounds (salts, acids) are dissolved in protic or aprotic solvents. In water, typically both cations and anions are solvated by, respectively, electron pairs and hydrogen bonds. This is expressed by the concept of donor numbers (DN) (electron pair donation from solvent to + charges) and acceptor numbers AN (propensity of the solvent to accept electron pairs from − charges), according to Gutmann [8]. The need to extend the electrochemical window beyond the reduction–oxidation domain of the protons has led to the use of molten salts at high temperatures (electrowinning of metals) and of *aprotic* organic solvents. In such cases, the main driving force for solvation is cation–electron pair interactions (high DN), whereas the anions are only poorly stabilized by dipolar orientation (low AN). In organic chemistry this property is currently used to enhance the basic–nucleophilic character of the negative charge ("naked anions"). Conversely, electrochemists make use of highly delocalized negative charges, as in covalent ClO_4^-, $CF_3SO_3^-$, or coordination anions such as BF_4^- or AsF_6^- to impart solubility to the corresponding salts. The relative indifference of these ions to their environment is expressed, for instance, by their free energy of transfer from water to aprotic solvents ($\Delta G[ClO_4^-]$ water/DMF = +0.4 kJ) [8]. It is noteworthy that species like BF_4^- or AsF_6^- are the result of an acid–base equilibrium:

$$AsF_6^- \Leftrightarrow AsF_5 + F^-$$

which is influenced by the chemical environment (electron-donating properties of the solvent) and in turn may influence its stability. Among alkali metal salts, the lithium derivatives are usually those for which the equilibrium is shifted to the right, because of the polarizing nature of Li^+ and the high lattice energy of LiF. Indeed, cascades of complex chemical reactions catalyzed by Lewis acids take place in organic liquid electrolytes [10] and the consequences for the behavior of batteries are often dramatic. The same applies to polymers, with possible depolymerization processes leading to loss of mechanical properties.

Usual organic anions such as carboxylates or nonperfluorinated sulfonates are "hard" bases in the sense given by Pearson [11] and require high AN solvents. They are thus poorly soluble and/or weakly dissociated in aprotic media; this also applies to most multiply charged anions. The same considerations apply to aprotic polymers.

More recently, new salts have been proposed, based on covalent bonds alone, with a highly delocalized charge [12,13] (see Fig. 2.3):

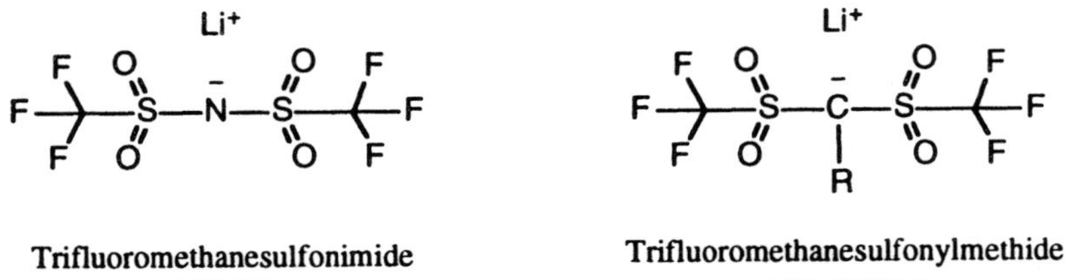

Trifluoromethanesulfonimide
LiTFSI

Trifluoromethanesulfonylmethide
LiTtriTFSM

Figure 2.3. Examples of anions with extensive charge delocalization; R = H, alkyl, acyl, $CF_3SO_2^-$.

the $CF_3SO_2^-$ group being the most electronegative known (inductive and resonant). These anions are thus quite resistant to oxidation; the anodic limit is found beyond +3.8 V versus $Li^+:Li^0$ [13,14]. In addition, the presence of a "soft" center (—N— or =C—) contributes to better dipolar solvation [15] and thus $Li(CF_3SO_2)_2N$ shows an extraordinary solubility in aprotic dipolar solvents. These salts are gaining increasing importance in liquid and polymer electrolytes.

2.3.2 Polymers as Solvating Media

Catenation (i.e., the possibility of associating atoms into long, chainlike molecules) is a distinct and spectacular property of carbon chemistry, though some limited examples are known based on other elements. In investigating how such primarily one-dimensional objects arrange themselves to fill a three-dimensional volume, the first consideration is the conformational flexibility of the macromolecule. Although "rigid rod" polymers expectedly offer relatively limited conformational choices, flexible bonds in the chains allow an unlimited number of spatial arrangements. The time scale for these rearrangements in bulk will be governed by the local energy–entropy considerations, expressed as a function of a temperature T_g (glass transition temperature) above which segmental motion involving $\approx 10^2$ atoms takes place. Solvent properties require chain flexibility to accommodate solute molecules and/or ions and are to be found in the class of flexible-backbone polymers.

In polymers the need to avoid hydrogen bonds is double: first, for the electrochemical reasons invoked earlier, a labile proton limiting the electrochemical stability; second, because the strong ≈ 15 kJ and directional H-bonds are most likely to form interchain cross links, appreciably stiffening the backbone. The consequence, as discussed in Section 2.3.4, is a reduction of the segmental *and* solute mobilities, parameters whose importance is basic to conductivity.

As expected, incorporation of a high concentration of electron-donating groups in flexible polymers leads to complex formations with salts, observed in solution or in the solid state. The study of polymer–salt complexes parallels that of host–guest chemistry whose development has been a major thrust recently [16]. Here the oxyethylene sequence forming an entropically favorable five-membered ring has been found to provide good solvation for most cations (Fig. 2.4):

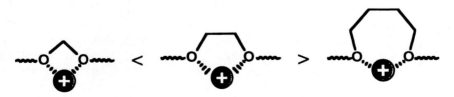

Figure 2.4. Chelate rings formed from cation–polyether dipolar interaction.

As a result of the preceding consideration, polyethers or polyamines, and to a lesser extent, thioethers containing the carbon–carbon–heteroatom sequence easily form complexes with most metal salts MX where the anion X^- fulfills the preceding criterion (i.e., a low solvation energy). Most host polymers studied today are based on (Fig. 2.5):

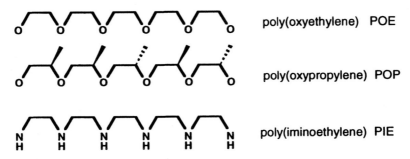

Figure 2.5. Principal solvating units used in polymer electrolytes.

2.3.3 Lithium Polymer Electrolytes

The interaction between solvating polymers and salts has mostly been studied using lithium derivatives. The pivotal role of lithium in intercalation chemistry, implying battery applications, has been one motivation. The Li cation differs markedly from its heavier homologues (Na, K, Cs), with higher solvation energies in various donor solvents. For instance, the gain resulting from Li^+/PEO interactions can overcome the lattice energy of LiCl, though Cl^- itself is considered to require high solvation energy; this is illustrated by the large free energy for transfer of chloride from water to propylene carbonate ($+38$ kJ [9]). NaCl shows no evidence of solubility ($<10^{-3}$ M) in aprotic solvents and polymers.

2.3.4 Phase Diagrams and Conductivity

The lower conductivities of polymer electrolytes, as compared with liquids, has been put forward as the main obstacle to their use in batteries. It is thus necessary to understand the mechanism of ion motion, the present achievements, and the prospects for improvement.

The two polyethers POE and PIE, having highly symmetrical repeat units, tend to crystallize. POEs with M_2 beyond $\approx 2.10^4$ reach the plateau melting point of 65°C. The behavior of the two polyethers, POE and PPO, is thus fundamentally different, as PPO, made by a nonstereoregular polymerization process, can be described as containing randomly distributed CH_3— groups, preventing short- or long-range order. PPO has thus served as a model for conductivity since it was recognized that conduction and diffusion processes take place in the amorphous phase only. However, the solvating power of the PO unit is weaker than that of OE, resulting, in particular, in a "ceiling" temperature (60–170°C) for PPO adducts, depending upon the dissolved salt, above which the solute precipitates out (i.e., the entropy of mixing the salt and the polymer is negative since complexation reduces the configurational freedom). PEO complexes do not show this trend but tend to form stoichiometric complexes, with 1:1, 3:1, and 6:1 EO units per cation. None of these compositions correspond to appreciable conductivities, especially the more concentrated ones. No complete structural data are available for these complexes, as lithium nuclei are too light to scatter X-rays effectively and they are very hygroscopic. These complexes seem, however, similar to their sodium homologues, whose structure consists of a helicoidal wrapping of the POE chains around the cation [17,18]. It is now customary to use phase diagrams to determine the domain of existence of the conducting amorphous phase as a function of content and temperature. For compositions between pure PEO and the first complex, or between successive complexes, a eutectic forms that has a lower melting point, above which the polymer becomes fully amorphous. In Fig. 2.6, the diagrams for the most interesting lithium salts with PEO are represented [19]. The existence of a 6:1 complex is to be taken as favorable to lower eutectic temperatures and a wider domain of amorphicity and, also, to better conductivities. Symmetry of the anion seems to play a role, as both ClO_4^- and $(CF_2SO_2)_2N^-$ form the 6:1 complex, whereas $CF_3SO_3^-$ does not.

It should be stressed that when dealing with polymers, considerable supercooling can be observed, the tendency increasing with molecular weight and chain entanglement, but also strongly influenced by the salt. Again, crystallization in PEO–$LiCF_3SO_3$ shows almost no delay, whereas perchlorates and thiocyanates can be kept supercooled for days. Thus conductivity data may sometimes be misleading unless experimental conditions are clearly specified.

The conductivities of selected examples of lithium salts–polymers complexes are shown in Fig. 2.7. Smooth conductivity curves are obtained for noncrystallizable polymers like the phosphazene comb polymer with short PEO solvating units attached (64). The PEO–$LiClO_4$ phase diagram is reflected for the two compositions O/Li = 8 and O/Li = 20 shown here.

The influence of the nature of the salt is apparent when using nitrogen or carbon anion-based salts, depicted in Fig. 2.3, which, because of to their extensive charge delocalization and low solvation requirement, show a more favorable phase diagram. These salts show the best reported conductivities in PEO systems, as shown in Fig. 2.8.

Since the exploration of polymer–salt complexes began, their ionic conductivity in the amorphous phase has been correlated to the segmental motion of the macromolecular strands above the glass transition temperature. Mainly, two expressions have been used to quantify this aspect, the Vogel–Tamman–Fulcher (VTF) [20] and the Williams–Landel–Ferry (WLF) [20]. These are shown in Table 2.1. Though very dissimilar at first glance, these two equations express the control of diffusion by the segmental motion of the chains above a critical temperature (T_0 and $T_{ref} - C_2$, respectively). Indeed, the values accepted for the universal number C_1 and C_2 (see Table 2.1) make the two approaches identical. However, T_0 is accessible only

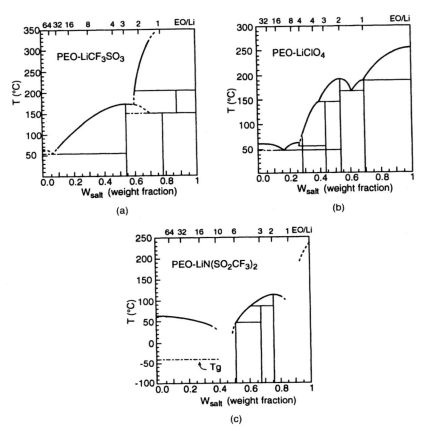

Figure 2.6. Phase diagrams for (a) the PEO–LiCF$_3$SO$_3$ system. The vertical boundary lines indicate the formation of 3:1, 1:1, and 0.5:1 crystalline compounds obtained by DSC; (b) the PEO–LiClO$_4$ system with 6:1, 3:1 and 2:1 and 1:1 crystalline compounds obtained by DSC and X-ray diffraction data; (c) the PEO–Li[(CF$_3$SO$_2$)$_2$N] system showing 6:1, 3:1, and 2:1 crystalline compounds obtained from DSC data. Mixtures EO/Li 8 to 10 are amorphous. (From Ref. 19.)

from data analysis and processing, whereas the reference temperature can be chosen as T_g, the glass transition value experimentally accessible through scanning calorimetry. It is currently acceptable to take T_0 either as $0.8T_g$ or $T_g - 25$ K.

VTF and WLF are phenomenological laws and do not correspond to classical notions of simple energies barriers and activated steps, despite the dimension of energy attributed to $T^{-1}B$. More descriptive models were later given by Doolittle [22] and Cohen and Turnbull [23]. The general picture is that a minimum free volume is needed for the motion of either the chain segments or the solute particles, which move in "cages" of neighboring atoms. The probability that this requirement will be met increases with T above T_0 according to the VTF and WLF laws. It is noteworthy that the temperature range over which the conductivity measurements are made does not allow us to infer or defer the $T^{1/2}$ term. Also, if the ion motion

POLYMERIC MATERIALS FOR LITHIUM BATTERIES

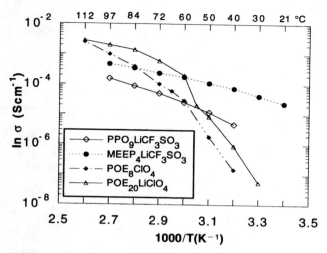

Figure 2.7. Arrhenius plots of conductivity versus temperature for PEO complexes with LiClO$_4$ for two compositions (O/Li = 8 V and O/Li = 20); poly(methoxyethoxyethoxyphosphazene)(MEEP)$_4$–LiCF$_3$SO$_3$ (monomer unit/Li = 4) from Ref. 64; PPO–LiCF$_3$SO$_3$ (O/Li = 9). Cooling cycles shown.

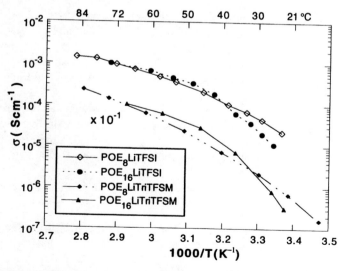

Figure 2.8. Arrhenius plots of conductivity versus temperature for PEO complexes with Li[(DF$_3$SO$_2$)$_2$N] (LiTFSI) for two compositions (O/Li = 8 and O/Li = 16) and Li[(CF$_3$SO$_2$)$_3$C] (TriTFSM) also for compositions O/Li = 8 and O/Li = 16. For readability, the curve is shifted by one decade. Cooling cycles shown.

Table 2.1 Comparison of the Origin and Expression of the Two Most Commonly Used Models (VTF and WLF) for Diffusion and Conductivity in Viscous Media and Their Correspondence

	Vogel–Tamman–Fulcher (VTF)	Williams–Landel–Ferry (WLF)
Expression	$\sigma = \sigma_0 \exp^{-}\left(\dfrac{B}{T - T_0}\right)$	$\sigma = \sigma_0 \exp \dfrac{-C_1(T - T_{ref})}{C_2 + T - T_{ref}}$
Typical values	$\sigma_0 = 0.4$ S cm^{-1} $B = 2.2 \times 10^{-3}$ K^{-1} $T_0 = 210$ K	$C_1 = 17.4$ $C_2 = 52$ K $T_{ref} = T_g = 240$ K
Derived from:	Variation of viscosity with T: $\eta \propto T^{1/2} \exp \dfrac{B}{T - T_0}$	Scaling factor a_T for relaxation: $\log(a_T) = \dfrac{-C_1(T - T_{ref})}{C_2 + T - T_{ref}}$
Valid relations needed	$D = \dfrac{kT}{6\pi r_1 \eta}$ Stokes–Einstein	$\sigma = \dfrac{ze^2}{kT}D$ Nernst–Einstein
Correspondence	$C_1 C_2 = B$	Neglect $T^{1/2}$ prefactor in σ_0 from VTF $C_2 = T_{ref} - T_0$

requires, as logically expected, a solvation–desolvation process, it is interesting to compare the size of an EO unit (60 Å^3), larger than that of any alkali ion (Li$^+$: 2.2 Å^3, K$^+$: 17 Å^3). The relative invariance of the B values may thus reflect the stepwise replacement of a coordinating oxygen from the EO units around the cation.

Addition of a salt always increases the glass transition temperature of the resulting complex. This is usually depicted in terms of "transient cross-links" following the formation of strong dipolar interactions. Indeed, this effect becomes dominant at high concentrations and completely offsets the advantage expected from a higher number of carriers when a relatively invariant salt dissociation is being considered.

Table 2.2 gives the following variation of T_g as measured by differential scanning calorimetry (DSC) for usual salts [19]:

Both salts based on the triflyl (CF$_3$SO$_2$—) group have the distinct advantage of a smaller increase of T_g for a given concentration. The impossibility of maintaining the triflate complexes in the amorphous state over a wide composition range, in contrast with the imide salt, highlights the advantage of using the latter as electrolyte.

Le Nest et al. [24], following an exhaustive study of urethane cross-linked

Table 2.2 Experimental Values of the Variation of Glass Transition Temperature with Concentration for the Most Studied Solute Salts (from ref. 19).

Salt	LiSCN	LiClO$_4$	LiCF$_3$SO$_3$	Li[(CF$_3$SO$_3$)$_2$N]
$\dfrac{\partial T_g}{\partial X_{salt}}$ (°C/mole%)	3.8	4.2	2.8	2.8

networks based on PEO of various chain lengths, have established the following relationship:

$$\frac{1}{T_g} = \frac{1}{T_g^\circ} - 2.7 \times 10^{-4} c_{salt} - 7.6 \times 10^{-4} c_{xlinks} \tag{2.8}$$

where T_g° is the glass transition temperature of salt-free network, c_{salt} and c_{xlinks} are the concentrations of salts and crosslinks respectively in mol dm^{-3}. This result implies that the effect on T_g of a cross-link is similar to that of three molecules of salt; a tempting interpretation is that the salt aggregates into clusters containing three salt units. Interestingly, theory predicts, when considering the dielectric constant of the solvent, such an order of aggregation [25].

More recently, the percolation theory [26] has been adapted to describe the situation in macromolecules above their T_g where lattice sites fluctuate with polymer motion [27]. Hopping probabilities renew their values on the time scale imposed by the polymer relaxation process. All these models apply only when the system is totally amorphous. In PEO systems, below the liquidus line imposed by the phase diagram, either crystalline PEO or stoichiometric complex coexists with the amorphous phase. The conductivity of these mixed phases is even more complex. However, the Arrhenius plots show an approximately straight line, reflecting the progressive dissolution of the crystallites in the amorphous phase [28].

2.3.5 Dissociation and Transport Numbers

The models (VTF and WLF) describe the motion of neutral particles in a macromolecular medium. When dealing with a salt, the question arises whether the dissociation of the salt and the activation energy for ionization have to be taken into consideration. Miyamoto and Shibayama [29] have modified the basic VTF law to include an Arrhenius term to account for both the dissociation energy of the salt and an activation:

$$\sigma_0 = \sigma_0 \exp\left[-\frac{\gamma v^*}{v_f} - \frac{\Delta E}{RT} \right] \tag{2.9}$$

where v_f = free volume per mole
v^* = critical volume per mole
γ = constant scaling factor allowing for free-volume overlap

The pseudoactivation term is derived from these parameters as

$$B = K\gamma v^* \bigg/ \left(\frac{\partial v_f}{\partial T} \right)_{T_0} \tag{2.10}$$

The activation energy ΔE was taken to represent two separate processes:

$$\Delta E = E_{hop} + \frac{W_{diss}}{2\epsilon} \tag{2.11}$$

where E_{hop} = energy barrier for ion transfer
W_{diss} = dissociation energy

The dissociation of salts dissolved in polymer electrolytes is still the subject of unresolved controversies. Simply taking into consideration the dielectric constant of polyethers ($\epsilon \approx 5$) leads to the conclusion that these systems are strongly associated at any concentration. Furthermore, aggregates of the type $[LiX_2]^-$ or $[Li_2X]^-$ or with a higher degree of condensation are expected to form. A thorough theoretical treatment has been given by Pettit and Bruckenstein [25] for liquid electrolytes. The increase of the molar conductance Λ with concentration ($10^{2-} - 1$ Ml^{-1}) has allowed MacCallum et al. [30] to estimate the concentation of free ions and triplets in the system LiCF$_3$SO$_3$/α,ω-dimethyl-PEO400. Writing the dissociation equilibria as

$$\alpha_1 = \frac{[LiX]}{[Li^+][X^-]} \qquad \alpha_2 = \frac{[LiX_2^-]}{[Li^+][X^-]^2} \cdots$$

implies a ΔG for the corresponding α_n, which in turn implies taking into account in the expression for the conductivity an Arrhenius exponential term containing the W_{diss}. In conventional liquids this energy is positive (i.e., $\alpha \Uparrow$ with $T \Uparrow$). An activation term of ≈ 20 kJ has been deduced in urethane-cross-linked PEO networks [31], using the Miyamoto and Shibayama model.

However, Raman spectroscopy applied to polyether–salt systems [32,33], including lithium derivatives, leads to opposite conclusions. The spectral signature of the CF$_3$SO$_3^-$ as free species or ion pairs shows a marked decrease of the dissociation with temperature from $\alpha \approx 75\%$ at 200 K to $\alpha \approx 0$ at 350 K. This result is in agreement with polymer thermodynamics where the entropy of mixing is usually positive. The polymer studied, PPO, loses its solvation properties above its ceiling temperature, and it is logical to suppose that ion association precedes salt nucleation. The rapid rise in conductivity with temperature is thus explained by a dramatic increase in mobility, compensating for the loss of carrier concentration. Along the same lines, Xu et al. predict [34] a universal decrease of α with temperature. PEO is, however, more solvating ($\Delta H \gg 0$, especially for Li$^+$); thus the ceiling temperature is expected to be much higher than that for PPO, and its crystallinity complicates similar studies.

In concentrated systems, at O/Li ≤ 12 (≥ 2 Mdm^{-3}), the situation seems different, as the effective dielectric constant of the system is markedly raised by the presence of the salt. The appellation "solvate melt" seems more appropriate here, as all particles are statistically close to each other, a favorable situation for the screening of coulombic interactions through quasi-spherical distribution. Of importance is the notion of the coordination number of the salt, that is, the number of solvating units present at the maximum concentration in the amorphous phase, which is surprisingly temperature invariant [35]. This number is itself the sum of the contributions of the cations (1 for Li$^+$, 5 for K$^+$) and the anions (2 for SCN$^-$, 6 for larger BΦ_4^-). The conductivities at constant $T - T_g$ are directly proportional to the cation coordination number, thus putting a penalty on lithium, which, because its number is smaller, gives lower performance than the heavier alkali metals.

Knowledge of the respective mobilities of the ionic species is equally important, since the operation of electrode materials corresponds to the ingress or egress of one type of ion. Determination of the transport numbers is also the source of unsettled questions. During the early development of polymer electrolytes, it soon became evident that both cationic and anionic species were mobile. This is in contrast with conventional solid electrolytes, such as β-alumina or alkali glasses, which show a transport number of unity for the cations. The possible situations are summarized in Table 2.3.

The existence of a transport number different from unity for the cationic charges results in the establishment of a salt concentration gradient Fig. 2.9, left, within the electrolyte, the total quantity remaining unchanged, when using an electrochemical cell where only lithium ions are exchanged at the electrodes (source and sink).

Lithium transfer is also possible even if anions are the sole charge carriers ($t_+ = 0$), on condition that neutral species LiX exist ($\alpha < 1$) *and* are mobile. The electrodes are then supplied with cation-generating species:

$$LiX_{interface} \Leftrightarrow Li^+_{inserted} + X^-_{electrolyte}$$

It is noteworthy that anionic aggregates such as $[LiX_2]^-$ or $[Li_2X]^+$ may serve the same purpose. In this case, if the transport number is zero, the transference number of the Li species is calculated as

$$T_{Li} = 1 - (t_{X^-} + 2t_{LiX_2^-} - 2t_{XLi_2^+}) \neq 0 \tag{2.12}$$

On the other hand, the use of an anion intercalation electrode against a Li^0 or Li intercalation electrode requires that both species be transported in the electrode, through direct (Li^+, X^-) or indirect (neutrals, triple ions). In addition, as discussed in Section 2.2, the average salt concentration in the electrolyte varies according to the quantity of current passed (reservoir-type electrolyte). This is shown in Fig. 2.9, left.

It is extremely difficult to measure the transport or transference numbers in polymer electrolytes, and the use of several techniques is necessary to reach valid conclusions. Table 2.4 summarizes the main reported results, the results being calculated for the apparent t_+ values.

Table 2.3 Predicted Salt Concentration Gradient Build-up in Electrolyte and Electrode Polarization for Different Salt Transference/Transport Numbers and Electrode Type Combinations.

Type of electrodes ⇒ electrolyte transport mechanism ⇓	⊖ Li source ⊕ Li sink (Li^+ intercalation)	⊖ Li source ⊕ Li sink (X^- intercalation)
$t^+ \neq 0$	electrolyte concentration gradient	concentration changes during operation
$t^+ = 0 \quad \begin{cases} D_{LiX} \neq 0 \\ D_{LiX} \neq 0 \end{cases}$	⊖ ⊕ electrode polarization concentration gradient	⊖ electrode polarization concentration gradient
$t^+ = 1$	no gradient	⊕ electrode polarization

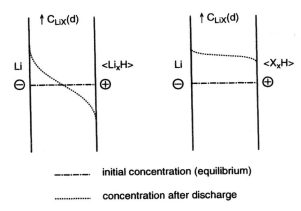

Figure 2.9. Schematic representation of salt concentration profile in electrolyte during battery operation. Left: source & sink type; right: electrolyte as salt reservoir.

It is generally agreed that the calculated transport numbers of Li$^+$ in salts like LiClO$_4$ as well as those of other alkali metals salts in the concentrated regime (O/Li $\leq$ 12) are close to 0.3. It is interesting to compare this value, obtained from both pulsed field gradient nmr, thus taking into account the ^{7}Li nuclei, irrespective of its charge, and various electrochemical techniques, accounting only for charged spe-

Table 2.4 Summary of Principal Studies and Techniques for the Measurements of Transport/Transfer Numbers of Lithium Species in Polymer Electrolytes. Sensitivity of the to Neutral (LiX) or Triplets ([LiX$_2$]$^-$, [Li$_2$X]$^+$ ...) Species is Given.

Technique	Sensitivity		Polymer Matrix	Concentration Range	Results (t_+)	Ref.
	[LiX]	[LiX$_2$]				
Radiotracer	Yes	Yes	PEO–Na*I*,	8	0.36	37
			PPO–Na*S*CN	8	0.38	38
PFG NMR	Yes	Yes	PEO–LiCF$_3$SO$_3$	20/1–6/1	0.4–05	39
			PEO–LiClO$_4$	20/1–8/1	0.3	40
			PEO–LiPF$_6$	100/1–20/1	0.0–0.3	41
Tubandt–Hittorf	No	Yes	PEO$_{XL}$–LiX^a	242/1–8/1	0.4–0.2	42
			PPO$_{Lw}$–LiCF$_3$SO$_3$		0.0	43
Concentration cells	No	Yes	PEO {LiClO$_4$	80/1–8/1	0.3	44
			PEO {LiCF$_3$SO$_3$	80/1–8/1	0.6	
DC polarization	No	Yes	PEO {LiClO$_4$	100/1–8/1	0.2–0.3^b	46
			PEO {LiCF$_3$SO$_3$	100/1–20/1	0.45–0.6^b	

^{a} Cross-linked PEO network; LiX = LiCf$_3$SO$_3$; LiClO$_4$.
^{b} I_+/I_0 cationic current fraction.

cies, which give the same result. This would indicate that undissociated ion pairs are immobile in polymer electrolytes, that is, that ionization of salt is a prerequisite for motion, as ion pairs reside in a stable coordination closed shell. This situation contrasts with that found in liquid electrolytes, where, although the transport numbers are the same ($t_+ \approx 0.35$), there is no restriction on the diffusion of uncharged species. The case of triflate salts appears quite special. The finding of $t_+ \approx 0.6$ is surprising and could suggest a preeminence of triplets $[Li(CF_3SO_3)_2]^-$ as charge carriers [44,46].

The recent results by Arumugam et al. [41] suggest that the diffusion of ^7Li-containing species becomes negligibly small compared to that of the anion-based ^{31}P in a dilute (O/Li $\geq$ 50) PEO–LiPF$_6$ system. The drop in either transport or transference number of the lithium species may precipitate the establishment of a salt-depleted zone close to the intercalation electrode during the reduction process. Before coming to conclusions about the generality of this behavior, attention should be given to data showing a t_+ relatively invariant with concentration [43] in PEO networks. However, the same group reports Li transport numbers close to zero in PEO–PPO block copolymer networks where phase separation takes place at the microscopic level. Cameron et al. also measured $T_+ \approx 0$ in low-molecular-weight copolymers of the same subunits [42]. The latter results should be considered together with the determination of the threshold $M_w \approx 3.10^3$, below which diffusion of the cations with the solvation shell (i.e., the net diffusion of the polymer strands) is faster than the solvation–desolvation process [46]. In this case, the quasi-totality of the electrolyte volume is carried with the cations, resulting in a moving boundary for the migration reference in Hittorf experiments. Beyond the threshold value, which also corresponds to the onset of chain entanglement, the net polymer diffusion falls to zero, resulting in "immobile solvent" behavior. Networks fall, of course, into this category independently of the molecular weight of the segments comprised between two cross-link knots.

Several attempts have been made to design and synthesize polymer electrolytes with transport numbers of unity for cations. These materials would represent the ideal case for lithium intercalation electrodes, as no concentration gradient appears during operation. The concept may be viewed as combining the principle of a solvating polymer with that of an ion exchange resin. Practical realization has, however, been hampered by two obstacles:

- The requirements that the anionic charges perform well in solvating polymers (i.e., charge delocalization and large overall electronegativity) is difficult to reconcile with a chemical tether destined to attach the molecule to the polymer backbone and immobilize it. The result is low conductivity due to low dissociation or microphase separation due to the incompatibility between the salt phase and the polyether segments.
- Careful removal of all foreign ionic contaminants is necessary to ensure that the conductivity is solely due to the cations. Ionic impurities are likely to be introduced during synthesis.

Earlier attempts to graft simple alkyl sulfonates, carboxylates, perfluorocarboxylates, and phenolates to a polyether backbone have been disappointing, yielding low conductivities [47,49]. The low dissociation or even incompatibility of the corresponding salts and the polyethers is responsible for this situation and evidenced by microphase separation. Again, the situation is less favorable for Li salts than for heavier alkali metals. More recently, Sylla et a. [50] have prepared PEO networks with attached perfluorosulfonate groups of the following type:

The $-I$ inductive effect of the carbonyl and fluorine on the $-SO_3^-$ results in improved conductivies over previous attempts. Conductivities $\geq 10^{-5}$ S cm^{-1} were obtained at 55°C. Careful removal of ionic impurities by washing was made easier by considering the cross-linked nature of the polymer–salt complex, which gave evidence for ^{7}Li as the only mobile species when checked by nmr. Interestingly, the conductivity of the salt after attachment to the polyether was ≈0.2–0.3 times that of the free salt at similar concentrations. This may indicate that the suppression of the anionic contribution leaves intact the σ_+ value ($\sigma_{total} \times t_+$). Also, no transport of ions via pairs or triplets is possible when the salt is immobilized.

2.3.6 Solvating Polymer Architecture

The ionic conductivity, as discussed earlier, is confined to the amorphous phase. On the other hand, among the polyethers, PEO has the best solvating capabilities and the highest conductivities when in the amorphous phase. Avoiding the limitations caused by crystallinity, regulation by the phase diagram has motivated most of the recent research on polymer electrolytes. Though it is possible to obtain quenched amorphous phases from high M_w PEO salt complexes, especially at compositions close to the eutectic, it is illusory to expect to maintain this metastable state indefinitely, especially when the materials are subjected to mechanical stress. Several strategies have been tried for the synthesis of intrinsically amorphous macromolecules.

In a first approach, EO units are kept as the majority component of the main chain. Disruption of the infinite sequence by other components whose geometry excludes them from the crystals progressively decreases the size, and, hence, the melting point of the crystalline phase according to [51]:

$$\frac{1}{T_m} = \frac{1}{T°_m} - \frac{R}{\Delta H°_{fus}} \times \ln p_{OE} \qquad (2.13)$$

T_m = melting point of the copolymer
$T°_m$ = melting point of the EO homopolymer
$\Delta H°_{fus}$ = enthalpy of fusion of the EO homopolymer
p_{OE} = propagation probability

In random copolymers, a comonomer is introduced during EO polymerization. In this case the propagation probability is given by the average composition. Since the kinetics of EO polymerization is appreciably faster than that of most epoxides, including the first homologue propylene oxide, the synthesis of a random EO–PO polymer requires care. Yet the copolymers effectively demonstrate low crystallinity at 15–20% OP and, hence, increased conductivities at room temperature.

In block copolymers or in cross-linked networks, the p_{OE} factor is much larger than that given by the average composition. These polymers can be made from commercial PEO glycols (PEGs) available in various molecular weights. The melting point of the PEO fraction is roughly equal to that of the starting oligomers. Thus amorphous materials are obtained from PEGs 400 or 600, whereas crystallinity is observed at higher values. Excellent conductivities are obtained from oxymethylene-linked PEG 400, through a simple Williamson condensation reaction that appears to be remarkably simple and efficient [52,53]:

$$\sim\sim OH + CH_2Cl_2 \xrightarrow{KOH,\ CH_2Cl_2} \sim\sim OCH_2Cl$$

$$\sim\sim OCH_2Cl + HO \sim\sim \xrightarrow{KOH,\ CH_2Cl_2} \sim\sim OCH_2O \sim\sim$$

A wide variety of networks have been obtained through the convenient urethane linkage, which allows control of the chain length through the starting oligomers and the cross-link density with bi- or trifunctional isocyanates [54–57]:

$$\sim\sim OH + OCN \sim\sim \longrightarrow \sim\sim OCONH \sim\sim$$

Pendant PEO chain polymers represent another route to low-crystallinity polymers. In this case the EO segments have one nonbonded extremity (usually end-capped by a methyl group to avoid the reactivity of the OH group), which gives greater conformational freedom. Such types of comb polymers are based on methyl-PEG acrylates, methacrylates [58,59], itaconates [60,61], or vinyl ethers [62,63]. The control by T_0 or T_g of the conductivity in the VTF–WLF models has led to the design of low glass-transition-temperature polymers based on a flexible backbone. The rotation barrier around a carbon–carbon or carbon–oxygen single bond is still 5–8 kJ. Lower values can be obtained only through the involvement in d–p π bonding with second-row elements. Effectively, —Si—O—Si— and —P=N—P— polymers have the lowest reported T_g's but no intrinsic solvating properties. Thus they require the attachment of side oligo-PEO segments. High conductivities in both siloxane and phosphazene comb polymers are then obtained [64–68]. As a drawback, the mechanical properties of the linear polymers are very poor, even at high molecular weight, a consequence of facile chain diffusion, and usually cross-linking is required.

An interesting and simple way to improve both conductivity and mechanical properties is to add a dispersed phase [79,70], such as $LiAlO_2$ particulates in a POE-based electrolyte. An optimum concentration of 10 w% was found [71]. The origin of the effect is not clear yet and may be related to limitation of the crystal growth in the polymer [72].

2.3.7 Plasticizers and Gel Electrolytes

As discussed earlier, the conductivity of binary polymer electrolytes (a salt dissolved in a solvating macromolecular network) is limited by the chain motion above T_g. Despite remarkable improvements due to better understanding of the conduction process and the design of new salts and polymers, the conductivity seems to reach a ceiling value of $\approx 10^{-4}$ S cm^{-1} around 25°C. To improve performance below room temperature, low-molecular-weight additives have been incorporated to enhance chain flexibility and decrease crystallinity. The following two main families of compounds have been selected.

2.3.7.1 Short PEO Chains (in the 200–600 M_w Range)

Methyl end capping is required to avoid the chemical reactivity of glycols [73,74]. Moderate improvements in conductivity and decrease in size, and thus in the melting point of the crystallites, are obtained at the expense of the mechanical properties, as a large weight fraction is required (>50%). It is interesting to note that for the same polymer, low and high M_w fractions are not fully compatible, and the miscibility is also governed by a phase diagram, as discussed by Fauteux [75]. The advantage of such plasticizers, despite their limited action, is that their chemical behavior is similar to that of PEO and they have low volatility.

2.3.7.2 Aprotic Dipolar Solvents

PEO–salt complexes are soluble in high-permitivity solvents, which can be used as additives. The most-studied molecules are ethylene carbonate (EC) and propylene carbonate (PC), which are widely used as organic solvents in electrochemistry [76]. Dramatic improvements in conductivity up to 10^{-3} S cm^{-1} at 25°C are obtained for ≈ 50 v% PC, but intermediate values are readily obtained with lower additive fractions. It is not clear in such cases if the enhanced conductivity is still due to segmental motion made faster by the plasticizer or if complete decoupling between polymer dynamics and ion motion is obtained. The concept can be extrapolated to gels in which a nonsolvating (but polar) polymer is solely used to impart mechanical and filmogenic properties to a liquid electrolyte based on PC–, THF–, or dioxolane–LiX (LiX = LiCF$_3$SO$_3$, LiClO$_4$, . . .). In fact, this idea preceded the discovery of solvating polymers [77]. The conductivities are marginally lower than those measured with the free liquid, and the polymer, as a minority component, is not required to provide solvation for the ions. Compatibility of the macromolecules with the liquid (i.e., comparable solvent parameters) implies that the monomer units are polar. Poly(acrylonitrile) or vinylidene fluoride have been selected for this purpose [78,79]. Cross-linked gels based on PC-ethylene glycol diacrylates with a low concentration of solvating sites fall in this category [80]. As predicted by thermodynamics, the activity of the liquid solution is very close to unity, since the macromolecule polymer as a solute, is very diluted on a mole fraction basis. The materials should be treated like immobilized liquids, with the corresponding stability and safety limitations and volatility, but keep their advantage for film processing, which is discussed in Chapter 4.

2.4 Polymers as Electrode Materials for Lithium Batteries

2.4.1 General Considerations

Finding semiconducting polymers that can compete with inorganic materials was the object of extensive and creative research in the 1950s, but with limited success [81]. It was not until the accidental discovery that oxidative insertion of iodine in poly(acetylene) leads to metallic behavior that the field became one of the most visible of the past decade [6,82,83]. This reaction, extended to various intercalants, would improperly be called "doping." The research on battery electrode material came as a corollary to the electrochemical insertion used to prepare some of these polymers or to bring them into the conducting state. The suggestion to use these redox polymers as electrode materials had been made much earlier, in 1973 [84], but little attention was paid at that time.

Redox polymers possess conjugated bonds providing a pathway for electronic conduction. The main families studied so far are shown in Fig. 2.10. Structurally, polyparaphenylene and the polyheterocycles from thiophene and pyrrole can be derived from *cis*-polyacetylene by substituting the two hydrogens by —CH=CH—, —S— and —NH— groups, respectively. Polyacetylene is the product of Ziegler–Natta polymerization, whereas most other conjugated polymers are prepared by a nonconventional oxidative coupling from the monomer via the radical cation. This type of polymerization, shown at its initial stage in Fig. 2.11, should not be confused with redox initiation of polymerization, where the chain growth is triggered by catalytic amounts of redox species. The vinylene-heterocyclic polymers have increased stability due to lower steric hindrance, permitting better π overlap.

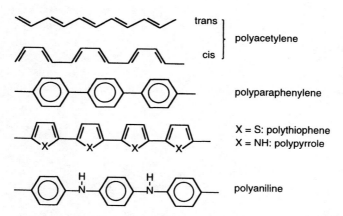

Figure 2.10. Most-studied families of conjugated polymers showing the structure units.

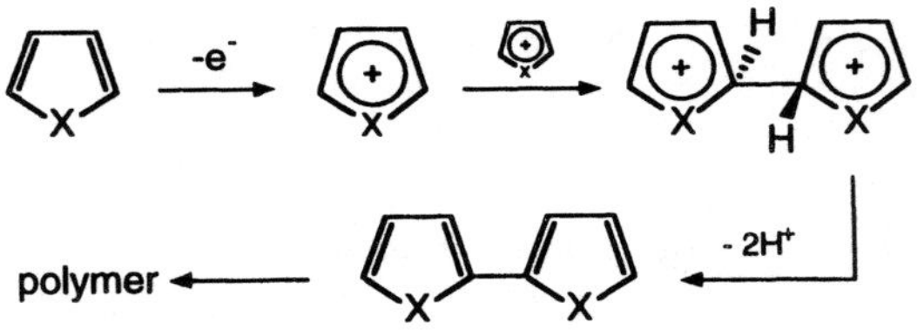

86   THE ELECTROCHEMISTRY OF NOVEL MATERIALS

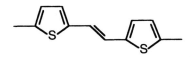

Figure 2.11. Initial electropolymerization stage proceeding via radical cation coupling; X = S, NH, (O, Se).

These could be considered alternating copolymers of the parent heterocycle and acetylene (Fig. 2.12):

Figure 2.12. Structure of polythienylene vinylene.

2.4.2 Redox Processes

The redox chemistry of conjugated polymers is possible via intercalation. All the described polymers can be oxidized without disruption of the backbone, the loss of electrons being compensated by the ingress of anions:

$$\langle r.u. \rangle + xX^- \Leftrightarrow \langle X_x r.u. \rangle \tag{2.14}$$

where r.u. represents the polymer repeat unit; for polypyrrole, $0 \leq x \leq x_{max} \approx 0.3$, which can be schematized in Fig. 2.13. In this sense these materials are the macromolecular equivalents of discrete molecules forming carbocations. In both cases electron-donating heteroatoms facilitate (by decrease in potential) and stabilize the positive charges. Yet the extraction of electrons occurs at relatively positive E-values, in the $+3.5$–4.5 V versus Li/Li^0 range. Anion intercalation, which is relatively rare in the inorganic world (occurring mainly in graphite), is almost a rule in redox polymer electrochemistry. The tendency for radical cations to associate in pairs to minimize the number of radicals in the structure (i.e., to have a conjugated canoni-

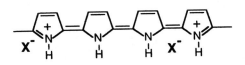

Figure 2.13. Canonical form of oxidized PPy.

cal form) has been the subject of extensive theoretical and practical studies. Usually, the physicist's terminology "polarons" for radical cations has been adopted:

2 polarons $\Leftrightarrow$ bipolarons

Polyacetylene (PA) and polyphenylene (PPP) are two examples where cation insertion is also known, including with lithium. In this case radical anions are formed at potentials not far above that of the lithium electrode, $+0.5-1.2$ V versus Li/Li^0. Cations reside in galleries left between the polymer strands (Fig. 2.14). The use of PA and PPP as low-voltage positive electrodes with a lithium anode is only anecdotal. These materials could be used in rocking-chair batteries as negative electrodes in place of carbon, which is presently in use, but their stability is far lower. Thus most lithium batteries employing a redox polymer in the positive electrode work through exchange of anionic species, requiring the use of electrolyte as a salt reservoir. The overall reaction can be written as follows:

$$\langle X_x \text{r.u.} \rangle + x Li^+ + x e^- \Leftrightarrow \langle \text{r.u.} \rangle + x(Li^+ X^-) \qquad (2.15)$$

In Table 2.5 the principal characteristics of redox polymer electrodes are summarized. The capacity of these materials as positive electrodes is relatively low if compared to inorganic intercalation materials (TiS_2: 2.4 Ah/kg; V_6O_{13}: 3.2 Ah/kg). This handicap is augmented if we consider the low density of the organic materials (≈ 1.2). In an electrode of fixed porosity (or a composite with a given volume fraction allowed for the electrolyte), the weight fraction of the active material depends on the specific weight of the components. Acceptable energy densities are in fact solely due to the very positive potential of the electrode which, as discussed later, poses stability problems.

2.4.3 Diffusion and Reaction Mechanisms

The electrode insertion reaction (2.14) is valid in most cases. It implies a mobility of the anion obeying kinetics compatible with the application in batteries ($D \geq 10^{-12}$ cm^2 s^{-1}). An exception is when the doping anion is affixed to a polymer, as in PAN/styrene sulfonate alloys or the polymers of sulfanilic acid. Here the cation must enter the poly–salt complex to compensate for the uptake of electrons during reduction; these are thus lithium intercalation compounds but with high equivalent weight.

As might be expected from the large number of parameters involved, the study of

Figure 2.14. Schematic representation of Li(Na) inserted in CH_x.

Table 2.5 Summary and Characteristics of the Most Studied or Conceptually Interesting Electrochemical Cells and Batteries Using Conjugated Polymers. r.u. = repeat unit

⊖ Electrode	E vs. $Li^+/Li°$	⊕ Electrode	Dopant	e^-/r.u.	E vs. $Li^+/Li°$	Capacity Ah/kg	Type	Electrolyte
Li°	0	(polyacetylene)	Li⁺	0.25	0.5–1.2	0.5	Source/sink	Liquid
Li°	0	(polyacetylene)	ClO_4^-	0.12	3.2–4	0.13	Salt reservoir	Liquid
Li°	0	(polypyrrole)	ClO_4^- BF_4^-	0.25	3–3.8	0.09	Salt reservoir	Liquid
Li° / LiAl	0 / 0.33	(polyaniline)	ClO_4^-	0.6	3.2–4.2	0.09	Salt reservoir	Liquid
Li°	0	(polypyrrole)	(phenyl-SO_3^-)	0.25	3–3.8	0.06	Source/sink	Liquid
Li°	0	(poly-aniline-SO_3^-)	Grafted	0.6	3.2–4	0.098	Source/sink	Liquid
Li°	0	(polydithiolene)	None	2	2.4	0.37	Souce/sink	Polymer
Li°	0	(polythiadiazole)	None	2	3	0.60	Source/sink	Polymer

ion mobility in redox polymers is still the subject of controversy and uncertainty. Three difficulties must be mentioned:

1. Redox polymers have, because of their growth process, a fibrillar morphology. The effective diffusion path is difficult to determine and the conditions for preparation often lack reproducibility.
2. In the presence of liquid electrolytes, swelling of the polymer takes place, with possible co-intercalation of solvents, thus altering the diffusion coefficients.
3. As for inorganic intercalation compounds, the domains of true nonstoichiometry are limited; distinct compounds exist that, in analogy with graphite insertion chemistry, are termed *stages*. Thus the voltage–composition curves show plateaus that represent boundaries between two pseudophases (stage$_{n-1}$/stage$_n$).

Mean diffusion coefficients in the presence of liquid electrolytes have usually been estimated to lie in the 10^{-9}–10^{-11} cm^2 s^{-1} range. These numbers are to be compared with the value of 10^{-17} cm^2 s^{-1} measured for the diffusion of oxygen in polyethylene single crystals. Measurement of Li diffusion in polyacetylene using polymer electrolytes led to a value of 10^{-14} cm^2 s^{-1} at 75°C [85]. In this case, no solvent co-intercalation takes place but the PEO electrolyte does not penetrate the entire fibrillar structure of PA.

In addition, studies using laser beam deflection at the polymer–electrolyte interface or a quartz microbalance to establish weight loss or uptake during electrode reactions have highlighted two possible mechanisms: At long time scales the reaction 2.15 is effective, and at short time scales, the reduction reaction corresponds to a cation insertion in the lattice, followed by salt ejection [86,88]:

$$\langle X_x \text{r.u.} \rangle + xe^- + x\text{Li}^+ \Leftrightarrow \langle \text{r.u.}(x\text{LiX}) \rangle \quad \text{fast} \quad (2.16)$$

$$\langle \text{r.u.}(x\text{LiX}) \rangle \Leftrightarrow \langle \text{r.u.} \rangle + x(\text{LiX})_{\text{solvent}} \quad \text{slow} \quad (2.17)$$

The role of the solvent in this mechanism is not yet clarified, but each ion incorporated is accompanied by two or three solvent molecules. These results are to be compared with solutions and polymers where the mobility of the anion is usually dominant.

2.4.4 Nonconjugated Redox Polymers as Electrode Materials

A novel and interesting principle for solid-state energy storage was proposed in 1987 [89,90] based on the duplicative oxidation of thiolate anions to form disulfide bonds:

$$\sim\!\!\text{S}^- + {}^-\text{S}\!\!\sim \rightleftharpoons 2e^- + \sim\!\!\text{S}\!-\!\text{S}\!\!\sim$$

If the same organic group bears at least two or more sulfur groups, a polymer is formed upon oxidation:

$$2n \text{ LiS}\!\!\sim\!\!\text{SLi} \rightleftharpoons 4ne^- + 4n\text{Li}^+ + {}^-\text{S}\!\!\sim\![\sim\!\!\text{S}\!-\!\text{S}\!\!\sim]_{n-1}\text{S}^-$$

The reaction, written with lithium ions, shows the similarity with an insertion reaction where a reversible polymerization–depolymerization process take place by electron plus ion injection/extraction. The macromolecule cannot be conductive, as extension of electron delocalization through the S—S bond is not possible. Yet complete electrode utilization is observed in solid-state cells [91].

The formation of an unactivated disulfide bond is chemically reversible but with slow kinetics. However, when the sulfur is conjugated with double bonds in the organic moieties, especially with nitrogen atoms, fast kinetics and higher potentials are observed. This is specially the case with the heterocycle dimercaptothiadiazole (DTZ) shown in Fig. 2.15.

The specific capacities are reported in Table 2.5. They are high, as are the corresponding energy densities, because of the low equivalent weight of the monomer units. Composite electrodes associating a conjugated polymer (PANi) and DTZ have been proposed, and they also show high utilization efficiencies [92]. Difficulties may arise from the nonnegligible solubility of the compounds at the final stages of reduction [i.e., of the lithium dithiolate or some oligomer (e.g., a dimer) salt in the electrolyte]. If this salt reacts chemically on the lithium, capacity will gradually be lost, but 500 cycles in a Li cell have already been obtained. Research is being undertaken to attach the salt to a polymer to avoid diffusion. The sulfide dimerization would then result in cross-linking or cycle formation [93,94], schematized in Fig. 2.16. This is analogous with biological processes where —S—S— bonds are used for X-linking proteins and also as two-electron-transfer mediators as in thioctic acid, a vitamin B cofactor.

2.5 Cell Geometry and Power Densities

As depicted, all electrochemical cells, including those of the lithium type, require a combination of ion migration and diffusion to provide for mass transport during operation. As discussed in Section 2.1, diffusion is inherent to the functioning of intercalation electrodes. Except in the case when $t_+ = 1$, and in lithium intercalation electrodes where only migration takes place, back-diffusion of the salt is needed to counter the establishment of the concentration gradient. Ion and electron transfer across interfaces will, at this point, not be considered limiting.

In a first approximation, the volumetric power of an electrochemical cell can be calculated by expressing the irreversible limitation due to mass transfer as series

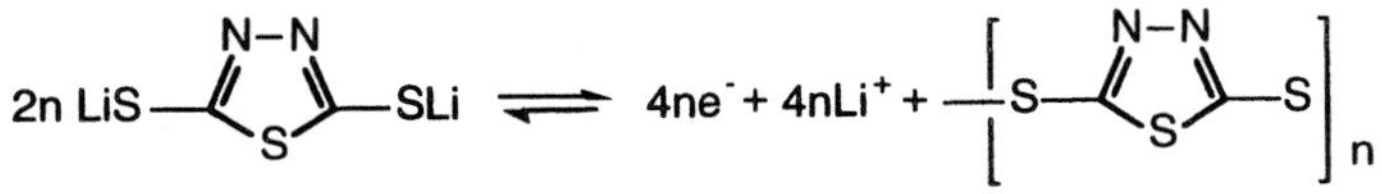

$$2n\, \text{LiS}{-}\underset{S}{\overset{N-N}{\diagup\!\!\!\diagdown}}{-}\text{SLi} \rightleftharpoons 4ne^- + 4nLi^+ + \left[{-}S{-}\underset{S}{\overset{N-N}{\diagup\!\!\!\diagdown}}{-}S{-} \right]_n$$

Figure 2.15. The redox polymerization chemistry of dimercaptothiadiazole (DTZ).

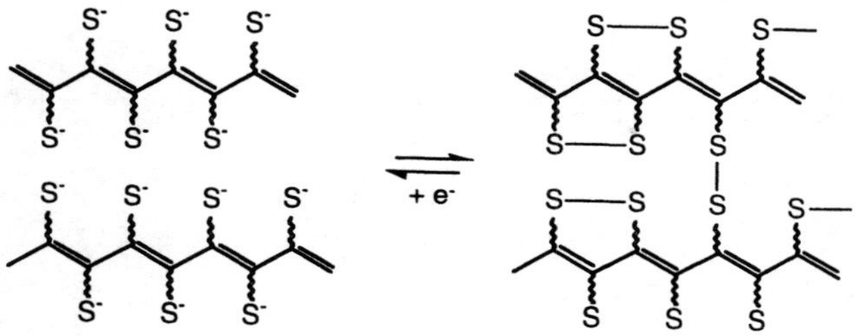

Figure 2.16. Principle of redox chemistry of polymer-attached sulfur.

impedances due to conductivity or diffusion. The volumetric power $\partial P/\partial V$ is then expressed as

$$\frac{\partial P}{\partial V} = \frac{\bar{E}^2\eta(1-\eta)}{d_\sigma + d_D}\left[\frac{d_\sigma}{\sigma} + \frac{kT}{e^2}\frac{d_D}{D}\right]^{-1} \qquad (2.18)$$

where $\bar{E}$ = average electromotive force
η = energy efficiency
d = average distances over which conduction d_σ or diffusion d_D takes place during operation

In a planar configuration, and neglecting the inert current collector thicknesses,

$$d_\sigma + d_D = d \approx \text{total cell thickness}$$

The modeling by Atlung of a composite electrode (intercalation material + electrolyte) [95,96] shows optimal utilization of electrode capacity when the diffusion and conduction impedance are of comparable values. Thus Eq. (2.18) can be more simply expressed as a function of a mean conductivity $\bar{\sigma}$ or diffusion coefficient $\bar{D}$ reflecting a balance between the materials involved in the electrochemical process:

$$\frac{\partial P}{\partial V} = \frac{\bar{E}^2\eta(1-\eta)\bar{\sigma}}{d^2} \qquad (2.19)$$

This equation reflects the interest in working with high-voltage couple systems in order to minimize the current flow. Also, the geometry of the cell, through the choice of the thickness, acting as the square power, permits compensation for the limitations due to mass transport, conductivity, and diffusional processes.

In Fig. 2.17, a schematic diagram illustrating these notions is drawn with $\bar{E}$ = 2.2 V, $\eta = 0.9$, to meet the criterion of 100 W/dm^{-3}. Three domains are apparent, determined by the available transport parameters:

1. *Conventional cells:* the average interelectrode distance is of the order of 1 mm and needed conductivities are in the 10^{-2}–10^{-4} S cm^{-1} range. Although this

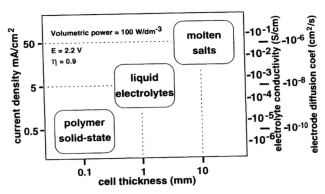

Figure 2.17. Zone diagram showing the electrochemical parameters required for sustained power of 100 W dm^{-3} assuming an average emf of 2.2 V as a function of cell thickness. Right scale: conductivity, diffusion coefficients; left scale: current densities.

requirement can be met with liquid electrolytes, the corresponding diffusion coefficients at 10^{-7}–10^{-9} cm^2 s^{-1} correspond to very fast solid-state processes, and only a few intercalation materials are known that have such kinetics (proton diffusion in H$_x$NiO$_2$, H$_x$LaNi$_5$; Li$^+$ in Li$_x$TiS$_2$). Other electrodes working at such rates usually involve dissolution–nucleation processes in the close vicinity of the electrode, available mainly in aqueous electrolytes.

2. *Molten salts:* these materials have higher conductivities than solution electrolytes, but the high temperatures at which they usually operate, posing materials problems in terms of corrosion and containment, imply the choice of a larger interelectrode distance (≈ 1 cm). Ionic conductivities reach the required levels, but again the diffusion characteristics are beyond those of most intercalation materials. Solid electrode materials are usually intermetallic compounds with liquid-like diffusion (e.g., LiAl alloys) or recourse to dissolution–nucleation processes. In sodium–sulfur batteries the electrodes are liquid and the anolyte (molten Na) and the catholyte (molten polysulfides) are separated by a β-alumina ceramic membrane, a superionic conductor with liquid-like Na$^+$ conductivity.

3. *Thin films:* the choice of cell geometries that are an order of magnitude thinner permits the use electrolytes and electrode materials with 100-fold slower mass transport parameters. Electrolytes in the 10^{-4}–10^{-5} S cm^{-1} range and electrodes 10^{-9}–10^{-11} cm^2 s^{-1} can be selected. An immediate advantage is the much wider choice of materials, solid electrolytes and oxide-type electrode materials (VO$_x$, MnO$_2$, ...) in which Li diffusion is of the order of 10^{-10}–10^{-11} cm^2 s^{-1}. These choices give far more opportunities to tailor the properties of the materials. In particular, the self-diffusion of Li in metallic lithium is also close to 10^{-9} cm^2 s^{-1} around room temperature. Adapting the transport parameters to this value is important for lithium redeposition, as discussed later, even if lithium stripping is limited, in theory, only by the conductivity and t_+ of the electrolytes. Figure 2.18 shows the schematic drawing of a solid-state thin-film battery.

POLYMERIC MATERIALS FOR LITHIUM BATTERIES

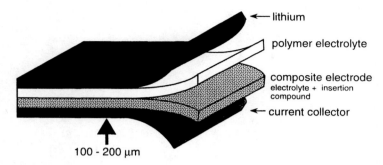

Figure 2.18. Schematic drawing of thin-film battery showing the superposition of electroactive layers.

Evidently, scaling down the cell thickness must be counterbalanced by larger surface areas for a given capacity. Thus the advantages foreseen in terms of the choice of materials and kinetic balance using thin-film configurations are decisive only if high-productivity, surface-intensive technologies can be used for fabrication. Polymer and coating technologies meet this criterion and are already in large-scale use for the paper, packaging, and information-recording industries. In this respect the advantage of plastic materials is evident because of their ease of processing.

2.6 EMFs and Stability Windows

Theoretical specific energies and power vary with, respectively, the first and second power of the cell EMFs. Most electrochemical systems, however, do function with electrode couples' potentials poised outside the thermodynamic stability window of their electrolyte. This is shown in Fig. 2.19.

Self-discharge of batteries based on an aqueous system is appreciable (1–10%/month) but the by-products of solvent decomposition (O_2, H_2) are not involved in secondary irreversible reactions. Organic compounds may offer the possibility of obtaining enough kinetic stability for the expected lifetime of a battery, but little or no reversibility (i.e., molecule reconstruction) is to be expected from the side-reaction products.

Metallic lithium corresponds to the most negative available potential and of course the lowest equivalent weight (3.86 Ah/g and 2.12 Ah/cm^{-3}), but Li^0 is thermodynamically unstable toward the majority of organic compounds, the main exception being aliphatic hydrocarbons, which have no value as electrochemical media. Similarly, multiatomic anions like ClO_4^-, $CF_3SO_3^-$, and $(CF_3SO_2)_2N^-$ are expected to be unstable at the potential of metallic lithium. Yet lithium presents a remarkable kinetic stability toward a variety of aprotic solvents. The corrosion products of the metal are probably very complex mixtures containing oxides, hydroxide, carbonate, fluoride, and so on, derivatives. The film formed reduces further growth but does not totally impede the electrochemical process. Most lithium

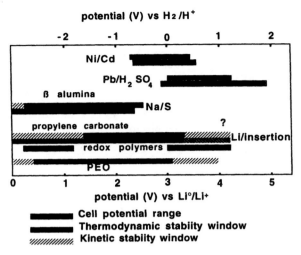

Figure 2.19. Stability window of various electrolytic media and operating potentials of electrode couples.

salts exhibit cationic mobility in the solid state. Although this property is generally observed at high temperatures, the crystallographic disorder inherent to the formation of the film and its thinness (10^2–10^3 nm) result in a nonblocking behavior (the appellation *passivating layer* is in this sense incorrect). Extensive studies have been undertaken to understand the growth and properties of this layer and several model have been proposed [97–102]. With liquid electrolytes, it is unlikely that the film can survive intact the volume variations inherent in battery operation, with the transfer of several micrometers of metal across the interface. More likely, a new film is formed after disruption, when scales are freed in the liquid electrolytes, progressively consuming the metal and the electrolyte irreversibly. The situation is more favorable with solid polymer electrolytes, which mechanically constrain the film at the interface. The growth of the reaction layer is also expected to be slower, thus resulting in a smaller fraction of the metal lost through nonelectrochemical processes. Complex impedance spectroscopy is usually the technique of choice for the study of interfacial phenomena [103,104].

PEO or PPO as ethers are chemically, thus electrochemically, inert with a window extending from 0 to about ≥ 3.7 V versus Li/Li^0 [13,14,105,106]. However, attention must be paid to the chemical bonds introduced to modify the architecture of the base linear polymer to improve the conductivity or mechanical properties. In particular, urethane linkages that result from isocyanates cross-linking with glycols contain the —CONH— amide group, with a labile hydrogen and a stability window close to that of water. Yet, because these reactive groups are immobile in the polymer network, the rate is slowed enough to allow a few tens of cycles of battery operation. Similarly, esters (as in oligo-EO acrylates) and siloxanes are expected to be reactive by analogy with small-molecule electrochemistry. There is a relative paucity of results concerning the long-term stability of polymers other than PEO, considering their recent emergence. Caution must then be used in analyzing all the results on liquid or polymer electrolytes, as foreign contaminants, especially water, may mask their intrinsic behavior. Interestingly, the addition of a dispersed phase ($LiAlO_2$) does diminish the interfacial resistance, which possibly is linked to a gettering of the water [72].

On the $\oplus$ electrode side, either aprotic solvents such as PC and ethers, hence polymer electrolytes, provide good kinetic stability with classical (TiS_2, VO_x, etc.) electrode materials in the 2–3.5 V versus Li/Li^0 range. Usually, the interface between polymer electrolytes and intercalation compounds appears reversible with low interfacial resistance; but increase in impedance with time at the POE–$LiCF_3SO_3$–V_6O_{13} contact has been reported [107].

The recently introduced high-voltage intercalation compounds Li_xMO_2, M = Mn, Ni, Co operate in the 3.8–4.4 V range [108,109], close to or even beyond the anodic limit of most organic solvents. Because the metastability of organic molecules is extremely temperature-dependent, operation above room temperature usually results in rapid capacity loss.

In all cases the interfacial film results in an added resistance per unit surface. Equation (2.19) is then modified as

$$\frac{\partial P}{\partial V} = \frac{\bar{E}^2 \eta (1 - \eta) \bar{\sigma}}{d^2 (1 + (R_s \bar{\sigma}/d))} \tag{2.20}$$

where R_s is the interfacial resistance. Again, the influence of R_s lowering the available specific power can be minimized through reduction of the interelectrode gap, as shown in Table 2.6, using the conductivity levels needed to reach ≈ 100 W/dm^3. Typical R_s values are in the $10^2 - 5 \times 10^2$ ($S^{-1}cm^2$) range.

2.6.1 Case of Conjugated Polymers; Self-Discharge and Lifetime

Conjugated polymers, when used as electrode materials, always show high self-discharge rates (percent per day). The simple explanation based on the highly positive operating voltage, close to the anodic limit of the stability window of the

Table 2.6 Influence of the Interfacial Resistance on Volumetric Power, as % Loss Compared to the $R_s = 0$ Situation, for Two Different Cell Thicknesses and Electrolyte Conductivities Situations

$d(\mu m) \Rightarrow$	10^3	10^2	10^3	10^2
σ (S cm^{-1})	10^{-3}	10^{-5}	10^{-3}	10^{-5}
R_s(S^{-1} cm^2)	10^2	10^2	5×10^2	5×10^2
$\Delta \frac{\partial P}{\partial V}$ (%)	-50%	-10%	-83%	-33%

electrolyte, is probably too expeditious, as inorganic compounds with voltages in the same range do not exhibit such rapid loss of capacity on standing. Several hypotheses can be proposed:

- The fibrillar morphology of the electrode materials greatly augments the surface exposed to the electrolyte, thus enhancing any parasitic chemical reaction.
- Conductive polymers may catalyze the oxidation of the electrolytes. This behavior is commonly observed with organic radical ions.
- The polymerization technique, like the electrochemical oxidative coupling, is likely to produce a distribution of molecular weights. Oligomers released in the electrolyte solution may diffuse to the negative electrode where they react. Back-diffusion would then trigger an effective shuttle to transport electrons across the electrolyte.

Inganäs and co-workers [110] have shown that, in fact, the self-discharge in solid-state cells of the type

$$Li/PEO-LiClO_4/\langle ClO_4 \rangle_{0.3}PPy \rangle$$

takes place only when the lithium electrode is present. The half-cell did not lose its stored capacity rapidly. A likely explanation is that the lithium surface, in addition to the passive film surface, produces soluble species that diffuse to the positive electrode. The chemistry of alkali metals suggest that these products (i.e., alkoxides) are basic and nucleophilic. The radical cations of conjugated polymers are prone to reaction with such species, by simple proton elimination in PPy and PANi or addition to the carbon skeleton. This fragility is obviously a major difficulty for the use of conjugated polymers in lithium batteries. Inorganic materials appear much less sensitive to such reactions.

2.7 Realizations and Prototypes

Scientific contributions proposing new materials and/or presenting a refined understanding of their properties have abounded in the last decade. There are studies of single-battery components focusing on properties such as conductivity, redox potential, interfacial resistance, and so on. Small experimental cells are assembled at laboratory scale to verify the viability of the materials under scrutiny.

Evaluation of prototype batteries requires large, long-term commitment of scientists from different disciplines to assess all the numerous parameters involved in practical devices. This is especially true with solid-state batteries, including polymers, where access to reproducible technologies is mandatory. Thus only a small amount of reliable data is accessible in the literature, sometimes for proprietary reasons. Table 2.7 summarizes the features of major programs that have been active in recent years.

Table 2.7 Principal R&D or Industrial Projects Concerning Polymer Electrolyte-Based Lithium Batteries

Groups Involved	Electrolyte	Electrode ⊖	Electrode ⊕	Date Started	Applications (temperature)	Stage
Hydro-Québec (Canada) Yuasa (Japan) (ACEP project)	Polymer	Li, LiAl	TiS_2, VO_x, MnO_2	1980	EV (60°C), electronics	10 Wh. from bench scale
Harwell–Dowty (U.K.)	Polymer Gel (RT)	Li	V_6O_{13}, MnO_2	1981	EV (130°C), electronics	?
Innovision (Denmark)	Gel	Li	V_6O_{13}	1981	Electronics	Transferred
Bridgestone (Japan)	Liquid	LiAl	PANi	?	Electronics	Commercial
Varta (Germany)	Liquid	Li	PPy	?	Electronics	?
Polyplus (U.S.)	Polymer	Li	DTZ	1988	EV (80°C), electronics	Development

2.7.1 Electronics: Established and New Applications

Electronics depends more and more on high-performance miniaturizable power supplies, with an average growth of about 10%/year. Very drastic requirements have been established in this field for operating temperatures, ranging to below −30°C, especially for the military market. Despite the progress in achieving better conductivities in solvent-free polymer electrolytes, these extreme temperatures are too close to the actual T_g of the complexes and ion motion is considerably slower at these temperatures ($<10^{-8}$ S cm^{-1}). Gel electrolytes are required if the advantages of solid-state technology are to be kept. There is especially a need for paper-thin power sources to handle memory backup or such new applications as "smart" credit cards.

Programs to develop batteries using inorganic intercalation oxides (V_6O_{13}) and lithium as electrodes and solvent-swollen polymers as electrolytes are active, as shown in Table 2.7. Preliminary disclosures indicate that excellent power densities, even at low temperatures, can be obtained in conjunction with respectable specific energies, ≥ 130 W L^{-1}. Cycle life seems, however, a weak point, with progressive loss in capacity over about 150–200 cycles. This is similar to the cycle life expectancy usually observed with free-liquid electrolytes, confirming the higher reactivity of the latter and the need to confine reaction products to the vicinity of the electrodes. Such systems would be in competition in this market segment with emerging new systems, such as the nickel–hydride couple or the rocking-chair battery based on lithium–carbon intercalation, characterized by lower energy densities but longer cycle life (see chapter 3 on intercalation materials and Ref. 7.)

In fact, most of the electronics applications for which a portable power supply is required are now associated with liquid crystal displays whose range of operation is −5 to +50°C. Thus systems meeting this less-stringent temperature requirement can use the improved polymer electrolytes without solvent additives. Modified amorphous PEO cells have been built and tested at Hydro-Québec in Québec, Canada, in conjunction with TiS_2, chosen because of its fast kinetics [111,112]. Cycle life with less than 50% loss in capacity extends to ≥ 600 cycles. At a C/10 discharge rate at 0°C, roughly half the capacity is available and still corresponds to the energy contained in a Ni/Cd cell of identical volume. Improvements can be expected in technology and optimization of composite electrode materials. Room-temperature operation operation with oxymethylene-linked PEO and DTZ as the redox polymerization electrode show a similar rate capability.

The only commercial devices using conjugated polymer electrodes are also aimed at the electronics market (Table 2.7). The coin-type battery produced by Bridgestone has a nominal capacity of 4 mAh, but is designed to undertake $>10^3$ shallow cycles at 1 mAh. It is used mainly as a supercapacitor, serving as an electrical buffer in electronic circuitry, often associated with solar cells. Here high capacity is nonessential and the possibility of directly producing the PANi films by electropolymerization is a substantial advantage. These devices have been very successful, as they fill a specific niche in the electronics market.

2.7.2 Electric Vehicles and Load Leveling

High-performance batteries are crucially needed for an application that has kept scientists and engineers at a stalemate for almost a century, electric vehicles. Growing concern for the environment has revitalized programs on "green" automobiles throughout the industrialized world. California requires that 2% of a manufacturer's sales be zero-emission vehicles by 1998, to reach 10% in 2003. No presently available system can meet these zero-emission requirements for the corresponding battery of an electric vehicle. The minimum requirements established by the U.S. Advanced Battery Consortium (USABC) are given in Table 2.8. Different driving habits and the smaller overall weight of vehicles would correspond to an even more stringent requirement for the European market.

The priority of energy density leaves a very small number of systems in competition, and concessions must then be made for the operating temperature, accepting

Table 2.8 Summary of the Battery Performances Requirements by the United States Advanced Battery Consortium for Implementations in Electric Vehicles

Gravimetric (80 Wh kg^{-1})		Volumetric (135 Wh kg^{-1})	
Sustained power	20 W kg^{-1}	Peak power	150–200 W$^+$ kg^{-1}
Life (years)	5	Cycle life	600 (80% DOD)
Self-discharge	<15% in 48 hours	Voltage	120–240 V
Recharge time	<6 hours	Efficiency: C/3 discharge, C/6 charge	75%
Production price: ($/KWh)	<$150	Operating environment (external)	−30 to 65°C

the added impediments of thermal management. For instance, the sodium–sulfur (350°C) and lithium–aluminum–iron sulfide (420°C) systems are still considered acceptable.

The polymer electrolyte battery can meet the USABC requirements at temperatures between 40°C and 80°C, when the conductivity of ether-based electrolytes become amply sufficient for the power demand. The interfacial resistance and the diffusion impedance in the electrode materials are both activated and lowered with temperature increase.

Battery tests undertaken at Hydro-Québec on the systems:

$$\text{Li} / ^M\text{POE} \begin{Bmatrix} \text{LiClO}_4 \\ \text{Li}[(\text{CF}_3\text{SO}_2)_2\text{N}] \end{Bmatrix} / \text{VO}_x \qquad (2.21)$$

have shown feasibility in terms of performance, as shown in the Ragone plot of Fig. 2.20. The diagram shows the energy–power relationship for the cell units, excluding or including packaging. The higher value obtained for an operating temperature of 80°C corresponds to better utilization of electrode materials, even at higher loading (positive electrode mass per unit surface), reflected in a higher fraction of active materials.

In all cases a threefold excess of lithium compared to the stoichiometry of the positive electrode is used. The better conductivities of either the perchlorate or imide, especially with modified PEO electrolytes (^{M}POE) compared to systems using LiCF_3SO_3 and pristine PEO for instance [115], explain the appreciably lower temperature of operation. These tests have been made on 10-Wh cells for a total surface close to about 5000 cm² comprising subunits connected in various series–parallel combinations. No scale-up effects have been observed when they are com-

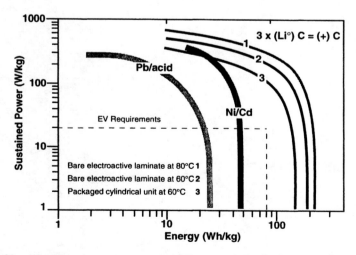

Figure 2.20. Specific power versus energy (Ragone plot) of thin-film solid-state battery Li/polymer electrolyte/VO$_x$. (From Ref. 115.)

pared with laboratory cells of a few tenths cm² [112]. This was confirmed more recently [116] on a 10-Wh cylindrical cell.

Cycle performances are shown in Fig. (2.21a,b). At least 70% of the initial capacities based on the tenth discharge are kept up to 500 or 1000 cycles. The 100% initial capacity is calculated on the basis of the stoichiometric intercalation in VO_x material. The best performances are obtained for low positive-electrode loading per unit surface (a thinner positive electrode) [117]. These results show how all transport parameters are linked in actual cells and point out the importance of finding an adequate balance between them, as discussed in Section 2.4.

The polymer electrolyte systems thus show a decisive advantage over those using liquid electrolytes in terms of lithium metal cycling. There are, indeed, very few reports in the literature of liquid electrolyte cells exceeding 300 deep cycles. Even then, most of the results were obtained solely with $LiAsF_6$ as a solute, which forms a passive film with favorable properties in relation to lithium but whose use is questionable (Section 2.8). On the contrary, good lithium–plating with polymer electrolytes have been obtained with a variety of solutes in different kinds of polyethers, demonstrating a general trend.

Usually the figure of merit (FOM), defined as

$$\text{FOM} = \frac{\sum Q_{\text{cycled}} - Q_{\text{initial}}}{Q_{\text{initial}}} \qquad (2.22)$$

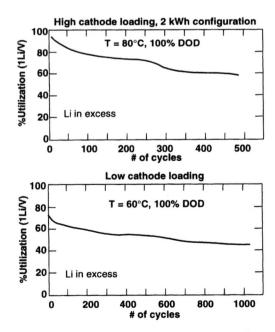

Figure 2.21. Cycle life (deep cycles) of solid-state batteries as in Fig. 20.

where $Q_{initial}$ represents the fresh electrode capacity and Q_{cycled} the total number of coulombs passed during the cycling. FOMs of 130 are currently obtained with polymer electrolytes, numbers which are exceptional when reached with liquid electrolytes [115].

Several reasons can be given for these remarkable results, though their relative importance is not known:

Low current densities allowing surface leveling from self-diffusion in Li^0
Mechanical resilience of the polymer electrolyte maintaining a even passivating layer on the Li^0 surface
Dendrite burnup through IR heating
Metal dissolution in the salt-depleted region in the vicinity of high local current density of the electrolyte (similar to liquid ether and ammonia), ensuring metal redistribution

Dendrites have been observed [112,115] in polymer electrolyte cells, but they do not subside on subsequent cycles unless long cycle life is reached.

As expected from the low reactivity of carefully chosen polymer electrolytes (i.e., they contain no chemical bond susceptible of attack by lithium), the shelf life is very good, with less than 2% capacity loss for three years storage at 60°C and no effect on further cycling utilization [113].

Scale-up of polymer batteries to large-size production is a technical challenge but has inherent advantages such as the fabrication of a continuous laminate that can be cut to desired surfaces for a given utilization with easy series–parallel connections, as shown in the schematic view of Fig. 2.22.

It is assumed that high-productivity polymer processing will largely compensate

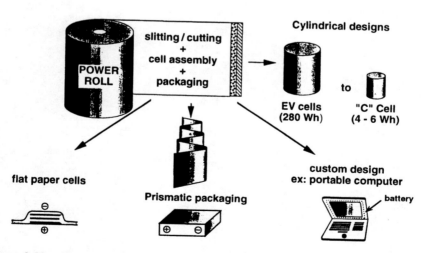

Figure 2.22. The "power roll" concept of utilization of continuous battery laminate for different end uses.

for the required tenfold increase in surface area. A schematic view of the bench-scale production line for 2-kWh batteries in the Hydro-Québec/ACEP program is shown in Fig. 2.23. Each battery requires a total of approximately 0.5 km of laminate for a 14-cm width.

As improvements already tested at the laboratory scale, in terms of either materials or technologies, are implemented the performances of batteries will improve from present values to reach packaged modules containing ≥ 150 Wh kg^{-1}, as shown in Fig. 2.24. Thin-film lithium foil is now produced at Hydro-Québec where the development of a molten-lithium printing process is also underway [115]. The ultimate performance of a solid-state battery using optimized components may reach 200 Wh kg^{-1}, corresponding to a quarter of the actual energy content of the electrode materials couple (Li/VO$_x$, Li/MnO$_2$).

2.8 Safety

The battery industry represents a sizeable segment of the economy. Major growth in the existing battery market, either primary or secondary, and the potentially gigantic outlook for development of electric vehicles suggest a careful evaluation of possible hazards and impact on the environment.

Primary batteries are presently the major source of diffuse mercury pollution worldwide. Consumer Ni/Cd systems are not recycled and are the principal contributor to contamination of landfills and air (via incinerators) by highly toxic cadmium. Legislature has so far been very lax, especially considering the absence of substitution products. This leniency is still sometimes taken as a warrant to extend this industry to larger-scale production (e.g., automobiles). This is not acceptable, considering that a 100% reclaiming efficiency of a product spread at the consumer level is unrealistic. It will, in any case, be necessary to recycle the materials involved in

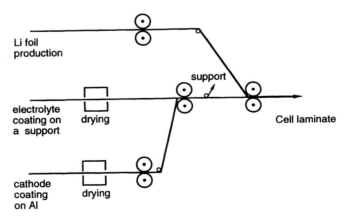

Figure 2.23. Schematic drawing of an automated line for continuous production of cell laminate; polymer and composite electrode cast from solvent.

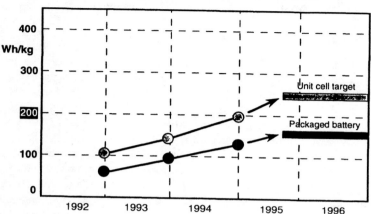

Figure 2.24. Evolution of the specific energy of the 2-kWh module destined for electric vehicle programs.

future automobiles (electrically powered or not), but the inevitable mishaps in the process should not turn into environmental incidents.

In addition, batteries may represent a danger because of their energy content, whose uncontrolled release after a short circuit or shock, for instance, may lead to catastrophic failure. Na/S batteries have been carefully redesigned to avoid immediate mixing of molten sodium and sulfur in case of cracking of the β-alumina ceramic membrane.

2.8.1 Toxicity Hazards

In terms of toxicity the components presently selected for polymer-based lithium batteries are less condemnable than their predecessors containing heavy metals:

Lithium ions have a low level of toxicity. Arsenic-based anions (AsF_6^-) cannot, of course, be considered. It has been argued that AsF_6^- itself is safe because of the complete shielding of the core metal by fluorine atoms. However, reactions with lithium in the interfacial layer, and of course incineration, liberate the arsenic in its toxic form. On the other hand, $(CF_3SO_2)_2N^-$ and its reaction products are acceptable.

Polymers, as electrolyte (PEO-based) or as electrode materials, are nontoxic. PEO is even biodegradable. The chemicals from which the macromolecules are made (EO) or some by-products of polymerization (benzidine for PANi) may pose a hazard in chemical plants.

Among the inorganic electrode materials, MnO_2 is benign, whereas moderate toxicity is attributed to vanadium derivatives. Nickel derivatives are recognized as allergenic and even carcinogenic.

Current collectors, either as metals or metallized plastic, are already in widespread use, especially for packaging applications.

2.8.2 Utilization Hazards

Despite its negative electrochemical potential, lithium is surprisingly stable in the presence of various chemicals because of the formation of a protective layer. This includes electrochemical solvents. However, this passivity no longer exists at the melting point of lithium, 180°C. Also, dendritic redeposition results in high-surface-area mossy or spongy lithium, which may unpredictably (via a local short circuit) trigger a chain reaction activated by the evolved heat. Perchlorate anion is considered as explosion-prone in contact with liquids. Several accidents have been reported for commercial systems, especially rechargeable ones. Safety with liquid electrolytes is claimed as one reason to abandon the metallic lithium electrode for a "rocking-chair" type using a negative electrode based on carbon materials [7].

The intrinsic safety of electrochemical devices using solid polymer electrolytes appears very good, for several additive factors:

Lower reactivity at the Li^0/solid interface
Absence of convection
No volatile component formed below the decomposition of the electrolyte
Thin-film configuration implying a limited mass of reactants available locally and the effectiveness of current collectors acting as heat sinks.

Safety tests have been performed on an experimental 10-Wh cell short-circuited when at the operation temperature of 60°C (Fig. 2.25). As shown, a temperature excursion brings the core of the cell to above the melting point of lithium, without

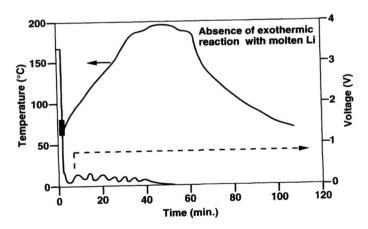

Figure 2.25. Short-circuit test on a 10-Wh battery; left scale: temperature versus time; right scale: lead potential versus time.

the onset of an exothermic reaction. Thus the safety characteristics of a solid-state battery, at the stage where no special feature or design has yet been taken into consideration for this special purpose, appear very promising.

2.9 Conclusions

Polymers, beyond the high volume of mediocre-quality structural materials that pervade our environment, are gaining enviable consideration in many fields of high technology. Electrochemistry appears to be one of these domains.

Moving ions or electrons in polymers was, two decades ago, considered either utopian or the object of a few scattered observations judged uninteresting. Ionically or electronically conducting polymers, still two distinct fields (they have respected the dichotomy attributing electrons to physicists and ions to chemists) are now mature in terms of understanding and new materials.

Electronically conducting polymers, which made headlines when they were expected to become a revolutionary new generation of batteries, have almost withdrawn from the field. Greater performance and reproducibility are found with inorganic intercalation compounds. Yet the new conductive polymers of today are more sophisticated, more stable, and processable from solution. They may be woven into the galleries of host materials, such as V_2O_5 xerogels, like a network of electric wires at the molecular level. A multitude of functionalities can be attached to the backbone, with catalytic effects, for instance, as in modified electrodes. There is a possibility that these materials will be used to prepare nanoscale composites for battery use. The domain of nonconjugated redox polymer (e.g., DTZ) is too recent to have received a complete assessment of its possibilities. The principle in itself is fascinating, and one should remember that nature routinely uses polymerization–depolymerization reactions to store its energy supplies (ATP, starch, or glycogen).

Polymers as electrolytes had a more modest start, but research now equals or exceeds the funding levels of the programs focused on liquid or molten salt electrolytes. Although significant progress was made in the understanding of these new materials, with subsequent improvements in performances, the conventional battery industry has been very resistant to the concept and to its technological implications. The argument about low current densities was and still is the base of opponents' complaints. With few exceptions, the polymer battery is being developed by newcomers in the field, with an acquired or historical expertise in film technologies.

For these reasons, the polymer electrolyte battery is neither a lithium nor a PEO battery. These components appear at present to provide the best performance for the goals of high energy density and cycle life: lithium because of its high voltage span and its specific role in intercalation chemistry, PEO for its combination of solvating power and chemical stability. Even if any of these components are changed, the basic concept of thin-film operation will remain, as it respects the intrinsic kinetic parameters of mass transport in matter.

References

1. Steele, B. C. H. In Fast Ion Transport in Solids, Gool, W. Van, ed. North Holland, Amsterdam, 1973, p. 103.
2. Armand, M. B. In "Fast Ion Transport in Solids," Gool, W. Van, ed. North Holland, Amsterdam, 1973, p. 665.
3. F. Wölher, *Pogg. Ann.* **1824**, *2*, 350.
4. Shaufhautl, *J. Prakt. Chem.* **1841**, *21*, 155.
5. Armand, M.; Duclot, M. French Patent 78 329 76 (1978).
6. Chiang, C. K.; Park, Y. W.; Heeger, A. J., Shirkawa, H.; Louis, E. J.; MacDiarmid, A. G. *J. Chem. Phys.* **1978**, *69*, 5098; Kaufman, J. H.; Mele, E. J.; Heeger, A. J.; Kaner, R.; MacDiarmid, A. G. *J. Electrochem. Soc.* **1983**, *130*, 571.
7. Scrosati, B. *J. Electrochem. Soc.* **1992**, *139*, 2776.
8. Gutmann, V. In "The Donor–Acceptor Approach to Molecular Interactions." Plenum, New York, 1978.
9. Cox, B. G.; Hedwig, G. R.; Parker, A. J.; Watts, D. W. *Austr. J. Chem.* **1974**, *27*, 477.
10. Abraham, K. M.; Goldman, J. L.; Natwig, D. L. *J. Electrochem. Soc.* **1982**, *129*, 2404.
11. Pearson, R. G. *J. Chem Ed. Soc.* **1968**, *45*, 581.
12. Armand, M.; Gorecki, W.; Andréani, R. "Proceedings of the Second International Symposium on Polymer Electrolytes," Scrosati, Sienna B., ed. Elsevier, London, 1992.
13. Benrabah, D.; Baril, D.; Sanchez, J. Y.; Armand, M.; Gard, G. G. *J. Chem. Soc., Faraday Trans.* **1993**, *89*(2), 355.
14. Baril, D.; Gauthier, M.; Lasia, A. *Electrochim. Acta* **1994**, in press.
15. Bordwell, F. G.; Branca, J. C.; Hughes, D. L.; Olmstead, W. N. *J. Organic Chem.* **1982**, *45*, 3305.
16. Lehn, J. M. In "Structure and Bonding," Springer-Verlag, Berlin, Vol. 16, 1973, p. 1.
17. Chatani, Y.; Okamura, S. *Polymer* **1987**, *28*, 1815.
18. Lightfoot, P.; Metha, M. A.; Bruce, P. G. *J. Mater. Chem.* **1992**, *2*, 379.
19. Vallée, A.; Besner, S.; Prud'homme, J. *Electrochim. Acta* **1992**, *37-9*, 1579.
20. Vogel, H. *Phys. Z.* **1921**, *22*, 645; Fulcher, G. S. *J. Am. Chem. Soc.* **1925**, *8*, 339; Tamman, G.; Hesse, W. *Z. Anorg. Allg. Chem.* **1926**, *165*, 245.
21. Williams, M. L.; Landel, R. F.; Ferry, J. D. *J. Am Chem Soc* **1955**, *77*, 3701.
22. Doolittle, A. K. *J. Appl. Phys.* **1951**, *22*, 1471.
23. Cohen, M. H.; Turnbull, D. *J. Chem. Phys.* **1959**, *31*, 1164.
24. LeNest, J.-F.; Gandini, A.; Cheradame, H. *Br. Polymer J.* **1988**, *20*, 253.
25. Pettit, L. D.; Bruckenstein, S. *J. Am. Chem. Soc.* **1966**, *88*, 4783.
26. Hammersley, J. M. *Proc. Camb. Phil Soc.* **1957**, *153*, 642; Zallen, R. "The Physics of Amorphous Solids." Wiley, New York, 1983.
27. Drudger, D.; Ratner, M. A.; Nitzan, A. *J. Chem. Phys.* **1983**, *79*, 3133; Drudger, D.; Ratner, M. A.; Nitzan, A. *Phys. Rev.* **1985**, *B 31*, 3939.

28. Berthier, C.; Gorecki, W.; Minier, M.; Armand, M. B.; Chabagno, J. M.; Rigaud, P. *Solid State Ionics* **1983**, *11*, 91.
29. Miyamoto, T.; Shibayama, K. *J. Appl. Phys.* **1973**, *44*, 5372.
30. MacCallum, J. R.; Tomlin, A. S.; Vincent, C. A. *Europ. Polym. J.* **1986**, *22*, 787.
31. Cheradame, H. In "IUPAC Macromolecules," Benoit, H.; Remps, P., eds. Pergamon, Oxford, 1982.
32. Kakihana, M.; Schantz, S.; Torell, L. M. *Solid State Ionics*, **1990**, *40/41*, 641.
33. Schantz, S.; Torell, L. M.; Stevens, J. R. *J. Chem. Phys.* **1991**, *94*, 6882; Petersen, G.; Jacobson, P.; Torell, L. M. *Electrochim. Acta* **1992**, *37-9*, 1495.
34. Xu, K.; Xu, Q.; Wan, G. *ISSI Lett.* **1992**, *3*, 1.
35. Besner, S.; Prud'homme, J. *Macromolecules* **1989**, *22*, 3029.
36. Spiro, M. In "Techniques of Chemistry," Weissberger, A.; Rossiter, B. W., Eds. Wiley, New York, 1970, Vol. 1, part 2A,
37. Chadwick, A. V.; Worboys, M. R. "Polymer Electrolytes Review," MacCallum, J. R.; Vincent, C. A., Eds. Elsevier, London, 1987, Vol. 1, p. 275.
38. Bridges, C.; Chadwick, A. V.; Worboys, M. R. *Br. Polymer J.* **1988**, *20*, 207.
39. Lindsay, S. E.; Whitmore, D. H.; Halperin, W. P.; Torkelson, J. M. *Polymer Preprints* **1989**, *301*, 442.
40. Gorecki, W.; Donoso, P.; Berthier, C.; Mali, M.; Roos, J.; Brinkmann, D.; Armand, M. B. *Solid State Ionics* **1990**, *28/30*, 1018.
41. Arumugam, S.; Shi, J.; Tunstall, D. P.; Vincent, C. A. *J. PHYS. C*, **1993**, *5*, 153.
42. Cameron, G. G.; Ingram, M. D.; Harvie, J. L. *Faraday Discuss., Chem Soc.* **1989**, *88*, 55.
43. Léveque, M.; Le Nest, J. F.; Gandini, A.; Chéradame, H. *J. Power Sources* **1985**, *14*, 1018.
44. Bouridah, A.; Dalard, F.; Deroo, D.; Armand, M. *Solid State Ionics* **1986**, *18/19*, 287.
45. Evans, J.; Vincent, C. A.; Bruce, P. G. *Polymer* **1987**, *28*, 2324.
46. Shi, J.; Vincent, C. A. *Solid State Ionics* **1993**, in press
47. Bannister, D. J.; Davies, G. R.; Ward, I. M.; MacIntyre, J. E. *Polymer* **1984**, *25*, 1291.
48. Liu, H.; Okamoto, Y.; Skotheim, T.; Pak, Y. S.; Greenbaum, S. G.; Adamic, K. J. In "Solid State Ionics," Nazri, G.; Huggins, R. A.; Shriver, D. F., Eds. Materials Research Society, Pittsburgh, **1989**, *135*, 337.
49. Shriver, D. F. *Macromolecules* **1988**, *21*, 2299; Tsuchida, E.; Kobayashi, N.; Ohno, H. *Macromolecules* **1988**, *21*, 96.
50. Sanchez, J. Y.; Sylla, S.; Armand, M. 4th European Polymer Federation Symposium on Polymeric Materials, Baden-Baden Germany, September 27 to October 2. Abstract O-60 132, 1992.
51. Booth, C.; Nocholas, C. V.; Wilson, D. J. *Polymer Electrolytes Review*, MacCallum, J. R.; Vincent, C. A., eds. Elsevier, London, 1987, Vol. 1, p. 275.
52. Craven, J. R.; Mobbs, R. H.; Booth, C.; Giles, J. R. M. *Macromol. Chem., Rapid Commun.* **1986**, *7*, 81; Craven, J. R.; Nicolas, C. V.; Webster, R.; Wilson, D. J.; Mobbs, R. H.; Morris, G. A.; Heatley, F.; Booth, C.; Giles, J. R. M. *Br. Polym. J.* **1987**, *19*, 509.
53. Nicolas, C. V.; Wilson, D. J.; Booth, C.; Giles, J. R. M. *Br. Polym. J.* **1988**, *20*, 289; Nekoomanesh, M.; Nagae, S. I.; Booth, C.; Owen, J. R. *J. Electrochem. Soc.* **1992**, *139*, 3046.

54. Bouridah, A.; Dalard, F.; Deroo, D.; Cheradame, H.; LeNest, J. F. *Solid State Ionics* **1985**, *15*, 233.
55. Cheradame, H.; Killis, A.; Lestel, L.; Boileau, S.; LeNest, J. F. *Polymer Preprints* **1989**, *30-1*, 420.
56. Lestel, L.; Boileau, S.; Cheradame, H. *Polymer Preprints* **1990**, *31*, 1154.
57. Watanabe, M.; Nagano, S.; Sanui, K.; Ogata, N. *J. Power Sources* **1987**, *20*, 327.
58. Xia, D. W.; Stoltz, D.; Smid, J. *Solid State Ionics* **1984**, *14*, 221.
59. Bannister, D. J.; Davies, G. R.; Ward, I. M.; MacIntyre, J. E. *Polymer* **1984**, *25*, 1600.
60. Kobayashi, N.; Ushiyama, M.; Tsuchida, E. *Solid State Ionics* **1985**, *17*, 307.
61. Cowie, J. M. G. In "Polymer Electrolyte Reviews," MacCallum, J. R.; Vincent, C. A., eds. Elsevier, London, 1987, Vol. 1, p. 69.
62. Pantaloni, S.; Passerini, S.; Croce, F.; Scrosati, B.; Roggero, A.; Andrei, M. *Electrochim. Acta* **1989**, *34*, 635.
63. Andrei, M.; Marchese, L.; Roggero, A.; Passerini, S.; Scrosati, B. In "Second International Symposium on Polymer Electrolytes," Scrosati, B., ed. elsevier, London, 1990, p. 107.
64. Blonsky, P. M.; Schriver, D. F.; Austin, P.; Allcock, H. R. *J. Am. Chem. Soc.* **1984**, *106*, 6854.
65. Blonsky, P. M.; Schriver, D. F.; Austin, P.; Allcock, H. R. *Solid State Ionics* **1989**, *18/19*, 258.
66. Nazri, G. A.; MacArthur, D. M.; Ogara, J. F. *Polymer Preprints* **1989**, *30-1*, 430.
67. Fish, D.; Khan, I. M.; Smid, J. *Makromol. Chem. Rapid Commun.* **1986**, *7*, 115.
68. Khan, I. M.; YuanI, Y.; Fish, D.; Wu, E.; Smid, J. *Macromolecules* **1989**, *21*, 2684.
69. Weston, J. E.; Steele, B. C. H. *Solid State Ionics* **1982**, *7*, 75.
70. Chen, L. In "Materials for Solid-State Batteries," Chowdari, B.; Radhakrishna, V., eds. World Scientific, Singapore, 1988, p. 69.
71. Capuano, F.; Croce, F.; Scrosati, B. *J. Electrochem Soc.* **1991**, *138*, 1918.
72. Croce, F.; Scrosati, B.; Mariotto, G. *Chem. Mater.* **1992**, *4*, 1134.
73. Kelly, I.; Owen, J. R.; Steele, B. C. H. *J. Electroanal. Chem., Interfacial Electrochem.* **1984**, *168*, 467; Kelly, I.; Owen, J. R.; Steele, B. C. H. *J. Power Sources* **1985**, *14*, 13.
74. Tsuchiya, J. In "Solid State Ionics," Nazri, G.; Huggins, R. A.; Shriver, D. F., eds. Materials Research Society, Pittsburgh, 1989, p. 357.
75. Fauteux, D. In "Polymer Electrolyte Reviews," MacCallum, J. R.; Vincent, C. A., eds. Elsevier, London, 1989, p. 69.
76. Cameron, G. G.; Ingram, M. D.; Sarmouk, K. *Eur. Polym. J.* **1990**, *26*, 197.
77. Feuillade, G. Ph. Perche, *J. Appl. Electrochem.* **1975**, *5*, 63.
78. Abraham, K.M.; Alamgir, M. *J. Electrochem. Soc.* **1990**, *137*, 1657.
79. Huang, H.; Chen, L.; Huang, X.; Xue, R. *Electrochim. Acta* **1992**, *37-9*, 1671.
80. Huq, R.; Koksbang, R.; Tonder, P. E.; Farrington, G. F. *Electrochim. Acta* **1992**, *37-9*, 1681.
81. Paushkin, Y. A.; Vishynakova, T. P.; Lunin, A. F.; Nizova, S. A. "Organic Polymeric Semiconductors." J. Wiley, New York, 1974.
82. Feast, W. J. in "Handbook of Conducting Polymers," Skotheim, T. A., ed. Marcel Dekker, New York, 1986, Vol. 1, p. 1.

83. Naarmann, H.; Theophilou, N. In "Electroresponsive Molecular and Polymeric Systems," Skotheim, T. A., ed. Marcel Dekker, New York, p. 1.
84. Jozefowitcz, M. In Fast Ion Transport in Solids, Van Gool, W., ed. North Holland, Amsterdam, 1973, p. 623.
85. Fouletier, M.; Degott, P.; Armand, M. *Solid State Ionics* **1983**, *8-2*, 165.
86. Okabayashi, K.; Goto, F.; Abe, K.; Yoshida, T. *J. Electrochem. Soc.* **1989**, *136-7*, 1986.
87. Servagent, S.; Vieil, E. *J. Electroanal. Chem.* **1990**, *280*, 227.
88. Bose, C. S. C.; Basak, S.; Rajeshwar, K. *J. Phys. Chem.* **1992**, *96*, 9899.
89. Visco, S. J.; Maillhé, C. C.; Liu, M.; Armand, M. B.; Dejonghe, L. C., Pacific Conference on Chemistry and Spectroscopy, Irvine, California, Abstract 175 October 28–30, 1987.
90. Visco, S. J.; Liu, M.; Dejonghe, L. C. *J. Electrochem. Soc.* **1990**, *137*, 1191.
91. Liu, M.; Visco, S. J.; Dejonghe, L. C. *J. Electrochem. Soc.* **1991**, *138*, 1891.
92. Sotomura, T.; Uemachi, H.; Takeyama, K.; Naoi, K.; Oyama, N. *Electrochim. Acta* **1992**, *37-10*, 1851.
93. Armand, M.; Degott, P. American Chemical Society Meeting, Annaheim, California, Extended Abstract 076, September 7–12, 1986.
94. Prasad, P. S. S.; Lee, H. S.; Xu, Z. S.; Skotheim, T. A.; Oyama, N. Electrochemical Society Meeting 92-2, Extended Abstract 45, Toronto, October, 1992.
95. West, K.; Jacobsen, T.; Atlung, S. *J. Electrochem. Soc.* **1982**, *129*, 1480.
96. Atlung, S. In "Solid State Batteries," *NATO ASI*, Nijhoff, Dordrecht, 1985, p. 129.
97. Peled E. In "Lithium Batteries," Gabano, J. P., ed. Academic, New York, 1983, Chapter 3.
98. Blomgren, G. E. In "Lithium Batteries," Gabano, J. P. ed. Academic, New York, 1983, Chapter 2.
99. Thevenin, J. *J. Power Sources* **1985**, *14*, 45.
100. Garreau, M. *J. Power Sources* **1987**, *20*, 9.
101. Nazri, G.; Muller, R. H. *J. Electrochem. Soc.* **1985**, *132*, 2050.
102. Aubarch, D.; Droux, M.; Faguy, P.; Yeager, E. *J. Electroanal. Chem.* **1991**, *297*, 225.
103. Fauteux, D. *Solid State Ionics* **1985**, *17*, 133.
104. Scrosati, B. In "Polymer Electrolyte Reviews," MacCallum, J. R.; Vincent, C. A., eds. Elsevier, London, 1987, Vol. 1, p. 69.
105. Armand, M. B. In "Lithium Non-aqueous Battery Electrochemistry," Yeager, E. B.; Schumm, B.; Blomgren, G.; Blankenship, D. R.; Leger, V.; Akridge, J., Eds. The Electrochemical Society, Pennington, New Jersey, **80-7** 1980, p. 261.
106. Sequiera, C. A. C.; North, J. M.; Hooper, A. *Solid State Ionics* **1984**, *13*, 175.
107. Bruce, P. G.; Krok, K. *Solid State Ionics* **1989**, *36*, 175.
108. Dahn, J. R.; Von Sacken, U.; Jukow, M. R. *J. Electrochem. Soc.* **1991**, *137*, 2207.
109. Tarascon, J. M.; Guyomard, D.; Jukow, M. R. *J. Electrochem Soc.* **1991**, *138*, 2864.
110. Novàk, P.; Inganäs, O.; Bjorklund, R. *J. Electrochem. Soc.* **1987**, *134-6*, 1341.
111. Vassort, G.; Gauthier, M.; Harvey, P. E.; Brochu, F.; Armand, M. B. "Proceeding of the Symposium on Lithium Batteries, Honolulu," Dey, A. N., eds. The Electrochemical Society, Pennington, New Jersey, **87-1** 1988.

112. Gautheir, M.; Bélanger, A.; Kapfer, B.; Vassort, G.; Armand, M. "Polymer Electrolyte Reviews," MacCallum, J. R.; Vincent, C. A., eds. Elsevier, London, 1989, Vol. 2, p. 285.

113. Choquette, Y.; Gauthier, M.; Bélanger, A.; Kapfer, B. 6th International Meeting on Lithium Batteries, Münster, Germany, May 10–15, 1992.

114. Hooper, A. In "Materials and Processes for Lithium Batteries," Abraham, K. M.; Owen, B. B., eds. The Electrochemical Society, Pennington, New Jersey, **89-4,** 1989.

115. Bélanger, A.; Gautheir, M.; Robitaille, M.; Bellemare, R. "Second International Symposium on Polymer Electrolytes," Scrosati, B., ed. Elsevier, London, 1990, p. 347.

116. Gauthier, M.; Duval, M.; Kapfer, B.; Vassort, G.; St-Amand, G.; Bouchard, P.; Laroche, G.; Ricard, S. "The Tenth International Seminar on Primary and Secondary Battery Technology and Application," Wolsky, S. P., Ansum Enterprises, Inc., Florida, March 1–4, 1993. (Available from the authors also.)

117. Belanger, A.; Kapfer, B.; Gauthier, M. "10th International Electric Vehicle Symposium, Proceeding EVS-10, Hong-Kong, p. 599, 1990. Available from the authors also."

CHAPTER 3

Insertion Compounds for Lithium Rocking Chair Batteries

Bruno Scrosati

3.1 Introduction

The rapid development of the present technology gives particular urgency to the need for new and efficient power source systems. This need comes from a series of crucial demands that range from the request of an efficient utilization of our energy resources to the constraints imposed by environmental protections. This has favored the research on advanced, high-energy, electrochemical systems capable of replacing the conventional batteries with a more efficient and less polluting operation. For instance, high-energy-density rechargeable batteries are needed today to replace bulky lead–acid batteries for the development of long-range electric vehicles, with consequent decrease in oil consumption and, most important, improvements in the air quality of large urban areas. Furthermore, reliable batteries are in demand for off-peak electric energy storage, as emergency power supplies in remote rural areas and as storage systems for intermittent energy sources, such as solar and wind. Finally, advanced and environmentally friendly batteries would be highly welcome in the electronic consumer market to replace the nickel–cadmium batteries or even the most common zinc–carbon dry cells, with the final goal of limiting the risk associated with their waste disposal.

For effective development of a high-energy-density battery, high-capacity electrode materials are essential. Alkali metals are the obvious choices; indeed, the most promising types of advanced batteries currently under production are based on these metals as negative electrodes. For example, the advanced system that appears to be closest to industrial production—namely, the sodium–sulfur battery—uses a molten sodium anode. A lithium–aluminum alloy is the anode of another system of present interest for advanced electrochemical storage (i.e., the lithium–iron sulfide battery). Both these batteries are today regarded as serious candidates for the large-scale development of efficient electric vehicles. In fact, these batteries have been included among the four most promising electrochemical systems to meet the criteria established by the so-called United States Advanced Battery Consortium (USABC), formed in early 1991 by three of the major car companies in North America—General Motors, Ford, and Chrysler—together with two U.S. government Institutions, the Department of Energy and the Electric Power Research Institute, with the purpose of promoting research and development work on advanced batteries for electric vehicle propulsion [1]. It is interesting to note that the USABC was established in response to the "air cleaning" project enacted in September 1990 by the state of California, according to which 2% of all new cars sold in the state in 1998 will be required to be "zero-emission" vehicles, which necessarily means electric vehicles.

However, both the Na–S and the Li(Al)–FeS$_2$ batteries operate at temperatures much higher than ambient—namely, at about 350°C and 450°C, respectively. Such high operational temperatures introduce some serious technological problems. First, the necessity of expensive materials that resist the highly corrosive molten electrode (e.g., Na and S) and electrolyte (e.g., LiCl–KCl) components and, second, the addition to the electrochemical system itself of a thermal control unit in order to keep the battery running. The latter does not appear to be a crucial problem for the off-peak (load-leveling) or even for the electric vehicle applications. However, it is unacceptable for a more versatile use, where operation may be expected at "ambient temperature," which, depending on the geographic location, means temperatures ranging from −10°C to 40°C. Therefore the development of ambient-temperature, high-energy batteries is today a major task; accordingly, many laboratories throughout the world are carrying out research aimed at reaching this important goal.

3.2 Lithium Batteries

The most promising results in the research of new, alternative power sources have been obtained so far with electrochemical systems using a lithium anode, a lithium-ion-conducting electrolyte, and a lithium-ion-accepting cathode material. The choice of the anode material is restricted by the already stressed need of the high-energy content, which is unavoidably linked to the use of an alkali metal as the main electrodic material. Lithium is generally preferred, since it is more easily handled than other alkali metals and, most significantly, is the lightest and the most electropositive of the family. In fact, lithium metal has an extremely high value of specific capacity—namely, 3.86 Ah g^{-1}.

The choice of the cathode is somewhat more flexible, since various materials can in principle assure the electrochemical balance for the overall lithium battery process. In this respect, one could select a given element A_y capable of reacting directly with lithium to form a lithium compound, LiA, according to a general electrochemical process of the type

$$yLi + A_y \Rightarrow yLiA \tag{3.1}$$

A practical example is the system where lithium is coupled with iodine, which is characterized by the following electrochemical reaction:

$$2Li + I_2 \Rightarrow 2LiI \tag{3.2}$$

Indeed, lithium–iodine batteries are currently produced as power sources in medical devices as, for instance, cardiac pacemakers [2].

Other possibilities involve the choice of compounds of the A_xB_z type, which are capable of assuring a displacement electrochemical process giving the corresponding Li_yB_z salt and the element A:

$$yLi + A_xB_z \Rightarrow Li_yB_z + xA \tag{3.3}$$

A typical example is the Li–CuS battery, which operates on the basis of the following process:

$$2Li + CuS \Rightarrow Li_2S + Cu \tag{3.4}$$

Again, this battery is presently of commercial interest for the electronic watch market [3].

However, these batteries, and many other examples of the same kind [4], are based on electrochemical processes that involve breakages and rearrangements of bonds. Consequently, all of them may suffer from poor reversibility at low temperature, thus restricting their use to the primary function.

Quite reasonably, the less the extent of bonding and structure modifications of the selected cathode material, the more likely it is that rechargeable, long-cyclable battery systems will be developed successfully. The most suitable materials in this respect are the so-called insertion compounds—namely, A_zB_y compounds having an open structure capable of accepting and releasing x number of lithium ions per A_zB_y mole, as the results of a reversible electrochemical reaction of the following type:

$$xLi^+ + A_zB_y + xe^- \Leftrightarrow Li_xA_zB_y \tag{3.5}$$

This reaction induces the formation of metastable phases accompanied with minor and reversible modifications in the structure of the host oxide A_zB_y. Such a topotactic or topochemical [5] reaction is usually called a *lithium insertion* or *lithium intercalation* process and is basically described as the interstitial insertion–extraction of both *mobile* lithium ions and *compensating* electrons into a rigid host structure. The guest Li^+ ions induce reversible modifications in the host structure, and the guest electrons induce reversible changes of the electronic properties of the host A_zB_y compound.

The insertion electrodes may then be classified as mixed-valence compounds, having a mixed ionic-electronic conductivity. A typical, well-known example is titanium disulfide (TiS_2), which has a structure formed by a sequence of S–Ti–S sandwiches, where the titanium atoms/ions are octahedrally surrounded by two layers of sulfur atoms in a hexagonally close-packed arrangement. The S–Ti–S sandwiches are stacked together by weak van der Waals bonds [6,7], and this provides easy access for the guest lithium ions, which can diffuse in the two-dimensional gaps between the dichalcogenide layers. The layered titanium disulfide structure provides one octahedral and two tetrahedral sites in the gaps per TiS_2 molecule. The lithium ions normally occupy the octahedral sites, since they are much too large to be accommodated into the tetrahedral sites.

The combination of a lithium metal anode with a titanium disulfide cathode in a Li^+-ion-conducting electrolyte gives rise to an electrochemical cell characterized by the process

$$xLi + TiS_2 \Leftrightarrow Li_xTiS_2 \tag{3.6}$$

The overall discharge process involves the dissolution of x lithium ions at the anode, their migration across the electrolyte, and their insertion within the crystal structure of the TiS_2 host compound, and the compensating electrons travel in the external circuit to be injected finally in the electronic band structure of the same hosting compound (Fig. 3.1). The overall charge process is just the opposite.

Therefore any liquid or solid conducting material characterized by fast lithium ion transport can be used as an electrolyte medium for these batteries. Common examples of liquid electrolytes are solution of lithium salts LiX (e.g., $LiClO_4$, $LiAsF_6$, $LiBF_4$, etc.) in aprotic organic solvent (e.g., propylene carbonate, PC, dimethoxyethane, DME, methyl-tetrahydrofurane, MeTHF, etc.) or their mixtures (e.g., PC–DME). Crystalline or glassy compounds having vacancy or interstitial defects energetically favorable for Li^+ transport may act as solid electrolytes. Typical examples are $LiI.Al_2O_3$ composite powders [8] or LiI–$Li_xP_yO_zS_w$ chalgogenide glasses [9]. Finally, an important class of electrolytes suitable for lithium batteries that have a compromised solid-to-liquid structure are polymeric membranes formed by the solvation of lithium salts in high-molecular-weight polymers, such as

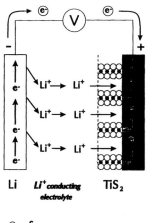

Figure 3.1. Scheme of the discharge process of a Li/TiS_2 cell.

poly(ethylene oxide), PEO (e.g., PEO–LiClO$_4$, PEO–LiCF$_3$SO$_3$, etc.) [10] or by liquid solutions (e.g., LiClO$_4$ in PC) trapped in a polymer matrix (e.g., a polyacrylonitrile, PAN, matrix) [11].

The insertion electrodes most commonly used as cathodes in lithium batteries are inorganic compounds, such as transition metal dichalcogenides and oxides, characterized by layered or tunneled structures capable of providing channels for the easy access and fast mobility of the guest lithium ions. In principle, these compounds can assure a very long cycle life to the battery. In practice, however, the life of the battery may be limited by the cyclability of the lithium metal negative electrode. In fact, corrosion side reactions may induce the growth of passivation layers on the electrode surface that greatly affect the uniformity of the plated lithium during the charge process, to an extent that may ultimately lead to a total cell failure (due to dendritic short circuiting) or even to serious safety hazard (due to local overheating and pressure buildup [12]). Indeed, a few incidents have occurred, with occasional fires in equipment powered by lithium batteries, even for prototypes assembled in industrial laboratories having recognized experience in lithium battery technology.

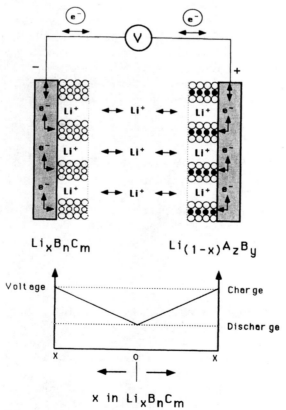

Figure 3.2. Scheme of the electrochemical process of a lithium rocking chair battery.

One way that has been proposed for overcoming this problem is to replace the lithium anode by another insertion compound—say, B_nC_m—capable of accepting and exchanging a large quantity of lithium ions. In this way, rather than lithium plating and stripping, as in the conventional systems, the electrochemical process at the negative side would be the uptake of lithium ions during charge and their release during discharge. Therefore the negative B_nC_m electrode acts as a "lithium sink," and a selected positive LiA_zB_y electrode acts as a "lithium source" and the total electrochemical process of the B_nC_m/LiA_zB_y cell involves the cyclic transfer of x equivalents of lithium ions between the two insertion electrodes (Fig. 3.2):

$$B_nC_m + LiA_zB_y \Leftrightarrow Li_xB_nC_m + Li_{(1-x)}A_zB_y \qquad (3.7)$$

These still uncommon electrochemical systems may be described as concentration cells where lithium ions "rock" from one electrodic side to the other; accordingly, these cells have been termed *rocking chair batteries* [13].

A large number of insertion compounds have been characterized and studied in view of their application in rechargeable lithium batteries. It would be impossible, and perhaps an unuseful exercise, to include here coverage of the preparation procedures and of the structural thermodynamic and kinetic properties of the many insertion electrodes that have been investigated in the latest decades, also considering that many extensive reviews [5–6,14–18] have been published on the subject. Therefore it seems more appropriate to restrict the content of this chapter to the description of the characteristics and of the behavior of selected types of insertion compounds—namely those that are presently considered as the most promising for the development of the novel types of rechargeable lithium batteries (i.e., the rocking chair batteries).

3.3 Criteria for the Selection of Insertion Electrodes for Rocking Chair Batteries

Considering the general design, which involves a positive lithium-source electrode combined with a negative lithium-accepting electrode and the related nature of the electrochemical driving process, a successful operation for a rocking chair battery and its effective competition with a metal lithium system requires some crucial conditions that should be fulfilled by the selected insertion electrodes. They are in essence the following:

1. The lithium activity in the negative electrode $Li_xB_nC_m$ must be close to 1 in order to assure open circuit voltages approaching those obtainable with the pure lithium.
2. The equivalent weight of both electrodes must be low in order to assure specific capacity values of practical interest.
3. The diffusion coefficient of Li^+ ions in both the lithium-source $Li_{(1-x)}A_zB_y$ positive electrode and in the lithium sink $Li_xB_nC_m$ negative electrode must be high to assure fast kinetics of the electrochemical process and thus fast charge and discharge rates.

4. The voltage changes upon lithium ion uptake and release must be small in both electrodes to limit fluctuations during charge and discharge cycles.
5. Both the ion-source and the ion-sink electrode must be easy to fabricate and based on nontoxic compounds, to assure low cost and environmental control.

To meet these conditions satisfactorily, insertion compounds must be selected with properties consistently different from those normally used for the conventional lithium batteries. For instance, conditions 1 and 2 can be achieved only by using innovative materials that could assure Li insertion voltages approaching zero (versus Li). The most popular in this respect are today graphite-type insertion compounds.

Furthermore, the achievement of the previously mentioned crucial conditions requires the selection of both a low-voltage negative and a high-voltage positive electrode. The most convenient are compounds having Li-insertion voltages around 4 V—namely, layered lithium metal oxides of the $LiMO_2$ type, where M = Co or Ni, and the three-dimensional, spinel-type lithium manganese oxides.

3.4 Carbon Insertion Materials

As already pointed out, one of the main requirements for a successful rocking chair battery is the proper replacement of the Li metal negative with a Li-insertion compound capable of maintaining the Li^+ activity as close to 1 as possible. In fact, only under this condition can the selected compound exhibit a potential approaching that of the metal and thus allow fabrication of rocking chair batteries with open circuit voltages approaching those of other conventional lithium batteries. In this respect, carbon electrodes, which have in their various graphite-type textures a well-established ability to accept large quantities of lithium, appear to be a very convenient choice. In fact, carbon is a common, low-cost material, and its use is familiar in the battery [19] and fuel cell [20] technologies. Many forms of carbon would be available in principle as electrode materials in lithium batteries. They can be obtained from a variety of sources, which range from natural graphite, pitches, cokes, pyrolitic carbon [21] to the C_{60} fullerenes of recent discovery [22].

All these forms, depending upon the starting raw material, the preparation method, and the heat treatment, differ in their crystalline organization and microtexture [23]. These differences influence consistently the characteristics of the lithium insertion process. In fact, from a mere structural point of view, a large degree of crystallinity would be desirable, since it provides the conditions for the insertion of lithium between organized graphite layers. In this respect, natural graphite with high crystallinity would appear the most suitable material. However, experimental results, obtained in liquid electrolyte cells [24,25], have shown that the intercalation process in highly organized crystalline graphite proceeds only to a very limited extent, or it is even totally unsuccessful, since its compact structure is unable to accommodate the solvated lithium ions. Indeed, large intercalation capacity of lithium in graphite was obtained only when using a solid, $(PEO)_8LiClO_4$, polymer electrolyte [26], where the Li^+ ions move free of salvation shells.

Therefore in common liquid electrolyte cells, a successful electrochemical insertion process requires the use of carbon forms that, still maintaining a basic graphite-type structure, have a less organized crystallinity than that of pure graphite. In synthesis, the crystallinity, which is necessary to promote the lithium insertion process, must be accompanied by free volumes to account for the cointercalation of solvent. In this respect, microtexture may become an important parameter in selecting the proper material.

Taking into account these considerations, many carbonaceous electrodes have been developed and characterized in the latest years for the use in advance, lithium rocking chair batteries. The most common of them is petroleum coke, which has a structure formed by disordered, randomly packed, graphite layers [27], where the spacing between the layers is generally the order of 3.47 Å. Successful lithium insertion has also been obtained in carbon fibers obtained by thermal decomposition of polymers [28], in phosphorus-containing films prepared by heat treatment of polyfurfuryl alcohol [29] and in alcohol mesophase-pitch-based carbon fibers [30]. As expected, because of their high crystallinity, attempts of lithium insertion in fullerenes have instead been unsuccessful [31], mainly because the reduced Li_xC_{60} fullerenes lose their crystallinity and dissolve into the most common liquid electrolytes [32].

The favorable form of carbon electrodes (e.g., petroleum coke) can insert lithium according to the following basic scheme:

$$xLi + 6C \Rightarrow Li_xC_6 \qquad (3.8)$$

where $0 < x < 1$.

The LiC_6 form, which can also be prepared chemically under adequate temperature and pressure conditions [21,33], has a golden color and belongs to graphite intercalation compounds (GICs) of stage 1, where the stage number corresponds to the number of graphite layers that separate two successive intercalated planes. The LiC_6 form has a hexagonal unit cell where the separation distance between the lithium and the carbon atoms is 4.3 Å.

Practical carbon electrodes can be prepared in a variety of morphologies, which include fibers, pellets, or films. The fabrication of the electrode requires the addition of a binder, the most used being ethylene propylene diene monomer (EPDM) [31,34]. However, other materials, e.g. polyvinyl chloride (PVC), have also been successfully used [35]. The final shape of the electrode is generally obtained by forming a slurry of carbon, the binder, and a suitable solvent. The electrode is then shaped in the desired form by a casting or pressing procedure. Pelletlike configurations have been realized by casting a cyclohexane slurry of powdered active carbon material and EPDM. The resulting blend was pressed at 10 tons cm^{-2} in the desired electrode shape and then annealed at 150°C for one hour [34]. Carbon electrodes can also be prepared in the form of membranes by dissolving in a suitable solvent [e.g., tetrahydrofuran (THF), a PVC former] and then slowly adding carbon in the desired concentration. The slurry is mechanically stirred and free-standing electrodic films can be obtained by a standard (e.g., Doctor Blade) lamination technique [35].

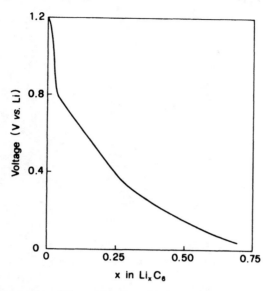

Figure 3.3. Voltage-composition curve for a Li_xC_6 electrode at room temperature determined in a cell having a Li counter electrode and a $LiClO_4$-PC electrolyte. (From Ref. 36.)

Figure 3.3 illustrates the voltage-composition curve for a carbon electrode, determined by monitoring the voltage during passage of small-intensity current in a cell using a Li metal counterelectrode and a $LiClO_4$-PC liquid electrolyte [36]. The flowing charge promotes the insertion of Li^+ ions within the layered, graphite-type structure of the carbon elecrode whose voltage decreases accordingly to finally reach at the maximum concentration ($x = 1$) a value approaching 0.02 V versus Li. Powder x-ray patterns have shown that upon Li^+ ions insertion, the interlayer spacing between the graphite layers of the Li_xC_6 electrode varies from 3.47 to 3.77 Å in the $0 < x < 0.7$ range [34]. This indicates that the electrochemical process does not induce formation of new staged phases but only the expected reversible Li^+ instertion and extraction process [34,37].

However, when submitted to charge–discharge cycles, the Li_xC_6 electrode inevitably shows an intial loss in capacity: Only a fraction of the charge consumed in the first "charge" (Li-uptake) is released and exchanged in the following discharge (Li-release). This effect, which is illustrated in Fig. 3.4, has been observed by various authors [36,31,35,28], and it is generally explained by assuming that part of the initial charge is dissipated by a side process involving the decomposition of the electrolyte. It is also generally accepted that this side process induces the formation on the electrode surface of a passivation layer that, being electronically insulating but ionically conducting, prevents further electrolyte decompostion while allowing ionic transfer with the solution [36]. In synthesis, the formation of the passivation layer seems to be an essential effect in assuring the stability and the cyclability of the Li_xC_6 electrode, since it provides the conditions for the desired electrochemical

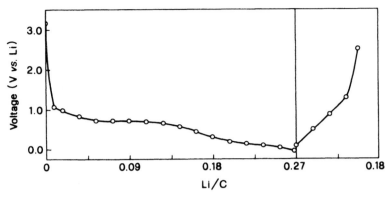

Figure 3.4. Charge balance between the first "charge" (lithium uptake)—discharge (lithium release) cycle of a Li_xC_6 electrode. (From Ref. 28.)

operation even at voltage levels that fall well below the stability window of the most common liquid electrolytes.

Figure 3.5 illustrates some recent results [35] obtained by running a slow-scan cyclic voltammetry of a Li_xC_6 electrode in the $LiClO_4$-PC electrolyte. The trend of the curves clearly shows that the amount of cyclable charge decreases consistently, passing from the first initial cycle to the second and third following cycles, after which a steady-state behavior is approached. Therefore these cyclic voltammetry

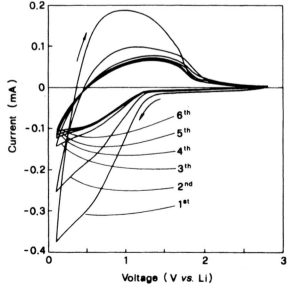

Figure 3.5. Cyclic voltammetry of a Li_xC_6 electrode in the $LiClO_4$-PC electrolyte at room temperature. Counter electrode: Li. Scan rate: 0.1 mV s^{-1}. (From Ref. 35.)

results also seem to support the commonly accepted theory that the initial loss of charge is followed by a stable and reproducible response for the Li_xC_6 electrode. However, the voltammetric curves do not provide conclusive evidence for the proposed passivation mechanism. In fact, by closely examining the shape of the first cycle of Fig. 3.5, one can detect a shoulder in the cathodic (Li-uptake) scan (at about 0.7 V versus Li) and no corresponding discontinuity in the following anodic curve. It is difficult to associate this irreversible effect just to a passivation phenomenon; other interpretations, more directly related to the feature of the lithium insertion process, may hold as well. In fact, considering that the structure of the carbon electrode may be described as a combination of distorted layered graphite stacks [27], it is reasonable to assume that different sites, with different coordinating energy, are available for accommodating the initial incoming lithium ions. Accordingly, the initial loss of charge could also be associated in part with those fraction of lithium ions stored in strongly screened lattice positions from which they cannot be easily removed by the electrochemical reverse process. In this connection it is important to point out that initial irreversible uptake of guest ions is a very common phenomenon in insertion electrochemical processes, which has been established in electrodes such as Li_xNiO_y, Li_xWO_3 and which in fact has been postulated to occur in the Li_xC_6 electrode of direct interest here [31,38].

Attempts to obtain some further information on the effective mechanism of the carbon electrode have been made using impedance analysis, a technique that has been successfully used to detect passivation and growth of film formation on lithium metal electrodes in contact with a variety of liquid [39] and solid [40] electrolytes. Figure 3.6 illustrates the response in the $-jZ'' - Z'$ plane of a Li_xC_6 electrode observed at $x = 0$, $x = 0.27$, and $x = 0.52$ compositions in a cell using $LiClO_4$-PC

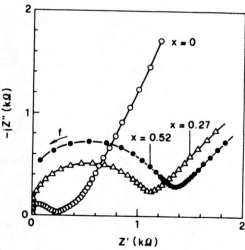

Figure 3.6. Impedance response of a Li_xC_6 electrode in the $LiClO_4$-PC electrolyte at various compositions ($x = 0$, $x = 0.27$, and $x = 0.52$) and at room temperature. Counter electrode: Li. (From Ref. 35.)

as liquid electrolyte. The response clearly reveals that the middle-frequency semicircle, which is associated with the interfacial resistance, increases upon lithium content. However, this result (which has also been observed in other liquid electrolyte media [33]) is not conclusive in fully clarifying the electrode mechanism, since if from one side the expansion of the semicircle could well be attributed to an increase of the interfacial resistance due to the formation of a passivation layer, from the other side the development of a second semicircle that could clearly distinguish the effect of this layer from other interfacial phenomena is not detected in the spectra of Fig. 3.6. The full clarification of the mechanism responsible for the initial loss of charge of the Li_xC_6 electrode is obviously a crucial requirement in view of the evaluation of the effective role that carbon electrodes may exert in the development of reliable rocking chair batteries, and further work should certainly be devoted to this problem.

Another point of concern in determining the conditions for the use of the Li_xC_6 electrode in replacement of the lithium metal electrode is in the respective values of the specific capacity, since that of the former ($x = 0.5$) is 0.186 Ahg^{-1} [i.e., more than one order of magnitude lower than that of the latter (3.86 Ahg^{-1})]. Admittedly, one has to recall that, because of the cited poor cyclability of the metal, an excess of lithium, generally estimated to be four times higher than the theoretical amount, is required to assure acceptable life to conventional Li/A_zB batteries. However, also under this consideration, the specific capacity of lithiated carbon remains lower than that of metal lithium, and this is reflected in lower attainable energy density values for carbon rocking chair batteries in comparison with lithium batteries. Unfortunately, there are no practical ways to overcome this drawback, since it is inherent in the choice of any carbon or other insertion anode electrodes used as an alternative to lithium metal. Therefore if priority is given to rocking chair batteries, one inevitably has to deal with comparatively low-energy-density systems and cope with the related penalties by emphasizing other advantages specific to the rocking chair concept, such as safety and cyclability [41].

To achieve these advantages, care must be taken to assure proper configuration of the rocking chair electrodes, since there is still the possibility that the practical electrochemical process may deviate from the expected eq. (3.7). In fact, depending on the type of negative electrode and the type of the cycling regime, conditions for promoting lithium plating rather than lithium insertion may be created unintentionally. For instance, in the example of the Li_xC_6 anode, one can reasonably consider that uncontrolled operating procedures may drive the voltage of the carbon electrode to values at which lithium plating may become the predominant process. In fact, this may very likely occur when the cell is anode limited: attempts to pull out the residual lithium stored in the cathode side will inevitably drive the potential of the carbon electrode to values favorable for Li plating, with the risk of building up regions of electrochemically deposited, highly reactive metallic lithium.

In addition to all the preceding stressed points, attention should be paid to fulfill condition 4 of the list in Section 3.3, which recommends small voltage fluctuations upon cell operation. One can notice from Fig. 3.3 that this requirement is difficult to obtain with the Li_xC_6 (coke) electrode, since its voltage varies as much as 1.5 V upon the exchange of the total removable and cyclable lithium ($\Delta x = 0.5$).

Under the most general cases, voltage composition curves for insertion electrodes characterized by processes similar to eq. (3.5) may be obtained by passing a known amount of charge in cell of the Li/A_zB_y type and then monitoring under open circuit conditions the voltage acquired by the $Li_xA_zB_y$ electrode until a constant value—namely, until equilibrium (i.e., when lithium ions have diffused uniformly throughout the entire electrodic mass) is reached. At this stage, the chemical potential μ_{Li}^+ of the lithium ions in the $Li_xA_zB_y$ medium is given by

$$\mu_{Li}^+ = \mu^o_{Li} + RT \ln a_{Li}^+ = \mu^o_{Li} + RT \ln (\gamma_{Li}^+ \cdot x_{Li}^+) \tag{3.9}$$

where x_{Li}^+ is the mole fraction of the Li ions, γ_{Li}^+ is their activity coefficient, and μ^o_{Li} is the chemical potential of the pure lithium at the same temperature (i.e., when $\gamma_{Li}^+ = 1$ and $x_{Li}^+ = 1$). The open-circuit voltage, OCV, is directly related to the difference in the chemical potential of lithium in the insertion electrode and in the lithium counter electrode. Thus the OCV of the $Li_xA_zB_y$ electrode at any composition x is given by

$$OCV = -(RT/F) \ln (\gamma_{Li}^+ \cdot x_{Li}^+) \tag{3.10}$$

It follows that large voltage fluctuations with concentration may be associated with large changes in the chemical potential of the electrochemically active ionic specie (e.g., of the Li+ ions in the host solid medium). In the case of Li_xC_6, one may then assume that the observed voltage fluctuations are associated with strong interactions between the Li+ ions and the carbon host structure. Therefore in future development of the rocking chair battery, some attention should be devoted to the characterization of anode materials where a lithium activity approaching one in the lithium-rich state is not the sole requirement, but also where a limited variation in Li+ chemical potential upon the insertion–deinsertion process is taken into consideration.

Another aspect of crucial importance in determining the performance of the carbon electrode, and of insertion electrodes in general, is related to the value of the diffusion coefficient of the Li+ ions throughout the solid framework of the host structure. As pointed out by condition 3 of the list in Section 3.3, the diffusion of the Li+ ions controls the kinetics of the electrochemical process and thus the power capability of the related battery.

The determination of ion diffusion coefficients in solid matrices has been the object of consistent attention in the latest years. Most of the measurements, however, have been based on the same experimental techniques and on the same interpreting equations that have been traditionally used and developed for diffusion processes occurring in liquid ionic media, and in certain cases this may lead to inaccurate results. In fact, quite different types of ion–ion and ion–solvent interactions are expected in the two media. In the solid matrices the ion mobility influenced not only by the number of available sites but also by their related energy. For instance, different crystallographic sites can have different lattice energies; thus they can influence the ion mobility differently. Furthermore, the ionicity of the lattice of the hosting compound can also play a specific role in influencing the mobility by shielding the effective lattice charge seen by the traveling ion. In synthesis, while in the liquid media the ions move in a "continuum" environment, thus permitting accurate diffusion models, in the solid, multiphase media the ions may travel along different routes having different lattice energies, thus favoring processes that cannot be analyzed directly by the same diffusion models. It is out of the scope of this chapter to discuss in more detail the implications of these differences, nor can we describe the various attempts that have been made to modify the classical diffusion equations in order to adept them to the study of the solid-state insertion processes. Here we can only remark that the values reported for the diffusion coefficient of lithium ions in various insertion electrodes may be affected by a certain degree of uncertainty and thus must often be taken with some caution.

Generally, one can measure two types of diffusion coefficients for mass transport in nonstoichiometric ionic medium, such as Li_xTiS_2—namely, the self diffusion coefficient D^* and the chemical diffusion coefficient D. The difference is in the fact that the former is measured in the absence of concentration gradient, so that the two are related through the following equation [42]:

$$D = D^* (d \ln a/d \ln c) \qquad (3.11)$$

which shows that the two diffusion coefficients become equivalent in ideal solid solutions where the activity coefficient is unity. However, for some of the reasons outlined earlier, the intercalated electrodes are generally far from approaching ideality, and thus the two values are different, sometimes by orders of magnitude.

Various experimental methods have been proposed for the determination of the chemical Li^+ diffusion coefficient in insertion electrodes. Most of the methods used in the early studies utilized modifications of the Cottrell transient techniques developed on the basis of the solution of the Fick's second equation for an instantaneous planar source of diffusing species in a semi-infinite geometry [43]. Basically, the technique consists of applying short (few seconds) galvanostatic (or potentiostatic) pulses to promote excess of concentration of the diffusing Li^+ species at the insertion electrode surface and then following the consequent voltage (or current) relaxation as a function of time after switching off the current [44,45]. Alternatively, the technique involves the application of a galvanostatic pulse of very short duration (<1 s) to the cell and the analysis of the resulting overvoltage-time transient during the pulse. Here the overvoltage is a function of the concentration of the diffusing lithium ions at the electrode–electrolyte interface.

A transient technique of these types, called *potentiostatic intermittent titration technique*, was used for the determination of the chemical diffusion coefficient of the Li^+ ions, D_{Li^+}, in the Li_xC_6 carbon electrode [34]. The technique consisted in applying to the electrode progressive voltage steps and monitoring the charge increment $q(t)$ versus time t, thus allowing the evaluation of the value of D_{Li^+} for the whole composition range of the Li_xC_6 electrode. The reader is referred to the cited paper [34] and to the references therein for more details on the equations used. Here we report the final results, as illustrated in Fig. 3.7. The figure shows that the room temperature lithium diffusion coefficient decreases linearly during insertion, remaining, however, confined between acceptable values. As a term of comparison, one may quote the value of $D_{Li^+} = 10^{-9}$ cm^2 s^{-1} in TiS_2—namely, in one of the most commonly used insertion electrodes [46].

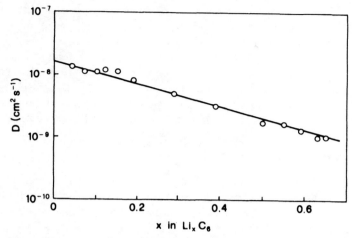

Figure 3.7. Li$^+$ ions chemical diffusion coefficient as function of the composition x in the Li$_x$C$_6$ electrode at room temperature. (From Ref. 34.)

3.5 Layered Lithium Metal Oxides

Layered lithium metal oxides, of the general formula LiMO$_2$ (where M = Co or Ni) are insertion compounds that have attracted considerable interest as electrodes for high-voltage lithium batteries. Indeed, the Li–LiMO$_2$ combination gives rise to cells with OCVs exceeding 4 V, and this has somewhat restricted the use of these highly oxidizing electrodes because of the lack of electrolytes having sufficiently high electrochemical stability windows.

The LiMO$_2$ compounds have a layered, rock salt structure where lithium and transition metal cations occupy alternate layers of octahedral sites in a distorted cubic close-packed oxygen-ion lattice. The layered MO$_2$ framework provides a two-dimensional interstitial space that allows for easy extraction of lithium ions. Therefore LiMO$_2$ compounds, being capable of releasing lithium ions, behave as lithium source electrodes and thus can be very conveniently coupled with a carbon electrode to form a C–LiMO$_2$ battery in its fully discharged state. The activation of this battery requires a "charging" process involving the removal of lithium ions from the LiMO$_2$ electrode and their insertion into the carbon electrode:

$$\text{LiMO}_2 + 6\text{C} \Leftrightarrow \text{Li}_{1-x}\text{MO}_2 + \text{Li}_x\text{C}_6 \tag{3.12}$$

The feasibility of this concept, although with a negative electrode different from carbon, was first demonstrated by Auborn and Barberio [47]. However, the system has gained great popularity after the recent announcement by a Japanese industry of its exploitation for large-scale production of batteries for consumer electronics [48].

The LiMO$_2$ compounds may be obtained by high-temperature synthesis from lithium oxide and the selected transition metal oxide. For instance, LiNiO$_2$ can be prepared [49] by annealing at 850°C an intimate mixture of Li$_2$O and NiO$_2$, and

LiCoO$_2$ is obtained [50] by heating a pelletized mixture of lithium hydroxide and cobalt carbonate in air at 850°C. However, regardless of the type of preparation, the LiMO$_2$ compounds may easily exhibit nonstoichiometry, generally due to an excess of M. Since M replaces structural sites otherwise available for Li$^+$ ions, the nonstoichiometry may ultimately affect the behavior of the LiMO$_2$ insertion electrodes. Therefore particular care must be devoted for controlling this unfavorable aspect.

Efficient electrode compositions involve mixtures of the LiMO$_2$ powder, carbon, and a binder to form a pellet or a film backed onto metallic substrates.

3.5.1 Lithium Cobalt Oxide

Among the LiMO$_2$ compounds, LiCoO$_2$ is the one that has attracted the largest attention because of its ability to give lithium cells with exceptionally high open circuit voltages—namely, on the order of 4.5 V [51,52], a value consistent with the high oxidizing power of the Co^{4+}/Co^{3+} couple [50]. Figure 3.8 shows the voltage-composition curve related to the process:

$$\text{LiCoO}_2 \Leftrightarrow \text{Li}_{1-x}\text{CoO}_2 + x\text{Li} \tag{3.13}$$

where $0.1 < x < 0.9$. Complete removal of lithium ($x = 1$) cannot be accomplished because of the nonstability of the highly reactive CoO$_2$ form. As shown by Plichta et. al. [51,52] and confirmed by recent x-ray diffraction study by Reimers and Dahn [53], the deintercalation of lithium in the $0.1 < x < 0.9$ range induces three-phase

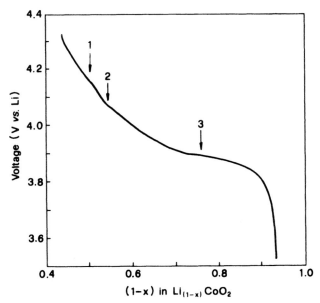

Figure 3.8. Voltage-composition curve at 30°C for a LiCoO$_2$ electrode determined in a cell having a Li counter electrode and a LiAsF$_6$-PC electrolyte. The three plateau are evidenced by arrows. (From Ref. 53.)

transitions in Li_xCoO_2, which are accompanied by lattice distortion. The occurrence of the phase transitions is revealed by the three plateau regions in the composition curve of Fig. 3.8.

Assuming an operating average voltage of 3.9 V, the *theoretical* energy density of the $Li/Li_{1-x}CoO_2$ electrochemical system is of the order of 1070 Wh kg^{-1} at room temperature. It is perhaps useful to remember that the theoretical energy density value includes the mass of the reactants only and thus provides a relative assessment of the energy content of the complete battery. The really significant value is the *practical* energy density—namely, the value that includes the weight of the electrolyte, separators, excess of electrode materials, and all the packaging hardware. A generally accepted figure of comparison assumes that the practical value falls between 20% and 25% of the theoretical value. Under this assumption, the practical energy density of the $Li/Li_{1-x}CoO_2$ battery would be around 200 Wh kg^{-1}, a still appreciable value, especially when compared with those of other lithium batteries of current interest, such as the Li/TiS_2 battery (about 100 Wh kg^{-1}).

The favorable energy content of the $LiCoO_2$ electrode material is somewhat opposed by the fact that the practical cyclability of process (3.13) is generally limited to $0 < x < 0.5$, since the reaction cannot be repeatedly extended over its full range because of the instability of the Co^{4+} oxidation state. Furthermore, because of the already mentioned difficulties of operating this system in the usual liquid electrolytes, the realization of practical $Li/Li_{1-x}CoO_2$ batteries has been confined to a few laboratory prototypes. Recent advances in the characterization of electrolytes with improved electrochemical stability [54,55] has opened the route for the exploitation of this high voltage couple. Plitchta et. al. [52] have used solutions of selected lithium salts in esters and alkyl carbonate-based mixed solvents (such as 1.8 M $LiAsF_6$ in MF/DMC, MF = methyl formate; DMC = dimethyl carbonate) which exhibited superior resistance toward oxidation, thus allowing the practical realization of rechargeable $Li/Li_{1-x}CoO_2$ cells. In fact, by limiting the charging voltage to around 4.3–4.5 V, these authors were able to operate the cells for 100 deep charge–discharge cycles, obtaining average theoretical (weight of active materials only) energy density of 500 Wh kg^{-1}. More recently, C- and D-size spirally wound $Li/Li_{1-x}CoO_2$ cells have been built and tested [55]. Practical energy densities meeting the expectations from the theoretical value—namely, of the order of 175 Wh kg^{-1} and of 145 Wh kg^{-1}, associated with a reasonably long cycle life—have been reported for these cells.

As already pointed out, the structure of $LiCoO_2$ may be described as layered arrangements of Co and Li in a rock salt–type lattice. Removal of Li^+ results in interlayer expansion (namely, from ~4.7 Å to 4.8 Å passing from $x = 1$ to $x = 0.5$) rather than the expected contraction; this is probably due to increasing electrostatic repulsion between O^{2-} centers upon delithiation [56]. The lattice expansion favors the extraction of the Li^+ ions; thus the diffusion kinetics of the electrochemical process is expected to be fast. The diffusivity of the Li^+ ions in $Li_{1-x}CoO_2$ has been investigated by various authors. However, the results are contradictory. Mizushima et al. [50], using a transient technique, obtained a value of chemical diffusion coefficient $D = 5 \times 10^{-9}$ cm^2 s^{-1} for x in the 0.2–0.8 range. A similar value was obtained by Kikkawa et al. [57], and Thomas et al. [58] reported a value of $D = 5 \times 10^{-8}$ cm^2 s^{-1}, as determined by impedance spectroscopy. (See the next section for a brief description of this technique.) Assuming an average value of $D = 10^{-9}$ cm^2 s^{-1}, one can consider that the lithium diffusivity in Li$_x$CoO$_2$ is high and thus that this compound may effectively operate as a high-rate lithium-source electrode for rechargeable rocking chair cells.

Indeed, LiCoO$_2$ has attracted considerable attention, and many laboratories are presently engaged in the realization of rocking chair batteries of the C/LiCoO$_2$ type. The charge–discharge process of these batteries is

$$6C + LiCoO_2 \Leftrightarrow Li_{0.5}C_6 + Li_{0.5}CoO_2 \quad (3.14)$$

and it evolves over an average voltage of 3.0 V. The theoretical energy density amounts to ~360 Wh kg^{-1}, and the life of the cell is reported to extend for many cycles [48,59]. AA and D size C/LiCoO$_2$ batteries are currently under large-scale production by Sony Japanese Industry [48]. Practical energy densities of 78 Wh kg^{-1} (AA size, 22 Ah kg^{-1} capacity) and of 115 Wh kg^{-1} (D size, 32 Ah kg^{-1} capacity) and cycle life of 1200 cycles have been demonstrated [48,60] for these batteries. These are very promising values that place the C/LiCoO$_2$ rocking chair battery in a highly competitive position with respect to other, more conventional power sources for consumer electronics, such as the D size Ni–Cd battery (25 Ah kg^{-1} capacity, 30 Wh kg^{-1} energy density) and the D-size portable sealed lead–acid battery (42 Ah kg^{-1} capacity, 25 Wh kg^{-1} energy density).

3.5.2 Lithium Nickel Oxide

The other well-known member of the LiMO$_2$ family is the lithium nickel oxide, LiNiO$_2$. Nickel is an abundant material and its electrochemical use in the battery industry has been largely acquired in the production of the common nickel–cadmium cells. All these facts make LiNiO$_2$ a very attractive electrode material for the new generation, rocking chair lithium battery development. The LiNiO$_2$ compound has a layered structure consisting of cubic close-packed O^{2-} ions with Ni^{3+} ions occupying alternate sheets of octahedral sites between adjacent close-packed oxide layers, while the Li$^+$ ions occupy the remaining sheets of octahedral sites between the oxide layers [49].

Similar to LiCoO$_2$, LiNiO$_2$ may be exploited as a lithium source electrode. The electrochemical delithiation process is:

$$LiNiO_2 \Leftrightarrow Li_{1-x}NiO_2 + xLi \quad (3.15)$$

where $0 < x < 0.5$ [61,62].

The variation of the voltage with lithium content in the Li$_{1-x}$NiO$_2$ electrode is shown in Fig. 3.9. Two plateaus are observed in two narrow composition ranges, around $(1 - x) = 0.55$–0.6 and $(1 - x) = 0.65$–0.75, respectively. However, x-ray diffraction studies [49] did not reveal evidence for discrete phases within the composition range reported in the figure. Short-range Li$^+$ ions and electrons ordering,

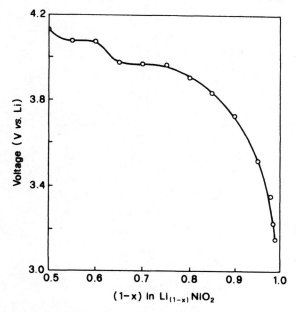

Figure 3.9. Voltage-composition curve for a Li_xNiO_2 electrode at room temperature determined in a cell having a Li counter electrode and a $LiClO_4$-PC electrolyte. (From Ref. 49.)

which might not be revealed by the diffraction pattern technique, have been postulated to account for the observed plateau in the voltage-composition curve [49].

The Li^+ chemical diffusion coefficient in $Li_{1-x}NiO_2$ has been determined by Bruce et al. [49], using impedance spectroscopy. This alternating current technique is particularly suitable for kinetic studies in thin-film insertion electrodes [63] where the semi-infinite diffusion boundary conditions, considered for the direct current transient technique, can no longer be assumed. The interpretation of the impedance results requires the definition of an electric circuit that is representative of the electrochemical system under study. The equivalent circuit generally used to interpret the impedance response of an electrode–electrolyte interface is the well-known Randles [64] circuit illustrated in Fig. 3.10a, where Re is the resistance of the electrolyte comprised between the given interface and the reference electrode; Cdl and Rct are the double-layer capacitance and the charge transfer resistance, respectively, of the interface; and Z_w is a complex impedance, called the Warburg impedance [65], representing the diffusion of the electroactive species within the interface. The Warburg impedance is given by the following equation:

$$Z_w = B_w \, \omega^{-1/2} - jB_w \, \omega^{-1/2} \tag{3.16}$$

where ω is the frequency of the applied signal, $j = (-1)^{1/2}$ is the imaginary factor, and B_w is a constant that may be described by the following equation:

$$B_w = [V_M \, (dE/dx)]/[n \, FA \, (2D)^{1/2}] \tag{3.17}$$

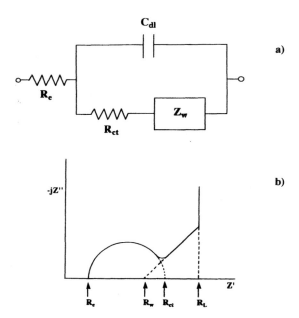

Figure 3.10. The equivalent Randles circuit for interpretation of the a.c. impedance response of an electrochemical interface (a) and ideal impedance response of a thin-film intercalation electrode (b).

where V_M and A are the molar volume and the surface of the electrode, respectively, dE/dx is the slope of the voltage-composition curve (e.g., the slope of Fig. 3.9 in the case of the $LiNiO_2$ interface), and the other symbols have their usual meanings.

The response of the insertion electrode over the entire frequency range can be represented by a plot of the imaginary $-jZ''$ part versus the real part Z' of the total impedance of the representative circuit (see Fig. 3.10b). The plot becomes representative of the diffusion kinetics at low frequencies where it assumes a linear trend with a 45° slope. The intercept of the 45° line with linear real axis gives the diffusional resistance R_w, and the chemical diffusion coefficient may then be calculated from the following equation:

$$R_w = B_w \cdot \omega^{-1/2} \tag{3.18}$$

where B_w is the constant expressed by eq. (3.17).

Using this impedance method, Bruce et al. [49] found for D_{Li^+} in $Li_{1-x}NiO_2$ a value of the order of $D = 2 \times 10^{-7} cm^2 s^{-1}$, namely, one of the highest observed in lithiated insertion electrodes. This makes $Li_{1-x}NiO_2$ a very promising and convenient electrode for high-rate rocking chair lithium batteries.

In fact, the nickelate compound is currently exploited [62] in alternative to the cobaltate one, as a lithium source positive electrode in combination with the negative lithium sink carbon electrode for the realization of $C/LiNiO_2$ rocking chair cells characterized by the following electrochemical process:

INSERTION COMPOUNDS FOR LITHIUM ROCKING CHAIR BATTERIES

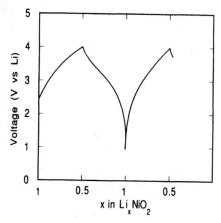

Figure 3.11. Voltage-composition curve at 21°C of a $C_6/LiNiO_2$ rocking chair battery using a 1 M $LiN(CF_3SO_2)_2$-PC/DME electrolyte. (From Ref. 62.)

$$0.5C_6 + LiNiO_2 \Leftrightarrow Li_{0.5}C_6 + Li_{0.5}NiO_2 \qquad (3.19)$$

where the cyclable lithium is confined within $\Delta x = 0.5$ [61,62].

The voltage profile of the associated charge–discharge cycle is illustrated in Fig. 3.11. Assuming a midvoltage of about 3.0 V, the theoretical energy of the $C/LiNiO_2$ couple is around 350 Wh kg^{-1}, which should lead to practical values around 80–100 Wh kg^{-1}. As expected this energy content value is lower than those generally associated with the parent lithium cells, since, as is always the case for rocking chair configurations, the loss in energy is directly related to the loss in specific capacity when passing from the lithium metal to the lithiated carbon. However, this shortcoming may be counterbalanced by a more extended cycling life and, particularly, by a safer operation, and this accounts for the present large interest in the rocking chair batteries of the general $C/LiMO_2$ type.

3.5.3 Lithium Nickel Oxide–Cobalt Oxide Solid Solutions

Besides the two nickelate and cobaltate compounds, their $Li_xNi_{1-y}Co_yO_2$ solid solutions have been taken into consideration as lithium source electrodes for carbon-based rocking chair batteries. These solutions exhibit an ideal layer structure formed by $(Ni_{1-y}Co_y)O_2$ sheets between which Li+ can be inserted in an octahedral environment [66]. This structure, which is observed for y larger than 0.3, allows an easy and efficient electrochemical insertion process. In fact, the main advantage of $Li_xNi_{1-y}Co_yO_2$ versus the pure $LiNiO_2$ is that in the former the occurrence of extra lithium in the intersheet gap of the structure can be avoided. Therefore although in $LiNiO_2$ some capacity may be lost during cycling due to Ni–Li repulsion, in the $Li_xNi_{1-y}Co_yO_2$ the presence of cobalt stabilizes the structure in a strictly two-dimensional fashion, thus favoring the full reversibility of the insertion–deinsertion process. This is an important aspect, especially for rocking chair battery applications, where the cyclability of the electrodes is a major requirement. Indeed, $Li_xNi_{1-y}Co_yO_2$ electrodes are currently of great practical interest for this type of battery [66].

3.6 Manganese Oxides

Manganese dioxide (MnO_2) is an inexpensive, nonpolluting, readily available material. There are various phase modifications of MnO_2, and by properly synthesizing the desired one, many electrochemical applications of this oxide can be envisaged.

Historically, manganese oxide has been used worldwide as the solid cathode in the popular Leclanché and alkaline–manganese dry cells [19,67]. Electrochemically prepared manganese dioxide (EMD), which has a γ-type structure and may be described as an intergrowth of rutile and ramsdellite-type structures [19], is predominantly used in these types of cells.

More recently, manganese dioxide cathodes have been used for the development of primary lithium batteries [4]. The efficient use of manganese dioxide in these nonaqueous power sources requires accurate preparation techniques to obtain the desired crystal structure characterized by the proper stoichiometry and the suitable water content. The most favorable procedure is the heat treatment at 350°C of EMD, which removes about 80% of water, thus favoring the transformation of the γ-MnO_2 phase to a γ-β MnO_2 phase [68].

The discharge process of lithium-manganese dioxide batteries consists in the insertion of Li^+ ions in the γ-β MnO_2 phase:

$$xLi + MnO_2 \Rightarrow Li_xMnO_2 \qquad (3.20)$$

where $0 < x > 1$. The batteries yield a remarkably high practical energy density [67]—on the order of 220 Wh kg^{-1}. However, process (3.20) is not reversible, and thus Li/γ-β MnO_2 batteries are confined to the primary use.

In view of the attractive high-energy, low-cost, and nonpolluting features of manganese dioxide, concerted efforts have been devoted to the synthesis of MnO_2 forms that could be suitable for fast and reversible lithium insertion and thus suitable for the achievement of rechargeable, widely accepted lithium batteries and, in particular, of environmentally friendly rocking chair batteries. In this respect, particular attention has been devoted to the synthesis of spinel-related manganese dioxide forms. Hunter [69] has reported the preparation of spinel-type λ-MnO_2 by lithium extraction from $LiMn_2O_4$. This author has shown that almost all lithium can be removed by a chemical reaction with sulphuric acid, thus yielding the λ-MnO_2 phase, which retains the $[Mn_2O_4]$ framework structure. This phase accepts reversible insertion of Li^+ ions, even if there is some controversy on the extent of this process when run electrochemically. Some authors [70] have reported that the delithiation cannot be extended beyond the $Li_{0.4}Mn_2O_4$ composition; others [71] claim that full removal of lithium, leading back to λ-MnO_2, is achievable. In any case, $LiMn_2O_4$ can easily accommodate up to one extra Li^+ ions per mole in its

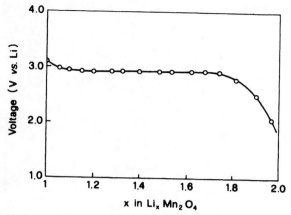

Figure 3.12. Voltage-composition curve for a $Li_{1+x}Mn_2O_4$ electrode at room temperature determined in a cell having a Li counter electrode and a 1 M $LiBF_4$ electrolyte. (From Ref. 72.)

spinel structure [72], and thus lithium may be reversibly cycled over a wide range of composition that extends at least from $Li_{0.4}Mn_2O_4$ to $Li_2Mn_2O_4$.

The voltage in the upper lithiated range evolves around 2.8 V in relation to the process:

$$xLi + LiMn_2O_4 \Leftrightarrow Li_{1+x}Mn_2O_4 \quad (3.21)$$

where $0 < x < 1$. The lithium insertion into $Li_{1+x}Mn_2O_4$ induces a Jahn-Teller type distortion that reduces the crystal symmetry from cubic in $LiMn_2O_4$ to tetragonal in $Li_{1+x}Mn_2O_4$, thus resulting in a two-phase electrode [72] characterized by a flat voltage-composition curve [72–74] (Fig. 3.12). The theoretical energy density of process [21], assuming an average voltage of 2.8 V, is 320 Wh kg^{-1} at 25°C. Practical energy densities of the order of 120 Wh kg^{-1} have been reported [56] for Li batteries using a 1 M $LiClO_4$ in PC–DME (PC = propylene carbonate; DME = dymethoxiethane) liquid electrolyte.

Figure 3.13 illustrates the voltage-composition curve for the delithiation–lithiation process:

$$LiMn_2O_4 \Leftrightarrow Li_{1-x}Mn_2O_4 + xLi \quad (3.22)$$

The curve, derived for a $LiMn_2O_4$ electrode prepared by reacting in air at 800°C stoichiometric amounts of Li_2CO_3 and MnO_2, is extended to the total removal of lithium [71]. One can clearly see that the process develops over two plateau, one extending between $1 > x > 0.4$ and the other between $0.1 > x > 0.4$. This may explain the controversy. In fact, some authors might be able to detect the first plateau only, and this may have brought them to the conclusion that electrochemical removal of lithium is limited to the $Li_{0.4}Mn_2O_4$ composition.

A loss of capacity of about 20% has been found in cycling the $LiMn_2O_4$ electrode [71]. Since practical electrode configurations involve blends of $LiMn_2O_4$ powder

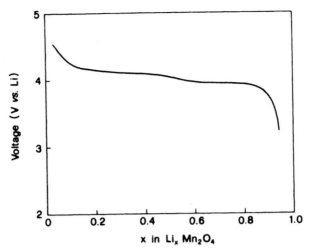

Figure 3.13. Voltage-composition curve for a $Li_{1-x}Mn_2O_4$ electrode at room temperature determined having a Li counter electrode and a $LiClO_4$-PC electrolyte. (From Ref. 71.)

with carbon black and a binder (e.g., EPDM), the loss of capacity has been attributed to lack of interfacial contact and of homogeneity in the electrode structure.

The $Li_{1-x}Mn_2O_4$ cycle evolves around 4 V versus Li, and this prevents the utilization of this electrode in lithium cells using common liquid electrolytes that have stability windows confined within 4 V. On the other hand, the high voltage makes $LiMn_2O_4$ a very promising lithium source electrode for carbon-based rocking chair cells. This type of cell, in fact, has been fabricated and tested by Guymard and Tarascon [34], using as preferred electrolyte a 1 M solution of $LiClO_4$ in an EC–DEE (50:50) mixture (EC = ethylene carbonate; DEE = diethoxyethane). This $C/LiMn_2O_4$ rocking chair cell can be cycled at 25°C with a voltage profile that varies between 4.4 V (full charge) and 2.4 V (full discharge) with a calculated (electrode materials only) energy density of about 230 Wh kg^{-1}. The power rate of this battery, as in general of all the rocking chair batteries, is controlled by the diffusion of the lithium ions within the electrode structure.

In the full composition range, the $[Mn_2O_4]$ framework possesses a three-dimensional space via face-sharing octahedra and tetrahedra, and this provides conducting pathways for the insertion and the extraction of Li^+ ions. This favorable structural situation should assure a high mobility of the Li^+ ions in the manganese spinel. However, the experimentally obtained values of D_{Li}^+ are somewhat controversial. Dickens and Reynolds [75] have reported a value of the order of 10^{-11} cm^2 s^{-1} at 25°C for the $Li_{0.4}Mn_2O_4$ composition, which would indicate a relatively low $Li+$ ion mobility and thus a low power capability for the $C/LiMn_2O_4$ cell. In contrast, Guyomard and Tarascon [34] found a value more compatible with the structural expectations—namely, a D_{Li}^+ on the order of 10^{-9} cm^2 s^{-1} at room temperature for the full composition range, leading to the provision that $LiMn_2O_4$-based rocking chair cells should be able to sustain high current rates.

More detailed study would certainly be welcome to obtain definite clarification of the diffusion kinetics in this, as in the generality of the insertion electrodes, and thus identify the most suitable technique for obtaining reliable and reproducible diffusion coefficient data. In the particular case of the $LiMn_2O_4$ electrode, a possible low Li^+ ion diffusivity should in principle limit the cycling rates and thus penalize $LiMn_2O_4$ in respect to other lithium metal oxide electrodes. However, the low cost and the nonpolluting characteristics contribute to counterbalance the kinetic limitations and thus to pose the $LiMn_2O_4$ compound among the most promising rocking chair electrodes, especially for the development of batteries designed for low-rate applications.

3.7 Other Types of Rocking Chair Configurations

It has already been stressed that carbon electrodes, despite many favorable characteristics, are affected by two major drawbacks in reference to their use as rocking chair electrodes—namely, an initial loss of capacity and a poor voltage stability upon cycling. Therefore some laboratories are attempting to replace the carbon with alternative negative electrodes with improved performance.

A recently proposed [76,77] interesting example is that of titanium disulphide. This compound, which is one of the common "cathodes" in conventional Li/A_zB_y ($A_zB_y = TiS_2$) lithium batteries, has been proposed as an "anode" in rocking chair batteries of the following type:

$$TiS_2 + LiCoO_2 \Leftrightarrow Li_xTiS_2 + Li_{(1-x)}CoO_2 \quad (3.23)$$

where TiS_2 acts as lithium sink negative electrode in replacement to carbon and $LiCoO_2$ has its usual function of lithium source positive electrode.

There are two main reasons to select TiS_2 as a possible negative electrode in rocking chair batteries. The first is related to its open, layered structure, which allows loose interactions with the guest lithium ions. As result of this, the insertion of these ions within the S–Ti–S layers is accompanied by a limited voltage variation (i.e., 0.7 V) passing from $x = 0$ to $x = 1$ [78]. Therefore the use of TiS_2 may effectively limit voltage fluctuation upon cycling the battery.

The second reason for using TiS_2 as a negative electrode is related to the fact that this compound intercalates with lithium at voltages about 2 V more negative than $LiCoO_2$, and thus the combination of the two compounds gives cell voltages that remain high enough to avoid solvent reduction. On the other hand, this second aspect is contrasted with a reduction of the overall energy density, a drawback that should be taken into account when balancing the performance of the $TiS_2/LiCoO_2$ rocking chair batteries.

Practical examples of these batteries have been fabricated and tested. Plichta et al. [76], by using a novel and interesting type of liquid electrolyte (i.e., solutions of lithium salts in acetonitrile), have shown that the $TiS_2/LiCoO_2$ electrodic system can be effectively cycled many times with reasonably high charge–discharge rates. Other work in the area includes the realization of a fully solid-state cell formed by combining the $TiS_2/LiCoO_2$ electrodic system with a polymer electrolyte membrane of the $PEO–LiClO_4$ type [77]. The resulting thin-layer, rocking chair batteries showed an excellent cyclability, even if the discharge process was affected by polarizations of the cobaltate electrode. However, preliminary indications provided by impedance spectroscopy suggest that this drawback may be solved by optimizing the design of the electrode configurations [77].

The choice of titanium disulfide as an alternative to carbon is certainly interesting, and although not conclusive, it indicates that the rocking chair design may indeed be exploited by a large variety of electrode pairs, and thus that the efforts in the field are far from being concluded and still open to a quite exciting variety of possibilities that, besides electrode pairs, involve also a list of alternative electrolyte materials.

3.8 Conclusions

The large availability of insertion electrodes capable of exchanging substantial quantities of lithium ions with relatively fast kinetics has promoted the development of various types of rechargeable lithium batteries with different design, size, capacity, power, and energy capabilities.

Sulphides and oxides of the transition metal family have been exploited for the realization of conventional batteries—namely, of batteries based on the combination of lithium metal and the selected insertion compound. Small Li/TiS$_2$ batteries are currently under development using both liquid [79] and solid [80] electrolytes. Large-scale production of AA size, liquid electrolyte, spirally wound Li/MoS$_2$ and Li/MnO$_2$ batteries has also been undertaken [81]. The lithium/vanadium oxides combination is presently exploited [82] for the realization of large-capacity polymer electrolyte batteries directed to the consumer market as well as to the development of electric vehicles.

All these lithium batteries offer a series of considerable specific advantages, such as high energy density and relatively low cost. However, their widespread utilization is still influenced by the high reactivity of the metal that, from one hand, assures the high energetic content, and from the other, induces safety hazards and limited cyclability.

Attempts to overcome this shortcoming have resulted in the resumption of the rocking chair concept with the development of batteries where the lithium metal is most commonly replaced by a carbon electrode. Penalties in lower voltage and energy density with respect to the Li systems are counterbalanced by an expected safer and longer operation. Although a very recent innovation, the rocking chair idea has already found enthusiastic response in many research laboratories that are presently involved in its investigation and development. Important achievements have been obtained with the characterization of new electrolyte compositions (e.g., solution of LiPF$_6$ in ethylene carbonate-dimethylcarbonate, EC-DMC mixtures) which have allowed the use of graphite electrodes for high-capacity, long-life batteries [83] or improvements in the performance of high-voltage cathodes [84]. As result of this, small lithium rocking chair batteries (or as otherwise named, "lithium ion batteries") are currently under development in Japan, the United States, and Europe.

However, rocking chair batteries are far from being totally accepted as safe, long-life, multipurpose batteries. Some unwanted characteristics, such as voltage fluctuations upon cycling, low power output, and chances of deviation from the expected electrochemical process, call for further systematic studies and improvements.

These studies should be mainly focused on the choice of the optimized insertion compounds; thus the investigation in these important electrode materials remains an essential task to be pursued in academic and industrial laboratories. This investigation is today highly motivated by the urgent need for new power sources that can combine favorable power and energy performance with an environmentally friendly structure. Rocking chair lithium batteries appear to approach these expectations, and thus this chapter has been written with the probably immodest hope of providing a basis of knowledge for those researchers who are planning to enter the important race for the development of clean and innovative power source devices.

References

1. Cairns, E. J. *ECS Quar.* **1992**, *1*, 25.
2. "Batteries for Implantable Biomedical Devices," Owens, B. B., ed. Plenum, New York, 1986.
3. Exnar, I.; Hep, J. *J. Power Sources* **1993**, *43–44*, 701.
4. "Lithium Batteries," Gabano, J. P., ed. Academic, London, 1983.
5. Murphy, D. W.; Christian, P. *Science* **1979**, *205*, 651.
6. Whittingham, M. S. *Progress in Solid State Chem.* **1978**, *12*, 1.
7. Frazer, E. J.; Phang, S. *J. Power Sources* **1981**, *6*, 307.
8. Liang, C. C.; Barnette, L. H. *J. Electrochem. Soc.* **1976**, *123*, 453.
9. Akridge, J. H.; Vourlis, H. In "Solid State Microbatteries," Akridge, J. R.; Balkanski, M., eds. Plenum, New York, 1990, vol. 217, p. 353.
10. Vincent, C. A. *Progress in Solid State Chem.* **1987**, *17*, 145.
11. Abraham, K. M.; Alamgir, M. *J. Electrochem. Soc.* **1990**, *137*, 1657.
12. Subbarao, S. In Ref. 3, abstr. THU-II.
13. Armand, M. In "Materials for Advanced Batteries," Murphy, D. W.; Broadhead, J.; Steele, B. C. H., eds. Plenum, New York, 1980, p. 145.
14. Whittingham, M. S. In "Fast Ion Transport in Solids," Vashista, P.; Mundy, J. N.; Shenoy, G. K., eds. Elsevier North Holland, New York, 1979, p. 17.
15. Murphy, D. W. In "Intercalation Chemistry," Whittingham, M. S.; Jacobson, A. J., eds. Academic, New York, 1982, p. 563.
16. Goodenough, J. B. In "Solid State Microbatteries," Akridge, J. R.; Balkanski, M., eds. Nato ASI Series, Plenum, New York, 1990, vol. 217, p. 213.
17. West, K.; Zachau-Christiansen, B.; Jacobsen, T.; Skaarup, S. In "Solid State Ionics," Nazri, G. A.; Shriver, D. F.; Huggins, R. A.; Balkanski, M., eds., "Mat. Res. Soc. Symp. Proceedings," Pittsburgh, 1991, vol. 210, p. 449.
18. Chippindale, A. M.; Dickens, P. G.; Powell, A. V. *Progress in Solid State Chem.* **1991**, *21*, 133.

19. Vincent, C. A.; Bonino, F.; Lazzari, M.; Scrosati, B. "Modern Batteries." Arnold Pu, London, 1984.
20. Kinushita, K. "Carbon." Wiley Interscience, New York, 1987.
21. Yazami, R.; Guérard, G., *J. Power Sources* **1993**, *43–44*, 39.
22. Heath, D. R.; O'Brian, S. C.; Zhang, Q.; Liu, Y.; Curl, R. F.; Kroto, H. W.; Trittel, F. K.; Sualley, R. E.; *J. Am. Chem. Soc.* **1988**, *110*, 1113.
23. Oberlin, A. *Carbon, 22* **1984**, 521.
24. Bashenard, J. O.; Fritz, H. P. *J. Electroanal Chem.* **1974**, *53*, 329.
25. Arakawa, M.; Yamamoto, Y. *J. Electroanal Chem.* **1987**, *219*, 273.
26. Yazami, R.; Touzin, P. *J. Power Sources,* **1983**, *9*, 365.
27. Ruland, W. *Acta Cryst.* **1985**, *18*, 992.
28. Imanishi, N.; Ohashi, S.; Ichikawa, T.; Yamamoto, O.; Kauno, R. *J. Power Sources* **1992**, *39*, 185.
29. Omanu, A.; Azuma, H.; Aoki, M.; Kita, A.; Nishi, Y. Fall Meeting Electrochem. Soc. Toronto, Canada, Oct. 1992, Abstr. 25.
30. Takami, N.; Sato, A.; Ohsaki, T. As in Ref. 29, Abstr. 28.
31. Huang, C. K.; Surmpudi, S.; Attia, A.; Halpert, G. "Proceedings of the IEEE 35th Intl. Power Sources Symp.," IEEE Customer Service Dept. Piscataway, N.J. USA, 1992, p. 197.
32. Seger, L.; Wen, L. Q.; Schlenoff, J. B. *J. Electrochem. Soc.* **1991**, *138*, L.81.
33. Yazami, R.; Guérard, D. *J. Power Sources,* **1993**, *43–44*, 39.
34. Guyomard, D.; Tarascon, J. M. *J. Electrochem. Soc.* **1992**, *139*, 937.
35. Rosolen, M. J.; Passerini, S.; Scrosati, B. *J. Power Sources,* **1993**, *45*, 333.
36. Fong, R.; von Sacken U.; Dahn, J. R. *J. Electrochem. Soc.* **1990**, *137*, 2009.
37. Dahn, J. R.; Fong, R.; Spoon, M. J. Phys. Rev. B, **1990**, *42*, 6424.
38. Bitthin, R.; Herr, R.; Hoge, D. *J. Power Sources,* **1993**, *43–44*, 405.
39. Peled, E. In "Lithium Batteries," Gabano, J. P., ed. Academic, London, 1983, p. 43.
40. Croce, F.; Gerace, F.; Scrosati, B.; *J. Power Sources,* in press.
41. Scrosati, B. *J. Electrochem. Soc.* **1992**, *139*, 2776.
42. Darken, L. S. *Trans. AIME* **1948**, *174*, 184.
43. Bard, A. J.; Faulkner, L. R. "Electrochemical Methods." John Wiley, New York, 1980.
44. Bonino, F.; Lazzari, M.; Vincent, C. A.; Wandless, A. R. *Solid State Ionics* **1980**, *1*, 311.
45. Bonino, F.; Scrosati, B. In "Solid State Batteries," Sequeira, C. A. C.; Hooper, A. eds., NATO ASI Series, Martin Nijhoff, Dordrecht, 1985, p. 119.
46. Vaccaro, A.; Palanisamy, T.; Kerr, R. L.; Maloy, J. T. *Solid State Ionics* **1981**, *2*, 337; **1981**, *6*, 363.
47. Auborn, J. J.; Barberio, Y. L. *J. Electrochem. Soc.* **1987**, *134*, 638.
48. Nagaura, T.; Tazawa, K. *Prog. Batteries Sol. Cells,* **1990**, *9*, 20.
49. Bruce, P. G.; Lisowska-Oleksiak, A.; Saidi, M. Y.; Vincent, C. A. *Solid State Ionics* **1992**, *57*, 353.
50. Mizushima, K.; Jones, P. C.; Wiseman, P. J.; Goodenough, J. B. *Solid State Ionics* **1981**, *3/4*, 171.

51. Plichta, E.; Salomon, M.; Slane, S.; Uchiyama, M.; Chua, D.; Lin, W. H. *J. Power Sources* **1987**, *21*, 25.

52. Plitchta, E.; Slane, S.; Uchiyama, M.; Salomon, M.; Chua, D.; Ebner, W. E.; Lin, H. W. *J. Electrochem. Soc.* **1989**, *136*, 1865.

53. Reimers, J. N.; Dahn, J. R. *J. Electrochem. Soc.* **1992**, *139*, 2091.

54. Dreher, J.; Haas, B.; Hambitzer, G. *J. Power Sources*, **1993**, *43-44*, 583.

55. Broussely, M.; Perton, F.; Labat, J.; Staniewicz, R. J.; Romero, A. In Ref. *J. Power Sources*, **1993**, *43-44*, 209.

56. Desilvestro, J.; Haas, O. *J. Electrochem. Soc.* **1990**, *137*, 5C.

57. Kikkawa, S.; Miyazaki, S.; Koizumi, M. *J. Power Sources* **1985**, *14*, 231.

58. Thomas, M. G. S. R.; Bruce, P.; Goodenough, J. *Solid State Ionics.*, **1986**, *18&19*, 794.

59. Mohri, M.; Yanagisawa, N.; Tajima, Y.; Tanaka, H.; Mitaie, T.; Nakajima, S.; Yoshimoto, Y.; Susuki, T.; Wadaet, H. *J. Power Sources* **1989**, *26*, 545.

60. Nagaura, T. 4th Intl. Rechargeable Battery Seminar, Deerfield Beach, Florida, 1990.

61. Thomas, M. G. S. R.; David, W. J. F.; Goodenough, J.; Groves, P. *Mat. Res. Bull.* **1985**, *20*, 1137.

62. Dahn, J. R.; von Sacken, U.; Jurzow, M. W.; Al-Janaby, H. *J. Electrochem. Soc.* **1991**, *138*, 2207.

63. Ho, C.; Raistrick, D.; Huggins, R. A. *J. Electrochem. Soc.* **1980**, *127*, 343.

64. Randles, J. E. *Discuss. Faraday Soc.* **1947**, *1*, 11.

65. Warburg, E. *Ann. Physik* **1899**, *67*, 493.

66. Delmas, C.; Saadune, I.; Auradou, H.; Menetrier, M.; Hagenmuller, P. In "Solid State Ionics: Materials and Applications," Chowdari, B. V., et/al.; eds., World Sci., Singapore, 1992, p. 255.

67. Kozawa, A. In "Comprehensive Treatise of Electrochemistry," Vol. 3, Bockris, J. O. M.; Reddy, A. K. N., eds., Plenum, New York, 1981, p. 207. "Handbook of Batteries and Fuel Cells," Linden, A., ed. McGraw-Hill, New York, 1984.

68. Pistoia, G. *J. Electrochem. Soc.*, **1982**, *129*, 1861.

69. Hunter, J. C. *J. Solid State Chem.* **1981**, *39*, 42.

70. Thackeray, M. M.; Jonhson, P. J.; de Picciotto, L. A.; Bruce, P. G.; Goodenough, J. B. *Mat. Res. Bull.* **1984**, *19*, 179.

71. Tarascon, J. M.; Wang, E.; Shokoohi, F. K.; MacKinnon, W. R.; Calson, S. *J. Electrochem. Soc.* **1992**, *139*, 2859.

72. Thackeray, M. M.; David, W. I.; Bruce, P. G.; Goodenough, J. B. *Mat. Res. Bull.* **1983**, *18*, 461.

73. Neat, R. J.; Macklin, W. J.; Powell, R. J. In "Second International Sympsiuim on Polymer Electrolytes," Scrosati, B., ed., Elsevier, London, 1990, p. 421.

74. Macklin, W. J.; Neat, R. J.; Powell, R. J. *J. Power Sources* **1991**, *34*, 39.

75. Dickens, P. G.; Reynolds, G. F. *Solid State Ionics* **1981**, *5*, 53.

76. Plichta, E.; Behl, W. Electrochemical Soc. Fall Meeting, Toronto, Canada, Oct. 1992, Abstr. 29.

77. Plichta, E. J.; Behl, W. K.; Salomon, M.; Schleich, D.; Croce, F.; Passerini, S.; Scrosati, B. *J. Power Source*, **1993**, *43-44*, 481.

78. Whittingham, M. S. *Progress in Solid State Chem.* **1978**, *12*, 1.

79. Andernan, N.; Lundqvist, J. T. Fall Meeting Electrochem. Soc., Honolulu, Oct. 18–23, 1987, Abstr. 47.
80. Akridge, J. R.; Vourlis, H. *Solid State Ionics* **1986**, *18–19*, 1082.
81. Laman, C.; Brandt, R. *J. Power Sources* **1988**, *24*, 195.
82. Scrosati, B.; Neat, R. In "Applications of Electroactive Polymers," Scrosati, B., ed. Chapman & Hall, London, 1993.
83. Tanaka K., Itabashi M., Aoki M., Hiraka S., Kataoka M., Fujita S., Sekai K., and Ozawa K., Fall Meeting Electrochem. Soc., New Orleans, Louisiana, Oct. 1993, Abstr. 21.
84. Guyomard, D. and Tarascon J. M., *J. Electrochem. Soc.*, **1993**, *140*, 3071.

CHAPTER

4

Thin Polymer Films on Electrodes: A Physicochemical Approach

Karl Doblhofer

4.1 Introduction

Polymer coatings on metallic substrates have become the subject of active interdisciplinary research, with a strong emphasis on electrochemistry [1–7]. Consider, for example, the *conducting polymers*, a field of great theoretical and practical interest. It relies on electrochemistry for preparing the films, for characterizing them, and also for possible applications (e.g., in batteries and displays) [1,4,5]. Second, polymer films are widely applied as protective coatings (paints, lacquers, moisture barriers). Electrochemistry plays a significant role in the preparation and characterization of the coatings and in investigations of the protection mechanism [8–11]. Similarly, information on the state of Langmuir-Blodgett films [12], self-assembled monolayers [13,14], and related thin organic films of current interest [15,16] can be obtained. Such studies yield thermodynamic data on partitioning equilibria of electroactive species between the aqueous and the polymer phases that provide a rational basis for developing ion-selective electrodes [17–19].

In all such applications the coated metal is immersed in an appropriate electrolyte and subjected to an electrochemical treatment. Three different types of behavior can be expected:

1. The film is an electronic conductor. Electrochemical charge-transfer reactions (e.g., $Ox + e^- \rightarrow Red$; see Fig. 4.1a) may proceed at the interface between the polymer and the electrolyte.
2. The film is an electronic insulator but an ionic conductor (i.e., it is permeable to species from the electrolyte). This is a film of the *membrane* type, as shown in

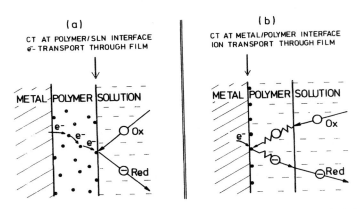

Figure 4.1. Schematic representation of two mechanisms of electrochemical charge transfer (CT, ox + e⁻ → Red) at polymer-coated electrodes.

Fig. 4.1b. Electrochemical charge-transfer reactions may proceed at the metal–polymer interface with the depolarizer partitioned into the membrane and transported across it [20].

3. The film is an electronic insulator and impermeable to solvent and solutes [10,21]. Such a film would be of the *barrier* type.

It turns out that the electron-conducting films are normally also ionic conductors; they constitute *mixed conductors* [22–24].

The polymer film shown in Fig. 4.1 is in an asymmetric arrangement; that is, it is in contact with an electronic conductor (metal) on the one side and with an ionic conductor (electrolyte solution) on the other side [24d]. Systems of great interest and importance arise when the same polymers are in corresponding symmetrical arrangements. When both contacting phases are electronic conductors, the polymer functions as a *polymer electrolyte* [25]. Polymer electrolytes have become an important independent field of research and development. Therefore they will not be reviewed in this work, despite the close fundamental proximity between the subjects. On the other hand, when the polymer film is in contact with (and separates) two electrolyte solutions, the polymer film is in the classical membrane arrangement. The behavior of membranes has been studied thoroughly because of their relevance in living systems as well as in technical devices, such as sensors and electrolysis cells. Basic concepts developed in this field are relevant for understanding the behavior of polymer-coated electrodes and will be reviewed and discussed here. However, the classical membranes as such are not the subject of this discussion.

The electrochemistry of polymer-coated electrodes is mainly governed by four concepts: first, the partitioning of solvent and ions between the polymer and the electrolyte; second, transport processes, in particular of ions, in the polymer phase; third, the nature of redox states in the polymer, and electronic charge transfer between these states; fourth, the state of the interface between the electrode and the

THIN POLYMER FILMS ON ELECTRODES

(solvated) polymer. It appears that in recent reviews the discussion of the fundamental relationships concerning the first two aspects and the consequences for the electrochemical behavior of the coated electrodes have been somewhat neglected. Therefore the emphasis in this work will be on these two themes. The work is organized so that after an introduction to basic concepts of polymer science and technology, in Section 4.2 simple nonionic polymers are considered; then (Section 4.3) fixed charges are introduced; and finally (Section 4.4) electron-conducting polymers are discussed. At every stage the relevance of the derived concepts for the electrochemical behavior of polymer-coated electrodes is pointed out.

4.1.1 Concepts for Characterizing Polymers

Polymers are among the most important natural substances (cellulose, starch, proteins, rubber, etc.), and synthetic polymers (polyvinyl chloride, polystyrene, polypropylene, etc.) constitute important commercial products.

Polymers are made up of monomer units. In the simplest case, one monomer species polymerizes (e.g., acetylene forms polyacetylene, Fig. 4.2a). This polymer is one of the *electron-conducting polymers* and is presently receiving considerable attention (discussed later). Another example is polystyrene (Fig. 4.2b). It can be prepared conventionally by a free-radical mechanism, producing a polymer with the phenyl groups randomly "up" or "down" (*atactic*). In a more modern process special metal-organic compounds are used as the catalyst (*Ziegler catalyst*). The produced polymer is then in a state of higher order, with all substituents either up or down (*isotactic*).

In more complex processes two different bifunctional monomers can react to form the polymer chains. Consider, for example, the formation of nylon 66 from

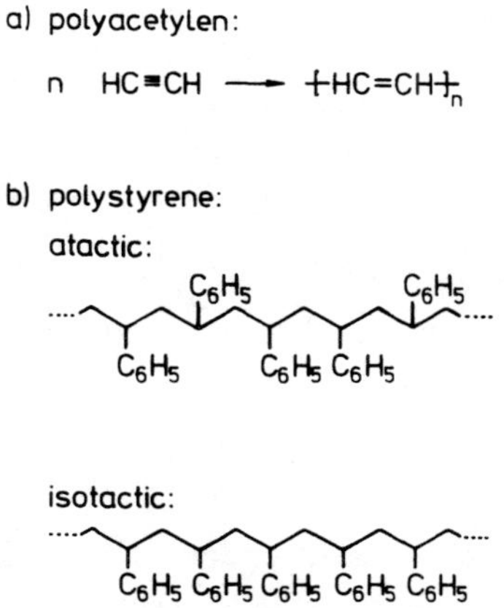

Figure 4.2. (a) Polymerization of acetylene to polyacetylene; (b) structure of atactic and isotactic polystyrene.

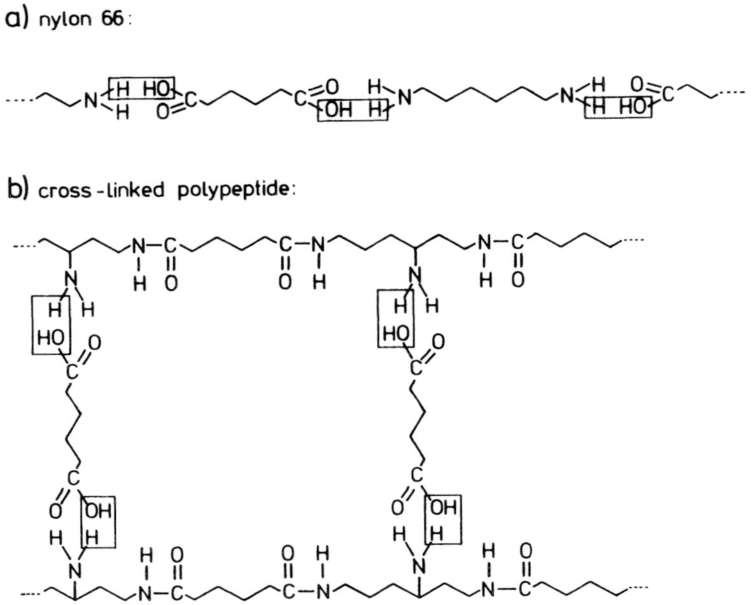

Figure 4.3. (a) Schematic representation of the formation of nylon 66 by polycondensation of adipic acid and hexamethylenediamine. (b) A trifunctional amine is used instead of the difunctional one; thus crosslinks are introduced into the polymer matrix.

adipic acid, $HOOC(CH_2)_4COOH$, and hexamethylene diamine, $H_2N(CH_2)_6NH_2$, as shown in Fig. 4.3a. The links between the monomer molecules are formed in the process of elimination of water (*polycondensation*). In this way, linear polymers of very high molecular weight may be formed. A wide variety of bifunctional species can be used as the monomers for such reactions. Furthermore, such substances can have numerous types of substituents (—OH, —CH_3, etc.). It is thus clear that polymers with widely varying properties can be prepared in a reasonably predictable fashion.

When a monofunctional compound is added to the reaction mixture characterized in Fig. 4.3a (e.g., a monocarbonic acid, R—COOH), at the position where this group reacts the polycondensation process is interrupted. In this way the average polymer-chain length can be adjusted to shorter values. Typical polymers consist of molecules of at least about 100 monomer units. Normally, a distribution of chain lengths exists and an average molecular mass can be specified. When the polymer chains consist of less than 10 monomer units, the term *oligomere* is used for the material.

The polymer chain length is an important parameter in determining the physical properties of the polymer. The solubility tends to decrease with increasing molecular mass. However, in general, linear polymers can be dissolved in suitable solvents. For example, starch (a polysaccharide containing four hydrdoxyl groups per

THIN POLYMER FILMS ON ELECTRODES

glucose unit) is hydrophilic and dissolves in water. On the other hand, atactic polystyrene is soluble in benzene or toluene.

4.1.1.1 Cross-Linking the Polymer Chains

Imagine that in the nylon 66 synthesis shown in Fig. 4.3a instead of bifunctional monomers a trifunctional monomer is used (e.g., 1,3,6-triaminohexane). Then in the process of polycondensation a three-dimensional polymer network is formed, as illustrated in Fig. 4.3b. The polymer chains are *cross-linked*. Depending on the ratio between the concentrations of the bifunctional and the trifunctional monomer, a more or less rigid polymer matrix forms. Cross-linked polymers do not dissolve in nondegrading solvents. They do not melt. Cross-linking is a most important technique for modifying the properties of a polymer. For example, vulcanizing rubber involves cross-linking the polyisoprene matrix by the formation of —S—bridges.

4.1.1.2 Crystallinity of Polymers

The van der Waals-type interactions between the polymer chains along with steric features and possibly ion-pair formation lead to a variety of self-organization phenomena in polymers. For example, polymers in biological systems are known to form extremely well-organized structures, such as the DNA helical rods (polypeptides). However, synthetic polymers can also be crystalline; even single crystals can form, for example, from polyethylene [26]. As shown in Fig. 4.4, isotactic polypropylene can form well-ordered helical coils that associate to form a crystalline state.

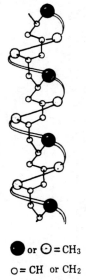

● or ☉ = CH₃
○ = CH or CH₂

Figure 4.4. Helical coil of polypropylene. (From Ref. 26.)

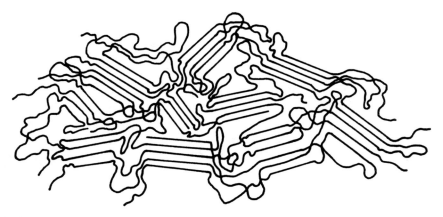

Figure 4.5. A two-phase model for a semicrystalline polymer. (From Ref. 26 by permission of John Wiley and Sons, Inc.)

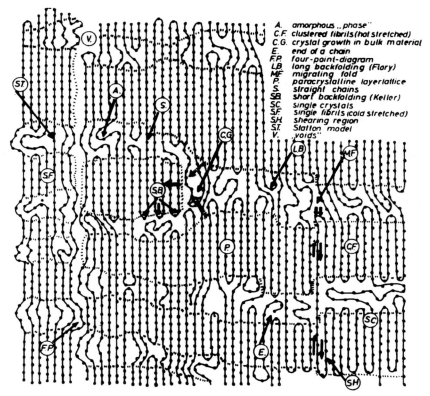

Figure 4.6. The Hosemann model of a semicrystalline polymer (stretched linear polyethylene) (From Ref. 27 by permission of the publishers, Butterworth Heinemann Ltd. ©.)

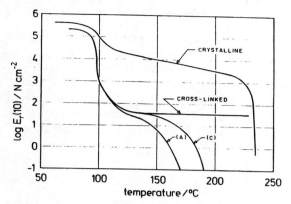

Figure 4.7. The temperature dependence of the elastic modulus [$E_r(10)$, characterizing the polymer stiffness] of four types of polystyrene. A and C are amorphous samples of different (weight-average) molecular mass (A: 210,000; C: 325,000). (From Ref. 26 with permission of John Wiley & Sons, Inc.)

Polypropylene polymerized in a conventional free-radical mechanism is atactic and contains side chains that disturb the ordered arrangement. Such polymers are in an amorphous or partially crystalline state. The structure of a partially crystalline polymer is represented schematically in Fig. 4.5, which shows the ordered, microcrystalline regions next to the polymer in the amorphous state. All stereochemically or structurally irregular polymers have a low degree of crystallinity or are completely amorphous. Of course, crystalline polymers become amorphous at temperatures above the melting temperature, and they can remain amorphous when quickly cooled from their molten state.

It turns out that even the well-crystallized isotactic polymers normally show some imperfections in the x-ray analysis. The picture given in Fig. 4.6 [26,27] is now widely accepted as representing a typical crystalline state of a polymer. However, the crystalline regions can also assume the form of fibrils (cf. [1, chapter 2, vol. 1]). In this case the polymer molecules are ordered parallel to one another over considerable distances. Feltlike structures may form that can assume dimensions visible in the light microscope. Such fibrillar structures can be oriented by stretching the polymer. Thus (e.g., in the case of polyacetylene) the electronic conductivity in the direction of stretching can be significantly enhanced.

The degree of both crystallinity and cross-linking affect the physical properties of polymers in a most significant way. Consider the temperature dependence of the viscoelastic properties of polystyrenes that differ in the degree of crystallinity and cross-linking (Fig. 4.7). $E_r(10)$ is the tensile–relaxation modulus determined by measuring the stress in a polymer sample that had been maintained at constant extension for 10 s. Note that the degree of cross-linking and thus the modulus of the polymer can be varied over a wide range.

4.1.1.3 Phase Transition in Polymers

The temperature-dependent local mobility of some chain segments in an amorphous polymer will differ from that of others. Furthermore, the thermal energy required to melt the amorphous segments in semicrystalline polymers may be lower than the one leading to melting the crystalline portions. Considering further that the crystallization equilibrium is normally not established, it is not unexpected that the melting process of a polymer usually extends over a considerable temperature range, with the polymer softening progressively (see Fig. 4.7), eventually to become a viscous fluid.

One usually defines the glass transition temperature T_g as the temperature range in which the viscosity of the polymer increases by several orders of magnitude within 10 K. The T_g values range typically between $-100°C$ (polyethylene) and about $+100°C$ (polystyrene). Values for common polymers have been collected [28]. Qualitatively, at constant-interaction energy high molecular flexibility decreases the absolute magnitude of T_g. Conversely, intramolecular stiffness raises T_g. Since cross-linking is a special case of structural immobilization, T_g rises rapidly with the cross-link density.

4.1.1.4 Adhesion of Polymer Films to Metallic Substrates

Considering the typical arrangement of polymer-coated electrodes (Fig. 4.1), it is clear that reliable attachment of the film to the substrate, in the presence of the bathing electrolyte, is important. The adhesion between polymer and metal is based on the following mechanisms.

First, specific interactions may be operative between the metal surface atoms and certain structures of the polymer. For example, thiol groups bind strongly to gold surfaces (bond energy of the order of 170 kJ/mol) [13].

Second, the approach of ionic or dipolar groups to metallic surfaces is associated with a decrease in electrostatic free energy. In the case of ionic polymers the electrode potential relative to the point of zero charge defines the electric field on the solution side of the electric double layer, which may be attractive or repulsive for the ions [29].

The third important mechanism of attachment is based on the hydrophobic (or amphiphilic) nature of the polymers. For organic substances it was first shown by Frumkin [29a] that adsorption at the metal–electrolyte interface corresponds closely to the amount of the same substance adsorbed at the air–electrolyte interface. This means that the surface excess of organic species at the metal–electrolyte interface is not so much defined by the metal–polymer interactions as by the surface activity of the organic phase, that is, by expulsion of the organic phase from the aqueous electrolyte.

4.2 The Permeability of Nonionic Polymers

4.2.1 Introductory Considerations

The transport of gases, vapors, liquids, and solutes in nonionic polymers and their partitioning into the polymer phase are important processes. For example, high

THIN POLYMER FILMS ON ELECTRODES

permeability for water makes a polymeric material suitable as fabric for clothings. In designing lacquers and varnishes for protective coatings, or polymer membranes for electrochemical sensors, it is important to control the permeability of the polymer. Ion transport is the process of central interest in the (normally nonionic) polymer electrolytes.

The concepts governing the partitioning and transport processes will be discussed next. They also constitute the basis on which the behavior of the more complex fixed-charge polymers and the electroactive polymers can be analyzed.

4.2.2 Partitioning and Transport of Electroneutral Species

Consider a neutral species i that is partitioned between the solution (S) and the polymer (poly) phases. In the polymer, contributions to transport by processes other than diffusion are considered negligible. The electronic conductivity of the polymer is considered negligibly small; thus electron transfer reactions can proceed (if at all) only at the interface between the metal (Me) electrode and the polymer phase. The situation corresponds to the "membrane-type" film shown in Fig. 4.1b.

The system can be analyzed on the basis of the classical picture (Fig. 4.8). It represents the partial molar standard (Gibbs) free energy, that is, the standard chemical potential of species i, μ_i^0, as a function of the spatial coordinate x extending across the interface. We consider first the equilibrium distribution of i across the interface, and then the diffusive transport in the polymer.

4.2.2.1 Partitioning Species i Across the Polymer–Electrolyte Interface

The equilibrium state is characterized by equal chemical potentials of i in the two phases:

$$\mu_i^{poly} = \mu_i^S \tag{4.1}$$

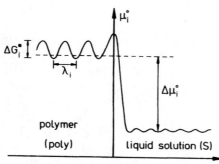

Figure 4.8. Free-energy (standard chemical potential, μ_i^0), on a molecular scale of a neutral species i partitioned across the interface between a solution and a polymer phase. $\Delta\mu_i^0$ is the free energy of transfer, ΔG_i^* is the activation energy for diffusive transport of i in the polymer.

The chemical potential can be divided into the standard chemical potential μ_i^0, and an activity a_i term:

$$\mu_i^{0(poly)} + RT \ln a_i^{poly} = \mu_i^{0(S)} + RT \ln a_i^S \tag{4.2}$$

Thus the fundamental equilibrium distribution equation results:

$$a_i^{poly} = a_i^S \exp\left(-\frac{\mu_i^{0(poly)} - \mu_i^{0(S)}}{RT}\right) \tag{4.3}$$

We now define the *free energy of transfer*, $\Delta\mu_i^0$ (often, at constant pressure and temperature, this is the same as the *work of transfer*):

$$\Delta\mu_i^0 = \mu_i^{0(poly)} - \mu_i^{0(S)} \tag{4.4}$$

and assume an ideal solution of an undissociated species i, so that the activities can be replaced by concentrations c_i:

$$c_i^{poly} = c_i^S \exp\left(-\frac{\Delta\mu_i^0}{RT}\right) \tag{4.5}$$

$$= c_i^S k_{part} \tag{4.6}$$

Equations (4.5) and (4.6) are representations of the well-known Nernst distribution law, where k_{part} is known as the distribution constant, the partition coefficient, or the medium-effect coefficient [24d,30,31].

As an experimental demonstration of the partitioning of water into polymer coatings, consider the electrochemical impedance results represented in Fig. 4.9. Two types of polymer films were prepared by plasma polymerization [21a], one from hexafluoropropene (HFP), the second one from acrylonitrile (AN). Such films are highly cross-linked, and the permeability for ions is very low. Thus one may assume that the polymer film constitutes the dielectric of a parallel-plate capacitor consisting of the metal electrode on one side and counter ions in the electrolyte (at the polymer surface) on the other side. The corresponding geometric capacitance C_g, which can be determined by impedance (Z) measurements [21], is given by:

$$C_g = A \frac{\epsilon \epsilon_0}{\delta} \tag{4.7}$$

where A is the electrode area, ϵ the relative dielectric constant, ϵ_0 the permittivity of free space, and δ the thickness of the film in the arrangement of Fig. 4.1b.

On the imaginary ["$Z(-90°)$"] axis of Fig. 4.9 the calculated impedance values for barrier films of $\epsilon = 3$ and 100-, 200-, and 250-nm thickness are shown. Clearly, the experimental impedance vector of the HFP film corresponds closely to the expected value, but the impedance of the AN film is smaller than expected by about an order of magnitude. This indicates that the HFP film is practically impermeable for water, whereas water enters into the AN film, raising its effective dielectric constant. Of course, impedance analyses can be conducted to obtain more detailed information about the state of a polymer film [9]; however, Fig. 4.9 shows the basic effect. From the change of impedance with time one can deduce the rate at which equilibration proceeds. It turns out that in the case of the HFP film the impedance

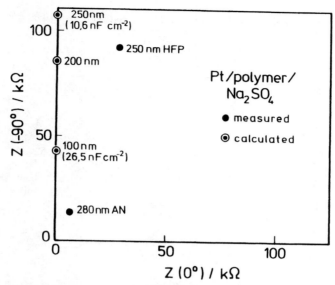

Figure 4.9. Experimental and calculated impedances of electrodes coated with plasma–polymer films of indicated thickness (in nm). Film prepared from hexafluoropropene (HFP) and acrylonitrile (AN). Experimental parameters: electrode area: 0.28 cm², ac frequency: 500 Hz, measurement 1 hour after immersion of the electrode.

was stable over periods of hours, whereas in the case of the AN film the impedance decreased over the first hour of immersion in the electrolyte by about one-half. The mass fraction of water can be estimated by assuming a linear combination of the dielectric constants of the polymer and water.

The reason for the more hydrophilic character of the AN polymer relative to the (Teflon-like) HFP polymer is the presence of the polar —CN groups, which interact with the water molecules. It turns out that the amount of water absorbed can be related stoichiometrically to the chemical structure of the polymer, when the matrix is open enough to permit the accumulation of water [32]. For example, at a relative humidity of 100% an aromatic ring in the polymer matrix has been found to bind 0.005 water molecules, a —C=O or —C≡N group binds 0.3, and an —OH or a —NH_2 group binds two water molecules [32]. Table 4.1 summarizes sorption equilibrium data of water in selected polymers [33], where p/p_0 is the partial pressure of water (the values in parentheses in the last column are the exact p/p_0 values).

Although water plays an outstanding role in permeability considerations, it should be noted that polymers are more or less permeable for a wide range of permeants [32,34]. It is a subject of great interest to modify the molecular design of the polymer, and thus change the permeability of the polymer for different diffusing species. In this way permselective membranes may be designed. For example, when the acrylonitrile content of an acrylonitrile–butadiene copolymer is increased, the solubility of CO_2 increases rapidly but that of H_2, N_2, and O_2 decreases [32].

Table 4.1 Smoothed Values for the Sorption of Water Vapor (g/100 g of polymer)

Polymer	\multicolumn{4}{c}{p/p_0}	T (°C)
	0.3	0.5	0.7	0.8–1.0	
Cellulose nitrate	—	1.9	2.92	3.5 (0.82)	40
Ethyl cellulose	0.44	1.22	2.08	7–9	50
Polyacrylic acid	5.0	7.0	~18.0	~70 (0.95)	26
Polyacrylonitrile	0.47	0.77	2.2	—	25
Polydimethylsiloxane	—	0.015	0.025	0.07 (0.95)	35
Polyethylene ($p = 0.923$)	\multicolumn{2}{c}{Linear isotherm}		0.0062 (1.0)	25
Polyethylmethacrylate	—	0.44	0.72	2.4	50
Polyisobutylmethacrylate	0.36	0.54	0.64	2.3 (0.95)	40
Polymethacrylic acid	3.42	6.12	9.72	~18.4 (0.95)	40
Polymethylacrylate	—	~0.99	—	—	—
Polymethyl methacrylate	0.42	0.75	1.18	—	50
Polymethylvinylketone	0.67	1.39	2.88	—	25
Polypropylene	\multicolumn{2}{c}{Linear isotherm}		0.0071	25
Polystyrene	0.019	0.032	0.048	—	25
Polytetrafluoroethylene	—	—	0.007	—	20
Polyvinylacetate	0.53	1.20	2.2	5.95	40
Polyvinylalcohol	4.0	8.5	18.0	>40 (0.9)	25
Polyvinylamine	14.0	16.0	30	—	26
Polyvinyl chloride	0.076	1.28	0.42	1.5 (0.9)	30
Polyvinylisobutyl ether	—	0.18	0.36	—	25
Rubber hydrochloride	0.26	0.45	0.93	—	25
Sodium polyacrylate	21.5	40	80	—	26

4.2.2.2 Diffusive Transport of Neutral Species

We consider the movement of particle i in the polymer (Fig. 4.8) from one position of minimum free energy, x_1, over the energy barrier of height ΔG_i^* to the next energy minimum at $(x_1 + \lambda_i)$. The rate of transport over the barrier can be characterized [35] by the frequency of transition, k_i^*:

$$k_i^* = \frac{kT}{h} \exp\left(-\frac{\Delta G_i^*}{RT}\right) \tag{4.8}$$

The flux J_i^* between two energy minima is given by the difference between all forward and backward jumps in unit time:

$$J_i^* = \lambda_i [k_i^* c_i(x_1) - k_i^{*(\text{back})} c_i(x_1 + \lambda_i)] \tag{4.9}$$

When all barriers are equal and symmetrical, one obtains from Eq. (4.9) Fick's law for the flux of species i in the polymer:

$$J_i^{\text{poly}} = -D_i^{\text{poly}} \frac{\partial c_i^{\text{poly}}}{\partial x} \tag{4.10}$$

where D_i^{poly} is the diffusion coefficient of i in the polymer. Note that in experiments with thin films on electrodes only a driving force across the film (direction x) is of interest. The barrier ΔG_i^*, which defines the diffusion coefficient, can be understood as characterizing the motion of polymer chain segments, which open the way for the particle to move to a new low-energy position. For a more thorough review of such ion-transport processes see Ref. 36.

Since the thermal motion of the polymer segments decreases drastically below the glass-transition temperature T_g, the temperature dependence of the diffusion coefficient has been described [37] by

$$D_i^{poly} = D_i^{0(poly)} T^{1/2} \exp\left(-\frac{\Delta G_i^*}{R(T-T_g)}\right) \qquad (4.11)$$

Crystallization of the polymer obstructs the movement of the permeating species; thus one may formulate for a homogeneous distribution of crystalline domains:

$$D_i^{poly} = D_i^{poly(a)} (1 - x_c) \qquad (4.12)$$

where $D_i^{poly(a)}$ is the diffusion coefficient in the amorphous polymer and x_c is the fraction of the polymer in the crystalline state. Cross-linking of the polymer has an effect similar to that of crystallinity. Obviously, the diffusion coefficient will be smaller for larger diffusing species. On the other hand, plasticizers tend to increase the value of D_i^{poly}. This is important, for example, in the field of ion-selective electrodes, to enhance the carrier mobility in polymers such as PVC [38]. But even water, dissolved in the polymer, can act as a plasticizer and enhance the rate of diffusive transport of such depolarizers as O_2.

The permeability of a polymer for a diffusing species is the directly measurable stationary flux across the polymer phase. Consider a film of thickness δ immersed between two solutions of concentrations $c_i^{S1} = 0$ and c_i^{S2}. The stationary situation is represented schematically in Fig. 4.10. The flux of i across the film is given by

$$J_i = - D_i^{poly} \frac{k_{part} c_i^{S2}}{\delta} \qquad (4.13)$$

Thus a possible barrier for i crossing the interfaces (see Fig. 4.10) as well as depletion of i in the liquid electrolyte near $x = \delta$ are neglected. Furthermore, it is

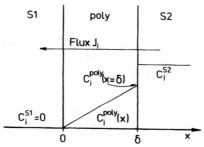

Figure 4.10. Stationary concentration (c) profile of species i permeating a polymer film of thickness δ from a solution containing i (S2) to a solution S1, which is free of it. The concentration changes caused by the flux of i in the solutions near the interfaces are disregarded.

assumed that k_{part} is not a function of the concentration of the diffusing species. This is in many cases unrealistic. In principle, it is therefore necessary to employ methods (for determining the permeability) in which the concentrations on both sides of the membrane are equal (see, e.g., Ref. 39).

In the arrangement of Fig. 4.1, it is experimentally straightforward to create the concentration gradient in the film electrochemically, when the species i is electroactive. When the overvoltage is large enough, the condition $c_i^{poly}(x = 0) \approx 0$ can be fulfilled by the interfacial charge-transfer reaction removing i. The observed current flow is a measure of the flux across the film (discussed later). Note that the determination of diffusion coefficients by such permeability measurements requires the knowledge of k_{part}.

4.2.3 Ionic Species in Polymers

In this section we consider the partitioning of ions into a polymer that itself is nonionic. The temperature is assumed to be above T_g [Eq. (4.11)] and the polymer is amorphous [Eq. (4.12)]. The ionic species should thus be mobile in the polymer. The electrostatic charge of the ions introduces new features into the partitioning and transport considerations.

4.2.3.1 Electroneutrality

Except at the interfaces, where electrical double-layer phenomena may prevail, the polymer phase will be practically free of excess electric charge. The partitioning of the compound K_mX_n (salt) between the electrolyte and the polymer phases can thus be formulated:

$$(K_mX_n)^{poly} \rightleftharpoons (K_mX_n)^S \tag{4.14}$$

and the partitioning equilibrium is described by Eqs. (4.2)–(4.6):

$$a_{salt}^{poly} = a_{salt}^S \exp\left(-\frac{\Delta\mu_{salt}^0}{RT}\right) \tag{4.15}$$

This compound is now considered to be dissociated according to

$$K_mX_n \rightleftharpoons mK_{z_K} + nX_{z_X} \tag{4.16}$$

where z_K and z_X are the ionic charges. Note that in this case the concentration of the salt is related to the activity in a way that is not straightforward [40]. It is thus preferable and more elucidating to consider the partitioning and transport of the single ionic species, which are formed in the dissociation process. Doing so is a common practice in electrochemical and ion-exchange work [40,54]. Thus the partitioning equilibria

$$(Kz_K)^{poly} \rightleftharpoons (Kz_K)^S \tag{4.17a}$$

$$(Xz_X)^{poly} \rightleftharpoons (Xz_X)^S \tag{4.17b}$$

will be considered. They establish in such a way that electroneutrality prevails in the polymer, as in the liquid phase. The concentrations of cations (K) and anions (X) in the polymer must thus adjust so that the condition

$$mc_K^{poly} = nc_X^{poly} \tag{4.18}$$

is fulfilled. This principle is known as *electroneutrality coupling*.

The equilibrium distributions of ionic species i are characterized by an equality (in the two phases) of the electrochemical potentials, $\bar{\mu}_i = \mu_i + z_i F \phi$, where ϕ is the electric potential in the considered phrase:

$$\bar{\mu}_K^{poly} = \bar{\mu}_K^S \tag{4.19a}$$

$$\bar{\mu}_X^{poly} = \bar{\mu}_X^S \tag{4.19a}$$

From these equations one obtains the partitioning equilibrium equation for each ionic species i:

$$a_i^{poly} = a_i^S \exp\left(-\frac{\Delta\mu_i^0}{RT}\right) \exp\left[-\frac{z_i F}{RT}(\phi^{poly} - \phi^S)\right] \tag{4.20a}$$

$$= a_i^S k_{part} \exp\left[-\frac{z_i F}{RT}(\phi^{poly} - \phi^S)\right] \tag{4.20b}$$

It is convenient and customary to choose as the reference states the solutions of i in the polymer and the liquid, both at infinite dilution [24,40a]. Then, $\Delta\mu_i^0$ is the molar free energy of transfer of i from the liquid solution into the polymer [Eq. (4.4)] at infinite dilution. The partitioning equilibrium is a function of both the partition coefficient $k_{part} = \exp(-\Delta\mu_i^0/RT)$, which characterizes the "medium effect" and of $z_i F(\phi^{poly} - \phi^S)$, the electric free-energy term; $(\phi^{poly} - \phi^S)$ is the Nernst–Donnan potential. For dilute solutions the activities of i may be replaced by the concentrations of i:

$$c_i^{poly} = c_i^S \exp\left(-\frac{\Delta\mu_i^0}{RT}\right) \exp\left[-\frac{z_i F}{RT}(\phi^{poly} - \phi^S)\right] \tag{4.20c}$$

$$= c_i^S k_{part} \exp\left[-\frac{z_i F}{RT}(\phi^{poly} - \phi^S)\right] \tag{4.20d}$$

Consider now the partitioning of both cations and anions. Electroneutrality coupling requires that $c_K^{poly}/c_K^S = c_X^{poly}/c_X^S$. Thus the system adjusts the interfacial electric-potential drop in such a way that at equilibrium the condition

$$\phi^{poly} - \phi^S = \frac{\Delta\mu_X^0 - \Delta\mu_K^0}{F(z_K - z_X)} \tag{4.21}$$

is fulfilled [it is assumed that Eq. (4.20c) is valid]. For example, if the anion is more hydrophobic than the cation (i.e., $\Delta\mu_X^0 < \Delta\mu_K^0$), the (hydrophobic) polymer in contact with the aqueous electrolyte will assume a negative electric potential rela-

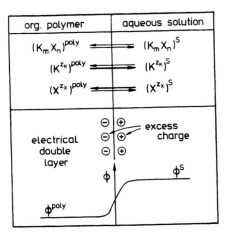

Figure 4.11. Equilibrium distribution of a dissociated salt $K_m X_n$ between its aqueous solution and a permeable polymer. The arrows of different thickness indicate that the cation is hydrophilic while the anion is hydrophobic. The resulting interfacial excess ionic charge and the corresponding potential difference $(\phi^{poly} - \phi^S) < 0$ are shown schematically.

tive to the electrolyte. At equilibrium, the electric work term $z_i F(\phi^{poly} - \phi^S)$ compensates the difference in the "medium effect" free energies of transfer. An interface of this type is shown schematically in Fig. 4.11. In extreme cases the equilibrium potential difference $(\phi^{poly} - \phi^S)$ (the *distribution potential*) can assume values of hundreds of millivolts, as is well known from the field of immiscible electrolytes; see, for example, Table I of Koryta and Vanýsek [41].

Note, finally, that the equilibrium distribution of the salt [Eq. (4.15)] can be expressed in terms of the $\Delta\mu^0$ values of the ions:

$$c_{salt}^{poly} = c_{salt}^{S} \exp\left(-\frac{m\Delta\mu_K^0 + n\Delta\mu_X^0}{RT(m+n)}\right). \tag{4.22}$$

where *salt* relates again to a substance dissociating according to Eq. (4.16).

4.2.3.2 The Electrostatic Free Energy of Ion Transfer

In the preceding discussion the important role of the free energy of transfer, $\Delta\mu_i^0$, in defining the partitioning equilibrium of species i was demonstrated. When i is a neutral species, $\Delta\mu_i^0$ is largely determined by the differences in van der Waals and donor–acceptor interactions of species i in the polymer and solution phases. When i is an ionic species, there is an additional electrostatic contribution to $\Delta\mu_i^0$, which is caused by the different electrostatic stabilization free energies of i in the two phases of different dielectric constant, ϵ. The difference in electrostatic free energy constitutes the *electrostatic free energy of ion transfer*, $\Delta\mu_i^e$. It has a dominating influence on the partitioning equilibrium of ions between aqueous solutions ($\epsilon \approx 80$) and organic polymers (normally $\epsilon < 10$ at room temperature), whenever the specific chemical interactions are not particularly strong in either phase. In such situations $\Delta\mu_i^e$ can be used as an estimate of $\Delta\mu_i^0$.

It was first pointed out by Born [42] that the electrostatic free energy of 1 mol of ions of radius r can be estimated on the basis of Coulomb's law, when one makes the assumption that the medium surrounding the ions (solvent or polymer) is a homogeneous dielectric [43,44]:

$$\mu_i^e = \frac{z_i^2 e^2 N_A}{8\pi\epsilon\epsilon_0 r} \tag{4.23}$$

where e is the charge of the electron and N_A is Avogadro's number. For a typical univalent ionic species i, of radius $r = 1. \times 10^{-10}$ m, one obtains with Eq. (4.23) a value of $\mu_i^e = 695/\epsilon$ kJ mol^{-1}. Thus in an aqueous electrolyte $\mu_i^e(\text{water}) = 8.7$ kJ mol^{-1}. On the other hand, in polyethylene ($\epsilon = 2.3$) $\mu_i^e(\text{PE}) = 302$ kJ mol^{-1}. The electrostatic work of transfer is thus $\Delta\mu_i^e = +293$ kJ mol^{-1}. If one assumes that in this system $\Delta\mu_i^0 \approx \Delta\mu_i^e$, and that the same value of $\Delta\mu_i^0$ is valid for the univalent anion X, then in contact with the dissociated aqueous KX electrolyte of concentration 1 mol dm^{-3} the equilibrium concentrations in the polymer $c_K^{poly} = c_X^{poly} = 4 \times 10^{-52}$ mol dm^{-3}. This value is certainly only a first approximation to the true value. However, even if one accepts that it may be incorrect by several orders of magnitude, one concludes that such ions will not enter into polyethylene, in agreement with the experimental experience.

The situation becomes quite different when the polymer is more polar. First, the dielectric constant is higher than for polyethylene (if the polar groups have sufficient local mobility that they can adjust their orientation to the electric field) [45]. Second, the polar groups interact specifically with water so that water is partitioned into the polymer matrix and raises the effective value of ϵ further (see Section 4.2.2). Third, the polar groups along with the incorporated solvent are likely to have specific interactions with the ions [43]. It may thus be concluded that polar polymers may well be permeable to a considerable extent to ionic species.

The ions may be partitioned into the polymer phase also as ion pairs. In this case the electrostatic free energy of the ion pair, say KX, in a homogeneous dielectric is smaller than the sum $\mu_K^e + \mu_X^e$. This is so because of the energy loss associated with the approach between the anion and the cation (Coulomb's law). The electcrostatic free energy of the ion pair can be estimated:

$$\mu_{KX}^e \approx \frac{e^2 N_A}{4\pi\epsilon\epsilon_0} \left(\frac{z_K^2}{2r_K} + \frac{z_X^2}{2r_X} + \frac{z_K z_X}{a} \right) \tag{4.24}$$

where a is the distance between the charge centers of the ions. As an example, consider again the uni-univalent salt KX where the anions and cations each have the radius $r_K = r_X = 1 \times 10^{-10}$ m, and $a = r_K + r_X$. The free energy of the ion pair is the same as that for one ion: $\mu_{KX}^e \approx 695/\epsilon$ kJ mol^{-1}. This is a remarkable result, indicating that the partitioning of an ion pair into the medium of lower dielectric constant is favored over the partitioning of the free ions. This preference for the ion pair is still more significant when the ionic charge separation in the ion pair is reduced by covalent bonding.

4.2.3.3 Dissociation and Association in the Polymer

In Eq. (4.24), the last term defines the electrostatic free energy of dissociation, μ_{diss}^e, of the ion pair. According to this classical model [30] the dissociation equilibrium constant, k_{diss}, again using the homogeneous-dielectric approximation, is given by

$$k_{diss} = \exp\left(-\frac{\mu_{diss}}{RT}\right) \approx \exp\left(-\frac{\mu_{diss}^e}{RT}\right) = \exp\left(-\frac{e^2 N_A z_K z_X}{4\pi RT \epsilon \epsilon_0 a}\right) \quad (4.25)$$

For the dissociation of the uni-univalent salt discussed earlier $\mu_{diss}^e = +695/\epsilon$ kJ mol^{-1}. Considering that at room temperature RT has a value of 2.46 kJ mol^{-1}, it is clear that in an aqueous electrolyte considerable dissociation takes place. However, even in a polymer phase with the relatively large dielectric constant of $\epsilon = 10$, a dissociation constant of only $k_{diss} = 5 \times 10^{-13}$ is expected. This result is important when the ionic conductivity is considered, which is proportional to the density of free ions. To have a high degree of dissociation the value of μ_{diss}^e should be small, that is, z should be ± 1, and a should be large. This is in agreement with the practical experience that salts consisting of large univalent ions tend to lead to ionic conductivity in media of low dielectric constant [43]. Of course, in actual systems specific interactions may exist, such as complexation of the ions [32,43]. This constitutes again a "chemical" free-energy term that has to be added to μ_{diss}^e to obtain a more appropriate value of μ_{diss} and thus of k_{diss}.

Note that ion pairs tend to associate with additional ions or ion pairs to form higher aggregates (triplets) [47]. It is intuitively plausible that two ion pairs, forming each an electric dipole, can associate to form a quadrupole (Fig. 4.12). This appears to be a process that can lead, upon further progress, to domains in the polymer, phase separation, and so on. It follows from the electrostatic model that such phenomena occur preferably in nonpolar polymers.

4.2.3.4 Ion Transport in Nonionic Polymers

The driving force for ion transport in a polymer phase may be either a concentration gradient (diffusive transport) or an electric field (migration), or both. The resulting

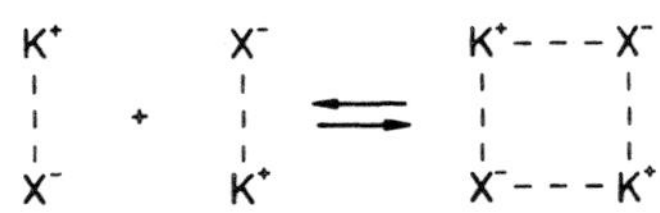

Figure 4.12. Formation of an ionic quadrupole from an ion pair.

flux of the ionic species i, J_i, is usually described by the Nernst–Planck equation [31,36]:

$$J_i^{poly} = -D_i^{poly} \left[\frac{\partial c_i^{poly}}{\partial x} - \frac{z_i c_i^{poly} F}{RT} \frac{\partial \phi^{poly}}{\partial x} \right] \quad (4.26)$$

Again, the transport in the film is considered to proceed only in the x-direction, that is, across the film. The diffusion coefficient can be correlated with the physical mobility u (in units of cm^2 V^{-1} s^{-1}) via the Einstein relation [19d,31]:

$$u_i = \frac{|z_i| D_i F}{RT} \quad (4.27)$$

The specific ionic conductivity σ_{ion} of the considered phase is given by

$$\sigma_{ion} = F \sum_i |z_i| u_i c_i \quad (4.28)$$

The situation is thus analogous to liquid electrolytes. In fact, in going from the organic liquid to the polymer by increasing the molecular mass, one can demonstrate the smooth transition of the ionic conductivity from the liquid to the solid electrolyte value [48]. Note that ions may cross-link the polymer matrix by interacting with sites of different chains, thus raising the viscosity and decreasing the ionic mobility. Further, the dissolved salt tends to increase T_g [48]. Otherwise, the same considerations as for the transport of neutral species in the polymer are valid.

Migrational transport proceeds normally by both the anions and the cations. The fraction of the ionic current transported, in the absence of concentration gradients, by a particular species i is the transference number t_i:

$$t_i = \frac{\sigma_i}{\sum_j \sigma_j} \quad (4.29)$$

where j symbolizes all the ionic species present, including i. Determining t_i is a subject of great interest in the field of polymer electrolytes [48].

Consider now the diffusion of the neutral salt KX. When the compound is undissociated, its treatment as a neutral species (cf. Section 4.2.2) is straightforward. However, an interesting situation arises when the salt is dissociated and the mobility of the anions differs from that of the cations. In this case, electroneutrality coupling requires that both anions and cations diffuse together. This is *ambipolar diffusion*, which is known to occur, for example, toward the walls confining electric plasmas, and which leads to the *diffusion potentials* well known in electrochemistry [40b]. The more mobile species advances ahead of the more immobile one. Thus an electric field is created that accelerates the slower and slows down the more mobile species. Eventually, in the stationary state, both the anions and the cations move at the same velocity (it is assumed that $z_K = |z_X|$). The situation can be analyzed on the basis of the Nernst–Planck equations for the anions and cations. A diffusion coefficient for the salt

$$D_{salt} = \frac{2D_K D_X}{D_K + D_X} \tag{4.30}$$

is obtained by eliminating the electric field from the two equations.

Consider, as a practical example for the preceding considerations, the common paints, lacquers, and varnishes that are used to prepare protective coatings. They are made up of a wide variety of organic compounds that contain polar groups [46]. It has indeed been long established that such organic coatings may be considered semipermeable membranes [46]. Along with such components as water and oxygen, ions permeate the coatings. In general, they provide sufficient ionic conductivity that electrochemical corrosion processes can proceed at the polymer–metal interface, that is, under the intact coating. It is therefore common practice to add inhibitors suppressing the interfacial charge-transfer reactions to such paints. In fact, it has been shown with model films that electrochemical charge-transfer reactions can proceed effectively at the interface between the metal and the polymer film, which is in contact with an aqueous electrolyte [9a].

4.2.4 Carrier Mediation and Ion-Selective Electrodes

4.2.4.1 Ionophores in Organic Phases

It was shown earlier that compounds dissociating into small ions, such as Na^+, Cl^-, and so on, are not partitioned to a significant degree into homogeneous media of low dielectric constant. However, numerous compounds (ionophores) exist that "mediate" the ion transfer into the organic phase. Consider, for example the natural ionophores nonactin and valinomycin (Fig. 4.13) [49]. They consist of a polar center in which the ion is complexed and a hydrophobic exterior that renders the complex soluble in organic phases. According to this principle, a large number of ionophores have been prepared synthetically [31,49]. Many of them are excellently

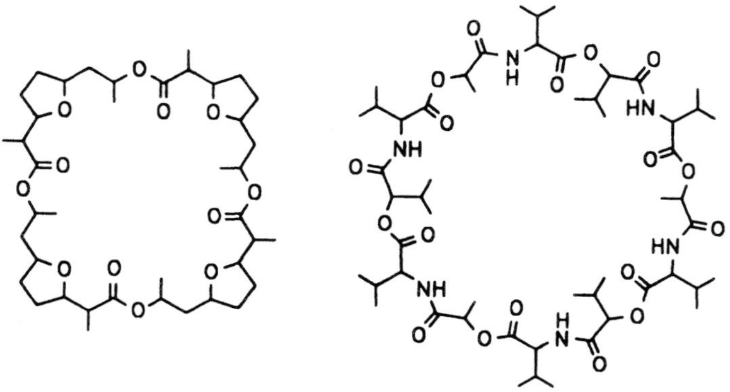

Figure 4.13. The ionophores nonactin (left) and valinomycin. (From Ref. [49], with permission of John Wiley & Sons, Ltd.)

selective for a particular ionic species. For example, valinomycin binds potassium ions, discriminating them from sodium ions by a factor of about 10^{-4} [31].

The carrier-mediated partitioning of an ionic species I of charge z_I and activity a_I can be analyzed as a two-step process:

$$I^S \rightleftharpoons I^{\text{poly}} \tag{4.31a}$$

$$I^{\text{poly}} + Y^{\text{poly}} \rightleftharpoons (IY)^{\text{poly}} \tag{4.31b}$$

where Y stands for the ionophore. Y is practically insoluble in aqueous electrolytes; that is, it is confined to the organic phase. The first step in the partitioning process is associated with the large positive free energy of transfer of I into the organic phase. The second step is the complexing reaction (*host–guest chemistry*) of the ion and the ionophore. The stability of the formed complex may be characterized as usual by the complex-stability constant K_{IY}:

$$K_{IY} = \frac{a_{IY}}{a_I a_Y} = \exp\left(\frac{\mu_I^0 + \mu_Y^0 - \mu_{IY}^0}{RT}\right) \tag{4.32}$$

In the following discussion, electroneutral carrier molecules, such as the ones shown in Fig. 4.13, are considered. The equilibrium distribution of the ionic species I is then characterized by the following equality of the electrochemical and chemical potentials [cf. Eq. (4.31)]:

$$\bar{\mu}_I^S + \mu_Y^{\text{poly}} = \bar{\mu}_{IY}^{\text{poly}} \tag{4.33}$$

From Eq. (4.33) one obtains immediately the equation describing the equilibrium partitioning of I between the solution and the polymer phase containing the carrier:

$$a_{IY}^{\text{poly}} = a_I^S a_Y^{\text{poly}} \exp\left(-\frac{\Delta\mu_I^0}{RT}\right) K_{IY}^{\text{poly}} \exp\left[-\frac{z_I F(\phi^{\text{poly}} - \phi^S)}{RT}\right] \tag{4.34a}$$

In Eq. (4.34a) the term $\exp(-\Delta\mu_I^0/RT)$ is the medium-effect coefficient k_{part} for partitioning into the carrier-free organic phase. One could combine the three constant pre-exponential terms to one *effective partitioning coefficient* $k_I = a_Y^{\text{poly}} k_{\text{part}} K_{IY}^{\text{poly}}$. Then, Eq. (4.34) reads

$$a_{IY}^{\text{poly}} = a_I^S k_I \exp\left[-\frac{z_I F(\phi^{\text{poly}} - \phi^S)}{RT}\right] \tag{4.34b}$$

Equation (4.34b) is of the form of the basic ion-partitioning equation, whereby $k_I > k_{\text{part}}$. The ionophore in the polymer has the function of rendering the considered ionic species I effectively more hydrophobic.

Consider now the partitioning of both cations K and anions X. If the ionophore Y has the desirable property of selectivity, it will complex either K or X. We assume for the following that: (1) Y is electroneutral and reacts with K to form KY while there is no significant specific interaction between Y and X; (2) the electrolyte is symmetrical ($z_K = |z_X| = z$); (3) activity coefficients may be disregarded; (4) the value of K_{KY}^{poly} is large enough that the concentration of free K ($=K^{z+}$) in the polymer is negligible compared with the concentration of KY ($=K^{z+}Y$). Electroneutrality coupling requires that $c_{KY}^{poly}/c_K^S = c_X^{poly}/c_X^S$. Combining this with the partitioning equations for K and X one arrives at an expression for the electric "distribution potential," that is, the equilibrium potential drop across the interface [cf. Eq. 4.21)]:

$$\phi^{poly} - \phi^S = \frac{1}{2zF} [(\Delta\mu_K^0 - \Delta\mu_X^0) + RT \ln(a_Y^{poly} K_{KY}^{poly})] \quad (4.35)$$

where $\Delta\mu_K^0$ and $\Delta\mu_X^0$ are the free energies of transfer of K and X into the carrier-free organic phase. Of course, deriving the corresponding equations for anionophores and unsymmetrical electrolytes is straightforward.

The following typical example should illustrate the preceding considerations. A simple uni-univalent electrolyte ($z = 1$) shall be partitioned between an aqueous electrolyte and a polymer–plastiziser system with the dielectric constant of polyvinyl chloride (PVC, $\epsilon = 4.55$). The polymer contains an electroneutral ionophore; it complexes the cation with a (typical [49]) value of $K_{KY}^{poly} = 10^6$ mol^{-1} dm^3. The concentrations of the salt in the electrolyte and of the carrier in the polymer are each 1 mol dm^{-3}. We assume that the free energies of transfer into the carrier-free medium can be approximated by the electrostatic free energies as shown earlier [Eq. (4.23)]; that is, $\Delta\mu_K^0 = \Delta\mu_X^0 = +144$ kJ/mol. Thus one obtains the interfacial equilibrium-potential difference according to Eq. (4.35): ($\phi^{poly} - \phi^S$) = $+0.176$ V. The concentrations of cations (as KY) and anions in the polymer containing the ionophore are $c_{KY}^{poly} = c_X^{poly} = 3.8 \times 10^{-23}$ mol dm^{-3}. In the absence of the ionophore the interfacial electric potential drop would be zero, and the equilibrium concentrations of both the anions and cations would be $c_K^{poly} + c_X^{poly} = 3.8 \times 10^{-26}$ mol dm^{-3}. This means that the presence of this carrier leads to an enhancement of the ion concentrations in the polymer by a factor of 10^3.

The preceding example illustrates the fact that, via the distribution potential, a fraction (in the considered case one-half) of the free energy of cation complexation is used to "pull" the required anions into the polymer. Therefore the effect of the ionophore on the partitioning of the salt is not as significant as it would be for the complexed ionic species alone. In this realistic example the presence of the ionophore raises the salt concentration to a value that corresponds to only two cations and two anions per cubic decimeter of the polymer phase!

4.2.4.2 Potentiometric Ion-Sensitive Electrodes

An ion-sensitive electrode (see, e.g., Buck [19], and Morf [31]) is a symmetric arrangement of a particular membrane between two electrolytes. The membrane potential, that is, the equilibrium potential difference between the two contacting solutions, is measured with reference electrodes. Thus the potentiometric signal $E \approx (\phi^S - \phi^{ref})$ is obtained. The situation is represented schematically in Fig. 4.14. Ideally, E is a Nernst-type function of the activity of the ionic species of interest, and it is not disturbed by the presence of other species in the sample solution.

THIN POLYMER FILMS ON ELECTRODES

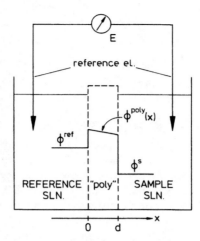

Figure 4.14. Illustration of the principle of an ion-selective electrode system. The polymer membrane ("poly") of thickness d separates the sample- from the reference solution. The potential (ϕ) distribution is shown schematically. The two reference electrodes produce the potentiometric signal E.

The membrane potential is the sum of the two interfacial potential drops and the diffusion potential that may develop across the membrane:

$$(\phi^S - \phi^{ref}) = (\phi^{poly}_{x=0} - \phi^{ref}) + (\phi^{poly}_{x=d} - \phi^{poly}_{x=0}) + (\phi^S - \phi^{poly}_{x=d}) \quad (4.36)$$

where $x = 0$ and $x = d$ mark the phase boundaries between the membrane and the electrolytes. The basic functioning of these sensors may be understood straightforwardly if it is assumed that concentration gradients in the membrane phase are small enough that the diffusion potential ($\phi^{poly}_{x=d} - \phi^{poly}_{x=0}$) is ≈ 0. This is a realistic assumption for relevant systems. It means that E is the algebraic sum of the two interfacial potential drops.

Note that the distribution potential for a single salt [Eqs. (4.21) and (4.35)] is not a function of the electrolyte concentration (except via the activity coefficients, which are disregarded in this consideration). The situation changes when two or more salts are involved in the partitioning process. Consider the partitioning of two uni-univalent salts with a common anion, KX and MX. In this case, in addition to Eqs. (4.19a,b) one has to consider the equilibrium

$$\bar{\mu}^{poly}_M = \bar{\mu}^S_M \quad (4.19c)$$

or, in the presence of the ionophore Y complexing M in the polymer

$$\bar{\mu}^{poly}_M + \mu^{poly}_Y = \bar{\mu}^{poly}_{MY} \quad (4.33a)$$

Neglecting activity coefficients [Eq. (4.20d)], and taking into account electroneutrality coupling, one derives the following equations for the equilibrium interfacial potential [50]:

$$(\phi^{\text{poly}} - \phi^S) = \frac{RT}{2F} \ln \left(\frac{k_{\text{part(K)}} c_K^S + k_{\text{part(M)}} c_M^S}{k_{\text{part}(X)} c_X^S} \right) \quad (4.37)$$

For ionic species that are complexed in the polymer, $k_{\text{part}(i)}$ should be replaced by the effective partition coefficient k_i. Clearly, depending on the concentrations and the partition coefficients, $(\phi^{\text{poly}} - \phi^S)$ depends more or less on the concentrations of the salts.

However, in order to obtain selective potentiometric response to one particular ionic species, it is required that this species be partitioned exclusively across the interface so that the interfacial potential is determined by the Nernst–Donnan relation [Eq. (4.20), (4.34)] for these ions. This may be achieved, for instance, with a pair of uni-univalent salts, KX and MZ, where the anion X is very hydrophilic (k_X or $k_{\text{part}(x)} \ll 1$), so that it is not significantly partitioned into the organic phase, whereas Z is hydrophobic (k_Z or $k_{\text{part}(z)} \gg 1$), so that it is confined to the organic phase. The ionic species Z might be attached alternatively via chemical bonds to the polymer matrix (*fixed-charge polymeres*, discussed later). For this system the interfacial potential drop is defined by [50]:

$$(\phi^{\text{poly}} - \phi^S) = \frac{RT}{F} \ln \left(\frac{k_{\text{part(K)}} c_K^S + k_{\text{part(M)}} c_M^S}{c_Z} \right) \quad (4.38)$$

Clearly, for establishing electroneutrality in the organic phase the condition $c_K^{\text{poly}} + c_M^{\text{poly}} = c_Z$ must be valid. If the (effective) partition coefficients of the ions K and M differ significantly, practically only the one cationic species with the larger partition coefficient will be present in the organic phase and can take part in the interfacial ion transfer processes. The Nernst–Donnan potential of this species determines the interfacial potential drop. Achieving such sensitivity is the aim of much research work in the field of ion-sensitive devices. One aspect is the use of ionophores, which raise the effective partition coefficient of the species of interest. For further aspects and more details on this important subject the reader is referred to the literature [e.g., 19c,31,50].

It has been a subject of considerable interest to build sensors by depositing the ion-selective membranes directly onto solid electronically conducting or semiconducting surfaces. One approach is the preparation of coated-wire ion-selective electrodes, in which the membrane is deposited as a thin coating on a metal electrode [18]. Another important field is the development of ion-sensitive field-effect transistors (ISFETs). In this case the membrane is deposited on the gate of the FET (usually a thin SiO_2 film on Si) through which the electric field (defined by the Nernst–Donnan potential generated at the membrane–electrolyte interface) affects the space-charge region at the gate–semiconductor interface, see, for example, Buck [19] and Sudhölter et al. [51]. In these systems the electric coupling of the membrane to the metal or semiconductor substrate is problematic [19a]. In principle, one should provide at this interface a liquid phase that defines the Nernst–Donnan potential (reference solution), and in addition, a system defining the electric potential of the substrate relative to this solution. Much work has recently been conducted with the aim of developing such well-defined interfaces; see, for example, Sudhölter et al. [51].

4.3 Ionic Polymers on Electrodes

4.3.1 Introductory Considerations

The subject of this section are films of electronically insulating polymers that contain immobile ionic groups and mobile counterions [54]. The emphasis will be on polymers that have the ionic groups attached to the polymer matrix by chemical bonds (*fixed-charge polymers*). The cross-linked quaternized poly[vinylpyridinium$^+$ X$^-$] (quat. PVP) shown in Fig. 4.15 is a typical example. Another important example of a fixed-charge polymer is NAFION [52], a polymer made up of a perfluorinated backbone and side chains with sulfonate ($-SO_3^-$) groups. It can be dissolved to form NAFION solutions [53]. Thin films on electrodes can be prepared from these solutions by evaporating the solvent.

When the ionic polymers are in contact with polar solvents, the fixed ions and their counterions are solvated. An osmotic pressure [54] develops that leads to swelling of the polymer. When the density of the fixed-charge sites (and consequently also the density of counterions) is large (typically of the order of 1 mol dm^{-3}), such polymers in contact with liquid electrolytes assume a solutionlike state; they constitute *polyelectrolytes* or *polyelectrolyte gels*. In fact, polymers such as quat. PVP tend to be soluble in aqueous phases. To render them insoluble they must be cross-linked. The cross-linking of the quat. PVP of Fig. 4.15 was achieved by heating a mixture of the partially quaternized PVP with bis(bromomethyl)benzene, which introduces the bridges [55,56]. Another way of cross-linking would be to subject the polymer to high-energy radiation [57,58] to introduce the type of cross-

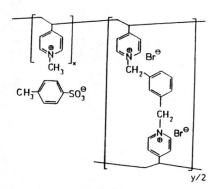

Figure 4.15. Quaternized poly[vinylpyridinium$^+$ *p*-toluene sulfonate$^-$] cross-linked with bis(bromomethyl)benzene.

linking reactions that are known from photoresist processing [59]. The counterions (toluene sulfonate and Br^- in Fig. 4.15) are generally mobile in the polymer matrix and can be exchanged for other ions of the same charge sign. The fixed-charge polymers are therefore termed *anion-* or *cation exchangers*, depending on the type of counterions [54].

In Fig. 4.16 the main points covered in the following discussion are shown schematically: (4.3.2) the metal–film interface; (4.3.3) the interface between the polymer film and electrolyte; (4.3.4) ion transport in the ionic polymer film (membrane transport); (4.3.5) electrochemical charge-transfer reactions (e^-) involving ionic redox systems. In particular, electroactive counterions (with respect to the fixed charges) will be considered. It will be assumed that these redox ions are mobile enough that they react at the metal–polymer interface, and that redox electron hopping may be disregarded. This latter process will be considered in Section 4.4 on electroactive polymers).

4.3.2 The Interface Between Polarized Electrodes and Ionic-Polymer Films

The structure of the metal–film interface depends crucially on the following factors:

1. Competition between water and polymer molecules for adsorption (or chemisorption) sites at the metal surface.
2. The polymer structure at the interface. It may be cross-linked by ions, and/or hydrophobic and hydrophilic domains may exist [26,27,60].

A powerful method of investigating the interface is to measure the double-layer capacitance as a function of electrode potential, film composition, and so on. Such investigations yield widely different results, depending on the system parameters. For example, platinum is known to bind organic π-systems [61] and form strongly

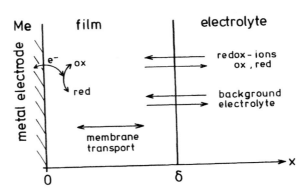

Figure 4.16. Schematic view of an electrode (Me) coated with a membrane-type film of thickness δ, in equilibrium with an electrolyte. An electrochemical charge-transfer reaction (e^-) proceeds at the interface between the electrode and the film.

bound adsorbates. Thiole groups bind strongly to gold; thus "self-assembled" monolayers may form that constitute dielectric layers covering the metallic substrate [62].

On the other hand, when the polymer does not contain groups that interact specifically with the metal, and when such polymers have a high density of fixed charges (and counterions), the polyelectrolyte differs relatively little from a concentrated liquid electrolyte. Thus the double-layer capacitance of a glassy carbon electrode was found to be changed very little by the presence of a film of quat. PVP [63]. Similarly, the double layer of a copper electrode was little affected by the deposition of a NAFION film [64], and the similarity of the cyclic voltammograms (Fig. 4.17) of a platinum electrode in the presence and absence of a NAFION film demonstrates the small effect of the film [65]. However, the NAFION layer interacts strongly with a gold surface, largely suppressing the anodic oxide formation [65].

Since the polymer film is considered to be an electronic insulator, charge-transfer reactions can proceed only at the metallic substrate. The depolarizer must thus be partitioned into the film. The equilibrium concentration in the film can thus be larger or smaller than in the electrolyte. For example, the cathodic oxygen reduction proceeds at a faster rate on the NAFION-coated than on the uncoated electrode [65], apparently because of the high affinity (negative work of transfer) of NAFION for O_2. The Cu/Cu^{2+} reaction rate is little affected by the NAFION film [64]. Since the mobility of the redox system in the polymer matrix is often orders of magnitude smaller than in the liquid electrolyte, one observes in many instances the rate of electron transfer to be governed by the rate of transport of the depolarizer crossing the polymer matrix toward the electrode [63].

Determinations of the rate of electron transfer with the depolarizer present in the

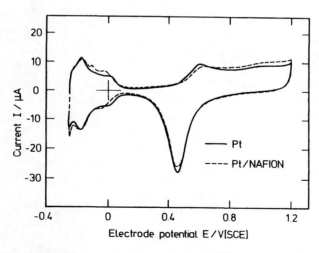

Figure 4.17. Cyclic voltammetry for a bare Pt disk electrode (———) and for a Pt disk electrode coated with a ≈ 2.5-μm Nafion film (- - - - -) in 0.1 M $HClO_4$ N_2-saturated solution. Scan rate 50 mV/s. Electrode area: 0.125 cm². (From Ref. 65, with permission of Elsevier Sequoia S.A.)

polymer matrix frequently show smaller heterogeneous rate constants than for the depolarizer in the aqueous electrolyte [20]. The reason for this is not quite clear, considering the fact that this observation was also made with polyelectrolyte films in which the electrical double layer can be established in a relatively normal way. Of course, specific adsorption of components of the film may constitute a substantial barrier for electron transfer. The decrease of the charge-transfer rate constants of simple depolarizers, such as $Fe(CN)_6^{3-/4-}$ and $Fe^{3+/2+}$ has therefore been taken as an indicator for the quality of organic monolayers [13].

4.3.3 The Interface Between Polyelectrolyte Gels and Electrolytes

4.3.3.1 The Interfacial Equilibrium State

The fundamental considerations for the nonionic polymers, Eqs. (4.14)–(4.20), are also valid for ionic polymers. This is so in particular for the partitioning-equilibrium equation for the ionic species i, Eq. (4.20):

$$a_i^{\text{poly}} = a_i^S \exp\left(-\frac{\Delta \mu_i^0}{RT}\right) \exp\left[-\frac{z_i F}{RT}(\phi^{\text{poly}} - \phi^S)\right] \qquad (4.20a)$$

However, now the density (concentration) of fixed charges, c_x of sign ω, enters into the electroneutrality considerations. For partitioning of the salt $K_m X_n$ [Eq. (4.14)] the electroneutrality condition is

$$z_K c_K^{\text{poly}} + z_X c_X^{\text{poly}} + \omega c_x = 0 \qquad (4.39)$$

Swollen fixed-charge polymers with a value of c_x of the order of ≈ 1 mol dm^{-3} will be discussed next. According to Eq. (4.39), they must contain at least the same concentration of counterions. Thus they constitute hydrophilic polyelectrolyte gels as opposed to the hydrophobic polymers and oils discussed in the previous section. In practice, for such polyelectrolytes the separation of the nonelectrostatic free energy of transfer into a "medium effect," characterized by $\Delta \mu_i^0$, and an activity-coefficient term, $RT \ln(\gamma_i^{\text{poly}}/\gamma_i^S)$, becomes ambiguous; consider Eq. (4.20a) written in an explicit form in terms of free energies:

$$z_i F(\phi^{\text{poly}} - \phi^S) + RT \ln \frac{c_i^{\text{poly}}}{c_i^S} + RT \ln \frac{\gamma_i^{\text{poly}}}{\gamma_i^S} + \Delta \mu_i^0 = 0 \qquad (4.40)$$

For describing partitioning equilibria of ions between liquid electrolytes and polyelectrolyte gels it is therefore customary and convenient [54,55,66] to set $\Delta \mu_i^0 = 0$. Any deviations from ideality of i will be reflected in the activity coefficients γ_i^S and γ_i^{poly}. The corresponding reference states are the (possibly hypothetical) infinitely diluted polymer and electrolyte, that is, $\mu_i^{0(\text{poly})} = \mu_i^{0(S)}$. The partitioning equilibrium of i between the electrolyte and such ionic polymers can consequently be formulated:

$$a_i^{\text{poly}} = a_i^S \exp\left[-\frac{z_i F}{RT}(\phi^{\text{poly}} - \phi^S)\right] \qquad (4.41a)$$

or, in terms of concentrations of i:

$$c_i^{\text{poly}} = c_i^S \frac{\gamma_i^{\text{poly}}}{\gamma_i^S} \exp\left[-\frac{z_i F}{RT}(\phi^{\text{poly}} - \phi^S)\right] \qquad (4.41b)$$

Thus the free energy of transfer of i from electrolytes to polyelectrolytes is defined by

$$\Delta\mu_i^T = RT \ln \frac{\gamma_i^{\text{poly}}}{\gamma_i^S} \qquad (4.41c)$$

where $\Delta\mu_i^T$ might be termed the *local-relaxation* free energy of transfer.

In order to estimate the equilibrium distribution of ions that do not specifically interact with the polymer, one may disregard the differences in the activity coefficients in the two phases; that is, the assumption is made that $\Delta\mu_i^T \approx 0$. Then one obtains the equation for the interfacial ionic equilibrium essentially as derived originally by Donnan [67]:

$$RT \ln \frac{c_i^M}{c_i^S} + z_i F \Delta\phi_D = 0 \qquad (4.41d)$$

where the equilibrium potential difference across the interface is now termed the *Donnan potential*, $\Delta\phi_D = (\phi^{\text{poly}} - \phi^S)$.

The dependence of the Donnan potential on the electrolyte concentration can be calculated using Eq. (4.41d), in combination with the electroneutrality condition within the membrane phase. For a $(1,-1)$-valent electrolyte of concentration c^S it is given by the equation:

$$(\phi^{\text{poly}} - \phi^S) = \Delta\phi_D = \frac{\omega RT}{F} \ln\left[\frac{c_x}{2c^S} + \left\{1 + \left(\frac{c_x}{2c^S}\right)^2\right\}^{1/2}\right] \qquad (4.42)$$

From Eq. (4.42) it can be seen that the sign of $\Delta\phi_D$ is determined by the sign of ω ($\omega = +1$, anion exchange membrane; $\omega = -1$, cation exchange membrane), and that the absolute value of $\Delta\phi_D$ is a function of the ratio (c_x/c^S). Equation (4.41d) and consequently Eq. (4.42) must fail whenever specific interactions or large differences in the [e.g., Born, Eq. (4.23)] solvation energy lead to strong differences of the activity coefficients in the two phases [54,55].

Despite these limitations, the preceding equations constitute the most practical basis for describing the equilibria across the interface between liquid electrolyte and fixed-charge polymer. They are used to describe membrane equilibria, including the partitioning of ions into the conducting polymers discussed later. It appears worthwhile to illustrate the interfacial situation by two typical examples.

Consider a polymer with fixed anionic groups, that is, a cation exchanger such as NAFION, $\omega = -1$. We assume that for the uni-univalent salt KX the local-relaxation free energy of transfer into the polymer is zero ($\Delta\mu_K^T = \Delta\mu_X^T = 0$. The interfacial situation is shown schematically in Fig. 4.18 [54]. Note the excess cations (⊕) that accumulate in the solution near the interface in correspondence to the electric-potential distribution shown schematically in the lower part of this illustration. The values of $\Delta\phi_D$, calculated with Eq. (4.42), and the corresponding concentrations of co-ions, c_-^{poly} and of counterions, c_+^{poly} are shown in Fig. 4.19 for two fixed-charge concentrations. The corresponding situation for an anion exchanger, $\omega = +1$ (e.g., quat. PVP), is represented in the following two diagrams. Figure 4.20 illustrates the concentration profiles of anions and cations across the interface, and in the lower part the distribution of the excess charge, σ, is shown, which corresponds to the interfacial potential drop. In Fig. 4.21 the ion concentrations and $\Delta\phi_D$ values are summarized, in a way analogous to Fig. 4.19, as a function of the concentration of KX in the solution (c^S).

The features of fixed-charge polymers illustrated in Figures 4.18–21 are relevant for an understanding of the electrochemistry of ionic redox systems on electrodes coated with polyelectrolyte gels (see Section 4.3.5). At electrolyte concentrations above the fixed-charge concentration, the polymer becomes "solutionlike." It contains both co- and counterions at substantial concentrations. The effect of $\Delta\phi_D$ becomes insignificant. On the other hand, as the electrolyte concentration decreases relative to c_x, the effect of the Donnan potential becomes significant. The concentration of counterions is then essentially constant and corresponds to the fixed-charge

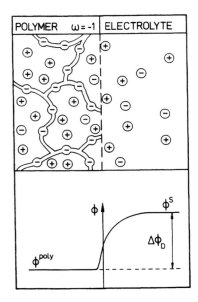

Figure 4.18. Illustration of the interface between a cation-exchange polymer and an electrolyte. The fixed charges on the polymer chains are shown as (-). ⊕: mobile cations, ⊖: mobile anions; ϕ is the electric potential, $\Delta\phi_D$ is the (negative) Donnan potential (cf. Helfferich [54]).

THIN POLYMER FILMS ON ELECTRODES

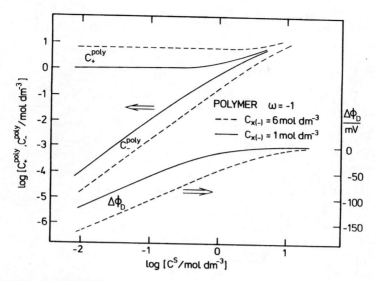

Figure 4.19. The equilibrium concentrations of mobile anions, c_-^{poly}, and cations, c_+^{poly}, in two cation-exchange polymers ($\omega = -1$) in contact with electrolyte of the K^+X^- type of concentration c^S. $\Delta\phi_D$ is the Donnan potential; c_{x-} is the concentration of fixed ions in the polymer.

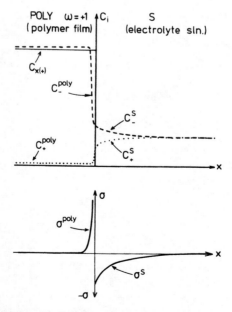

Figure 4.20. Illustration of concentration (c) and excess charge (σ) profiles across the interface between an anion-exchange polymer and a uni-univalent electrolyte solution. The sub- and superscripts have the same meaning as in Fig. 4.19.

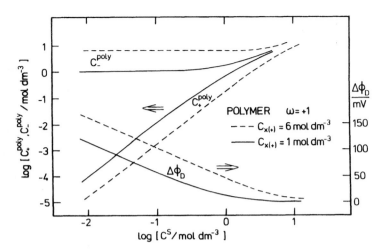

Figure 4.21. The equilibrium concentrations of anions, c_-^{poly}, and cations, c_+^{poly}, in two anion-exchange polymers ($\omega = +1$) in contact with electrolyte of the K^+X^- type of concentration c^S. $\Delta\phi_D$ is the Donnan potential; c_x is the concentration of fixed ions in the polymer.

concentration, whereas the co-ion concentration becomes very small and varies for 1 decade per 59-mV change in $\Delta\phi_D$. This is the condition known as *Donnan exclusion* of co-ions.

4.3.3.2 Selectivity for Counterions

Consider two ionic species, i and j, which both constitute counterions with respect to the fixed charges. When they compete for incorporation into the ionic-polymer phase, an ion-exchange equilibrium is established. It may be characterized by

$$m\bar{\mu}_i^S + n\bar{\mu}_j^{poly} = m\bar{\mu}_i^{poly} + n\bar{\mu}_j^S \qquad (4.43)$$

where m and n are the stoichiometric coefficients of the exchange reaction. By arguments like those given earlier, for polyelectrolyte gels Eq. (4.43) can be transformed into the convenient form:

$$\left(\frac{c_j^{poly}}{c_j^S}\right)^n \left(\frac{c_i^S}{c_i^{poly}}\right)^m = \left(\frac{\gamma_j^S}{\gamma_j^{poly}}\right)^n \left(\frac{\gamma_i^{poly}}{\gamma_i^S}\right)^m \qquad (4.44)$$

The selectivity of the membrane for a species is thus characterized by the activity coefficients of the competing species and by their charge, which defines n and m. It is found that in the absence of considerable specific interactions (e.g., the alkaline metal ions in polystyrene sulfonate) the right-hand side of Eq. (4.44) is near unity [68]. The well-known electrostatic preference of ion-exchange membranes for the more highly charged ionic species can be seen immediately. Note that in the case of hydrophobic polymer or oil phases Eq. (4.44) would be more appropriately formulated in terms of $\Delta\mu_i^0$ and $\Delta\mu_j^0$, or the single-ion partition coefficients; this would lead to results as discussed in Section 4.2.4 [50].

The local-relaxation free energy of transfer may be divided into two terms [69]. The "chemical" free energy of transfer, $\Delta\mu_i^{TC}$ is the result of stronger interactions (solvation, complexation, ion pair formation, etc.) in the solution or the membrane. The "osmotic" work term, $\Delta\Pi(V_i - V_j)$ relates to the pressure difference, $\Delta\Pi$, between the two phases; V_i and V_j are the molar volumes of i and j. Thus, Eq. (4.44) can be rewritten (for the $m = n$ case):

$$\frac{c_j^{poly} c_i^S}{c_j^S c_i^{poly}} = \exp\frac{(\Delta\mu_i^{TC} - \Delta\mu_j^{TC}) + \Delta\Pi(V_i - V_j)}{RT} \quad (4.45)$$

With increasing cross-linking of the polymer the effect of the osmotic term tends to become determining for the ion-exchange equilibrium [70]. However, in cases where specific interactions are significant, $\Delta\mu_i^{TC}$ tends to dominate the selectivity behavior of the system [54].

It should be pointed out again that the single-ion activity coefficients are not accessible with purely thermodynamic methods. However, nonthermodynamic experimental methods for estimating the value of $\Delta\phi_D$ are available [71]. This opens the way for determining the γ_i^{poly} values and thus $\Delta\mu_i^T$ via Eq. (4.41b), see also Braun et al. [55].

4.3.4 Ion Transport in Ionic Polymers

It was pointed out earlier (see Fig. 4.16) that with the electronically insulating coatings electrochemical charge-transfer reactions take place only at the interface between the electrode metal and the polymer. Such electrochemical processes are therefore associated with transport of the depolarizer and the reaction products across the polymer film.

With uncoated electrodes in electrolytes containing an excess of background electrolyte one may normally assume that the transport of ionic redox species to/from the electrode proceeds by diffusion and convection only. This means that one assumes that the electric field is suppressed. This assumption may also be justified in ionic polymers when the ion concentration of the bathing electrolyte is very large (see Figs. 4.19 and 4.21). However, under typical conditions of electrochemical experiments the redox ions may be practically the only mobile species in the film. This is true, for instance, for the reduction of $Fe(CN)_6^{3-}$ or $IrCl_6^{2-}$ in quaternized polyvinylpyridinium films up to concentrations of background electrolyte in the contacting solution of about 0.1 mol dm^{-3}. The situation was characterized experimentally with ring-disk electrodes [57,72,73]. The disk was coated and the ring was uncoated. Redox species leaving or entering the coating could thus be detected and compared with the disk currents. Experiments were conducted with the oxidized forms of redox systems, $Fe(CN)_6^{3-}$ or $IrCl_6^{2-}$, present in the electrolyte and partitioned into the quat. PVP coating. It was shown that, upon reduction of these species following an applied large potential step, the oxidized form of the redox system is initially ejected from the film into the electrolyte, so that electroneutrality is preserved in the film. Only after a delay, corresponding to transport across the coating, the produced reduced form of the redox system appears in the electrolyte and can be detected at the ring (after the additional intrinsic disk-ring transit time; cf. Fig. 4.22).

The physical picture corresponding to the experiment of Fig. 4.22 is represented schematically in Fig. 4.23 [72]. When the potential step is applied to the coated disk electrode, $Fe(CN)_6^{3-}$ is essentially the only mobile charged species in the matrix; that is, its transference number $t_{ox}^{poly} \approx 1$. Therefore, during cathodic reduction the $Fe(CN)_6^{3-}$ ions transport the charge across the film and leave the film. This process is detected at the ring electrode as I_{ring} ($E = 0.2$ V/SCE). The flux of $Fe(CN)_6^{3-}$ out of the film will decrease when the produced $Fe(CN)_6^{4-}$ ions reach the film–solution interface. The arrival of the produced $Fe(CN)_6^{4-}$ in the electrolyte is detected with the ring electrode set to an oxidizing potential as I_{ring} ($E = +0.4$ V/SCE).

It should be pointed out that the migrational flux of the redox ions out of the polymer (across the membrane–solution interface) constitutes a source of error in experiments in which the amount of redox ions is quantitatively determined by integration of faradaic current transients. For example, in the experiment of Fig. 4.22 the $Fe(CN)_6^{3-}$ ejected into the solution is not available for reduction at the electrode surface; that is, the current integral does not correspond to the correct

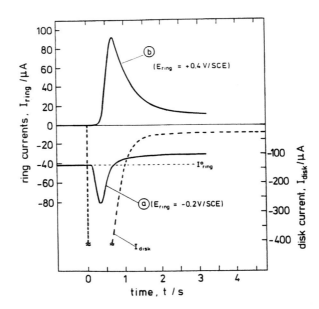

Figure 4.22. Current transients on the coated disk, I_{disk}, and the corresponding ring-current transients, I_{ring}, detecting $Fe(CN)_6^{3-}$ (a), and $Fe(CN)_6^{4-}$ (b). Electrode rotated at 81 s^{-1}. Collection efficiency $k = 0.29$; transit time disk ring: 0.08 s. Electrolyte 2×10^{-4} M $K_3Fe(CN)_6$ + 0.1 M KCl.

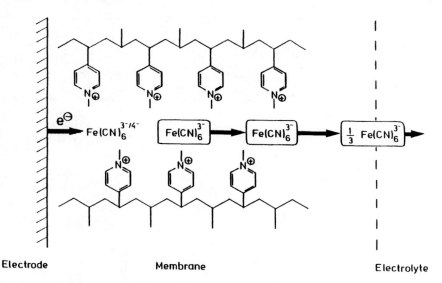

Figure 4.23. The initial stage of cathodic $Fe(CN)_6^{3-}$ reduction in an electrode coating of quat. PVP. The transference number of $Fe(CN)_6^{3-}$ is unity. The arrows symbolize the transport of negative charge across the film ("membrane").

equilibrium amount of $Fe(CN)_6^{3-}$ in the coating. For a more realistic evaluation of the amount of redox ions in the film the ring-current integral should also be considered [72].

The decrease of the transference number t_{0x}^{poly} with increasing concentration of background electrolyte is observed (in experiments as represented in Fig. 4.22) as a decrease of the cathodic ring current. As an example, Fig. 4.24 shows such ring currents obtained during cathodic reduction of $Fe(CN)_6^{3-}$ on an electrode coated with quat. PVP in electrolytes of varying concentration. Clearly, when the electrolyte contains 1 mol/dm³ KCl the ring-current peak has disappeared. This indicates that the contribution of $Fe(CN)_6^{3-}$ to the flux of negative charge across the film (which is about the same as before) has become insignificant (i.e., $t_{0x}^{poly} \approx 0$). A good estimate of the transference number of redox ions such as $Fe(CN)_6^{3-}$ is possible from such ring–disk experiments [74].

Figure 4.22 includes the disk-current transient following the large-negative-potential step, which is applied at time $t = 0$. Note that at the background electrolyte concentration of 0.1 mol dm⁻³ this current transient is not correctly described by the Cottrell equation [75], because of the contribution of migration to the transport of the redox ions (the complication caused by the "thin-layer cell" behavior of the coated electrode, which leads to lower than expected currents at the longer times [76], is disregarded). An analysis of such transients, which takes migration into account, has been conducted for the general electrode reaction $A \pm ne^- \rightarrow B$ on the basis of the Nernst–Planck equation [Eq. (4.26)] [77]. For this analysis it was

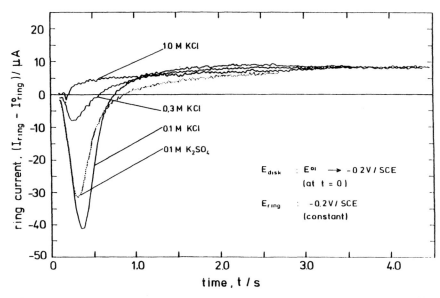

Figure 4.24. Ring current transients, as in Fig. 4.22, (a), but with varying background electrolyte.

assumed (1) that only the redox ions transport the charge in the film ($t_A^{poly} + t_B^{poly} = 1$); (2) at time $t = 0$ the applied potential step leads immediately to the surface concentration of the depolarizer $c_A^{poly(x=0)} = 0$ (Cottrell condition); (3) the film is infinitely thick. In Fig. 4.25 results are represented for the one-electron reduction of a two-valent anionic depolarizer (A) in an anion exchanger. This corresponds, for example, to cathodic reduction of $IrCl_6^{2-}$ in a quat. PVP matrix [73].

The most remarkable conclusion of these calculations is the fact that the diffusion–migration current transients show straight-line behavior on the $t^{-1/2}$ scale, as do the purely diffusional "Cottrell" transients [77,78]. This conclusion was confirmed experimentally; see, for instance, Lange and Doblhofer [74]. Migration only has the effect of changing the observed diffusion coefficients. One may define an effective diffusion coefficient, D_{eff}, which includes the contribution of migration to the current flow [77]. Figure 4.26 summarizes values of D_{eff}/D_A for experiments as represented in Fig. 4.25 ($z_A = -2$, $z_B = -3$) and for other redox systems of common valences involved in one-electron transfer reactions. Two-electron transfer reactions, such as $Tl^{+/3+}$, lead to more significant deviations from Cottrell behavior than the one-electron transfer reactions. Note that the situation for reduction of anions is equivalent to the corresponding oxidation of cations of the same valence, the only difference being the signs of the current and the electric field in the film. However, the oxidation of anions and reduction of cations lead to different values of D_{eff}.

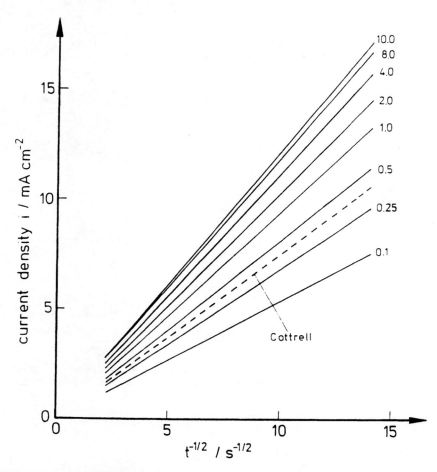

Figure 4.25. Calculated diffusion–migration current transients associated with the one-electron reduction of species A with $z_A = -2$, $D_A = 1 \times 10^{-9}$ cm^2 s^{-1}, and $c_A = 5 \times 10^{-1}$ M. The charge is transported only by A and the product B. For comparison, the pure diffusion case ("Cottrell") is included.

4.3.5 Electrochemistry on Electrodes Coated with Ionic Polymers

Consider for the following discussion an inert electrode coated with a fixed-charge polymer. The coated electrode is immersed into an electrolyte containing an ionic redox system (ox/red) and usually an inert salt. The redox ions are counterions with respect to the fixed charges in the polymer. The redox system $Fe(CN)_6^{3-/4-}$ in quat. PVP discussed earlier is a typical example. The situation with a cation exchanger and a cationic redox system (e.g., NAFION/$Fe^{2+/3+}$) would be analogous.

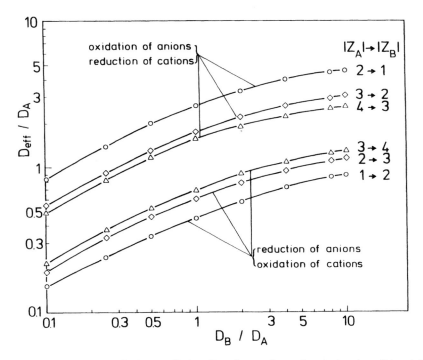

Figure 4.26. Effective diffusion coefficient D_{eff}, for an electrode reaction $A \rightarrow B$ involving both diffusion and migration of the redox ions of charge z_A, z_B. The redox ions are assumed to be the only mobile charge carriers. The data are derived from simulated experiments as shown in Fig. 4.25.

4.3.5.1 The Equilibrium Potential

At the interface between the metal electrode and the solvated polymer phase the electrochemical charge-transfer equilibrium can be established. The equilibrium potential $E^{eq(poly)}$ of a redox system in the polymer can be defined via the electrochemical potential of electrons, $\bar{\mu}_e$ [79]

$$-FE^{eq(poly)} = \bar{\mu}_e^{Me} = \bar{\mu}_e^{poly} \qquad (4.46a)$$

$$= \bar{\mu}_{red}^{poly} - \bar{\mu}_{ox}^{poly} \qquad (4.46b)$$

With the definition of the electrochemical potential one obtains:

$$E^{eq(poly)} = \frac{1}{F}(\mu_{ox}^{0(poly)} - \mu_{red}^{0(poly)}) + \frac{RT}{F} \ln \frac{a_{ox}^{poly}}{a_{red}^{poly}} + \phi^{poly} \qquad (4.47)$$

This is, of course, the Nernst equation giving $E^{eq(poly)}$ relative to a reference electrode that defines ϕ^{poly}.

Consider, for comparison, the metal electrode in direct contact with the redox electrolyte. One can formulate in an analogous way:

$$-FE^{eq(S)} = \bar{\mu}_e^{Me} = \bar{\mu}_e^{S} \qquad (4.48a)$$

$$= \bar{\mu}_{red}^{S} - \bar{\mu}_{ox}^{S} \qquad (4.48b)$$

and one obtains

$$E^{eq(S)} = \frac{1}{F}(\mu_{ox}^0 - \mu_{red}^0) + \frac{RT}{F} \ln \frac{a_{ox}^S}{a_{red}^S} + \phi^S \qquad (4.49)$$

which is the Nernst equation for the electrolyte.

When the coated electrode is in equilibrium with the redox electrolyte, then according to Eq. (4.19) the conditions $\bar{\mu}_{ox}^{poly} = \bar{\mu}_{ox}^{S}$ and $\bar{\mu}_{red}^{poly} = \bar{\mu}_{red}^{S}$ must be valid. Considering the Eqs. (4.46) and (4.48), it follows immediately that $E^{eq(poly)} = E^{eq(S)} = E^{eq}$. This means that when the redox ions can establish their equilibrium distribution across the interface, one obtains the same equilibrium electrode potential in the presence and absence of the polymer film. The presence of a Donnan potential ($\Delta\phi_D = \phi^{poly} - \phi^S$) corresponds to changes of the activities of the redox ions such that the electrostatic work of transfer is just compensated:

$$RT \ln\left(\frac{a_{ox}^{poly} \, a_{red}^S}{a_{ox}^S \, a_{red}^{poly}}\right) + F\Delta\phi_D = 0 \qquad (4.50)$$

The preceding formula refers to the case of a polyelectrolyte gel; that is, the same reference state is chosen for ions in the polymer and the solution. For hydrophobic polymers it would be preferable to include the single-ion free energies of transfer or the partition coefficients [80].

Note that one can establish experimental conditions so that the redox system is partitioned irreversibly into the polymer, for example, $Os(bipy)_3^{2+/3+}$ in NAFION [81]. The bonding of such redox systems to the polymer matrix can be so strong that they are retained over long periods of time even when in contact with an electrolyte free of the redox system. In this case the condition of Eq. (4.50) cannot be fulfilled. However, Eq. (4.47) is still valid when the redox system can establish a charge-transfer equilibrium with the metal. Consequently, when $E^{eq(poly)}$ is measured versus a reference electrode in the bathing solution, the measurable equilibrium potential is a linear function of the Donnan potential. This means that the change in $\Delta\phi_D$ that results from varying the electrolyte concentration can be observed as a shift of the equilibrium potential of the confined redox system [81,82].

The preceding polymer, containing irreversibly bound electroactive sites, belongs in principle to a large group of thin-film materials that are both electronically conducting or semiconducting and electroactive. Other examples are the "conducting polymers" [22,23], organic radical-ion salts [83], and even inorganic polymers, such as Prussian-blue type hexacyanoferrates [84]. Their electroactivity defines an equilibrium-redox potential according to Eq. (4.47) whereby the "polymer" phase contains counterions required by the electroneutrality condition. At the interface between the polymer and the contacting electrolyte a measurable Donnan-type potential drop develops. This means that such systems "respond" to the activity of the counterion species in the electrolyte, and they have thus been proposed as more or less ion-selective sensors [83,84].

4.3.5.2 Partitioning of Redox Ions: An Example

The equilibrium concentrations of ox and red (the components of a redox system) in the polymer phase may differ significantly from their solution values, depending on the activity coefficients and the Donnan potential [Eq. (4.40)]. As an example, consider again $Fe(CN)_6^{3-/4-}$ partitioned between an aqueous electrolyte and quat. PVP. In the liquid phase the concentrations of ox and red are adjusted to be equal. The equilibrium potential is the formal potential, $E^{eq} = E^{0'} \approx 0.18$ V/SCE. In the experiment represented in Fig. 4.27, at time $t = 0$ the overvoltage ($E - E^{eq}$) is stepped to ± 0.5 V. Thus the diffusion–migration current transients discussed earlier are obtained. The experimental situation is such that the contribution of redox-ion diffusion from the electrolyte to the current is insignificant. The current integrals give thus an indication of the equilibrium amounts of redox species that are present in the polymer film at $t = 0$.

It is qualitatively obvious that the charge associated with $Fe(CN)_6^{3-}$ reduction is significantly larger than the charge associated with oxidation of $Fe(CN)_6^{4-}$. One concludes that $Fe(CN)_6^{3-}$ is partitioned more strongly into the polymer matrix than $Fe(CN)_6^{4-}$. This conclusion is substantiated by the IR spectroscopic results [85] shown in Fig. 4.28. At $E^{0'}$, the film contains practically exclusively $Fe(CN)_6^{3-}$.

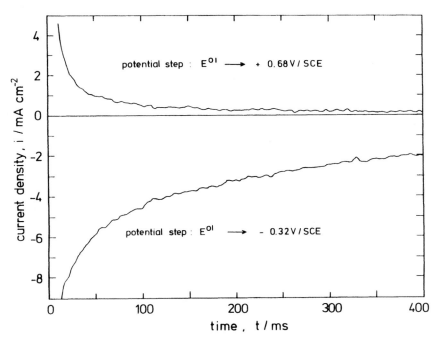

Figure 4.27. Current-time transients obtained with a platinum electrode coated with a quat. PVP film. The electrode potential (E) was stepped from the equilibrium potential, $E^{0'} = 0.182$ V/SCE, to the indicated value. Electrolyte: 2×10^{-4} M each $K_3Fe(CN)_6$ and $K_4Fe(CN)_6$, and 0.1 M KCl.

THIN POLYMER FILMS ON ELECTRODES

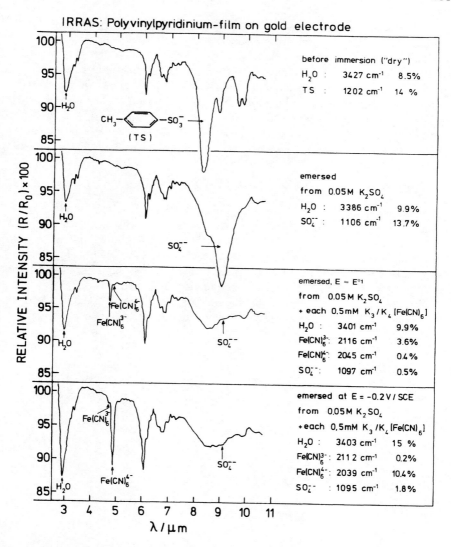

Figure 4.28. Typical IR reflection-absorption spectra of quat. PVP coatings in four states. (a) (= upper spectrum): After preparation of the film; note the strong SO peak from the toluene sulfonate (TS), which had been introduced in the quaternizing reaction (Fig. 4.15). (b): In an aqueous electrolyte TS was exchanged by SO_4^{2-}. (c): Spectrum obtained after equilibration with a solution containing the redox system $Fe(CN)_6^{3-/4-}$. The CN absorption bands indicate that the film contains $Fe(CN)_6^{3-}$, but practically no $Fe(CN)_6^{4-}$. (d) (= bottom spectrum): The coated electrode was withdrawn ("emersed") from the redox electrolyte while an overvoltage of -0.39 V was applied, i.e., during cathodic current flow. The film contains nearly only $Fe(CN)_6^{4-}$. For more details on the experiments see Niwa and Doblhofer [85].

After reduction at -0.2 V/SCE the absorption band corresponding to the CN vibration in Fe(CN)$_6^{4-}$ is clearly noticeable.

Such experiments may be conducted at electrolyte concentrations high enough to justify the assumption that $\Delta\phi_D \approx 0$. Then for each of the redox ion species the partitioning equilibrium equation, Eq. (4.41b), can be applied. On this basis one can obtain an estimate of the single-ion activity coefficients in the polymer (and the local-relaxation free energy of transfer) from the ratio c_i^S/c_i^{poly} and the known activity coefficient in the electrolyte. Certainly, one can derive an order of $\Delta\mu_i^T$ values, that is, an order in which one ionic species replaces the following one [85].

4.3.5.3 Cyclic Voltammetry with the Coated Electrodes

Consider again an electrolyte containing the redox ions Fe(CN)$_6^{3-}$ and Fe(CN)$_6^{4-}$ at equal concentration. The electrode is coated with a PVP that has been quaternized with n-dodecyl groups (Q12) [55]. This coating is insoluble without cross-linking and more hydrophobic than the methyl-quaternized PVP shown in Fig. 4.15. To simplify the following analyses, the background electrolyte concentration was large enough (0.5 M NaCF$_3$COO) that the assumption $\Delta\phi_D \approx 0$ was justified.

With this system the cyclic voltammogram of Fig. 4.29 was taken in the conventional way [75]. It is clearly seen that in the voltammogram obtained with the Q12-

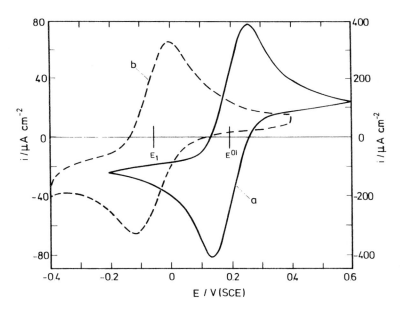

Figure 4.29. Cyclic voltammograms obtained with a bare glassy carbon electrode (a), and a Q12-PVP coated electrode (b). Film thickness 260 nm. Electrolyte: 2.5 mM K$_3$Fe(CN)$_6$, 2.5 mM K$_4$Fe(CN)$_6$, and 0.5 M CF$_3$COONa. Scan rate 100 mV s^{-1}. Curve b was obtained after 30 min continuous cycling.

PVP coated electrode the positions of the current peaks are shifted relative to the ones obtained on the bare electrode. The measurable equilibrium potential (open-circuit potential) was found to be the same on the coated and the uncoated electrodes.

To understand this result consider the Nernst-type diagram in Fig. 4.30. It was prepared under the following assumptions: (1) As the electrode potential changes, the Nernstian equilibrium concentration ratio of the redox ions is established (under the condition of cyclic voltammetry this is assumed to be true at the interface between the metal electrode and the polymer phase); (2) when this ratio between the concentrations of ox and red changes, the activity coefficients of the redox ions remain constant; (3) the total charge in the film remains constant, that is, $z_{ox}c_{ox}^{poly} + z_{red}c_{red}^{poly} + \omega c_x = 0$ ($c_x = 1$ mol dm^{-3}). Now the electrode potential halfway between the anodic and the cathodic current peaks obtained with the coated electrode is termed E_1, and it is assumed that at E_1 the equilibrium concentrations of ox and red in the coating are equal [75]. Starting from E_1, with the preceding assumptions the equilibrium concentrations of the redox ions in the polymer are calculated for all E values.

In particular, the equilibrium concentrations of ox and red prevailing at $E^{0'}$, $c_{ox}^{eq(poly)}$ and $c_{red}^{eq(poly)}$ are defined, considering that $c_{ox}^S = c_{red}^S$, by

$$\frac{c_{ox}^{eq(poly)}}{c_{red}^{eq(poly)}} = \exp\left[-\frac{F}{RT}(E_1 - E^{0'})\right] \quad (4.51)$$

and, considering Eqs. (4.50) and the definition of the local-relaxation free energy of transfer, one may formulate [55]:

$$\Delta\mu_{ox}^T - \Delta\mu_{red}^T = -F(E_1 - E^{0'}) \quad (4.52)$$

The shifts of the voltammetric peaks can thus be used conveniently in combination with an analytical technique to analyze the thermodynamics of the system. It is

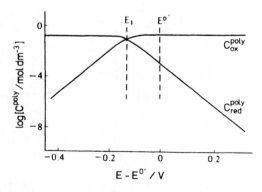

Figure 4.30. The equilibrium concentrations of a redox system ($|z_{ox}| = 3$; $|z_{red}| = 4$) in an ion-exchange film as a function of the electrode potential, E applied vs. a reference electrode in the electrolyte. Electroneutrality is provided by the redox ions only; $E^{0'}$ is the equilibrium potential of the redox system.

normally no problem to determine experimentally the concentration of the species that is strongly accumulated in the polymer, and thus its local-relaxation free energy of transfer or the activity coefficient in the polymer. When conditions prevail so that $\Delta\phi_D \approx 0$, $\Delta\mu_i^T$ of the second species, which may be present at an inconveniently low concentration, can be determined via Eq. (4.52). Of course, the preceding equations can be modified on the basis of Eq. (4.50) to include $\Delta\phi_D$. On this basis a more general analysis can be conducted, when $\Delta\phi_D$ is known.

In an investigation of the system of Fig. 4.29 with a Q12-PVP coating [55], a shift of the $Fe(CN)_6^{3-/4-}$ voltammetric current peaks of -0.24 V was found. The concentration of $Fe(CN)_6^{3-}$ in the polymer at $E^{0\prime}$ was found analytically to be $c_{ox}^{eq(poly)} = 1.8 \times 10^{-1}$ mol dm^{-3}. Since $c_{ox}^S = c_{red}^S$, one obtains immediately for the equilibrium concentration of $Fe(CN)_6^{4-}$ a value of $c_{red}^{eq(poly)} = 1.3 \times 10^{-5}$ mol dm^{-3}. When in such experiments the polymer is rendered less hydrophobic by quaternizing with shorter alkyl groups, the value of $E_1 - E^{0\prime}$ becomes less negative. The local-relaxation free energy of transfer of $Fe(CN)_6^{3-}$ was found to be little affected by changes of the type of quaternization; the value is $\Delta\mu_{ox}^T \approx -11$ kJ mol^{-1}. The change in $E_1 - E^{0\prime}$ must thus result from changes in the local-relaxation energy of transfer of $Fe(CN)_6^{4-}$. Indeed, a value of $\Delta\mu_{red}^T = +13$ kJ mol^{-1} is found in the case of Q12-PVP, and a value of -1 kJ mol^{-1} with the methyl-quaternized quat. PVP.

4.3.5.4 Electrochemical Rectification by Selective Partitioning

The equilibrium distribution of the $Fe(CN)_6^{3-/4-}$/quat. PVP system discussed earlier is shown schematically in Fig. 4.31. In the following example, stationary reduction and oxidation currents are produced on the coated rotated disk electrode. This is

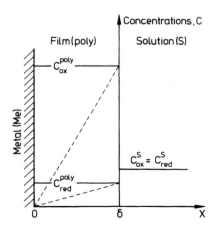

Figure 4.31. Distribution of the redox system $Fe(CN)_6^{3-/4-}$ (ox, red) between electrolyte and polymer film of thickness δ. The superscripts poly and S symbolize the film (polymer) and solution phases, respectively. The limiting diffusion currents are determined by the concentration gradients across the film (dashed lines).

done by adjusting the applied overvoltage to values at which the surface concentration (at $x = 0$) of either ox or red is near zero. We assume that this condition is fulfilled, and that the effect of migration is negligible. Then the observed diffusion current density, i_d, will be defined by [57]:

$$\frac{1}{i_d} = \frac{1}{i_d^{poly}} + \frac{1}{i_d^S} \tag{4.53}$$

where i_d^{poly} is the limiting diffusion current density across the film:

$$i_d^{poly} = \frac{F c_i^{poly} D_i^{poly}}{\delta} \tag{4.54}$$

and i_d^S is the limiting diffusion current density across the Nernst diffusion layer in the solution, δ^S:

$$i_d^S = \frac{F c_i^S D_i^S}{\delta^S} \tag{4.55}$$

The index i stands for ox or for red [e.g., $Fe(CN)_6^{3-/4-}$]. For the rotating disk electrode δ^S is given (in centimeters) by:

$$\delta^S = 1.61 (D_i^S)^{1/3} \omega^{-1/2} \nu^{1/6} \tag{4.56}$$

where ω is the angular frequency of rotation and ν the kinematic viscosity of the solution [75].

The partitioning equilibrium affects i_d^{poly} and thus i_d, because it defines the limiting concentration gradients (c_i^{poly}/δ) (the dotted lines included in Fig. 4.31). The experimental results represented in Fig. 4.32 demonstrate indeed a significant asymmetry between the anodic and the cathodic currents. Note that in these experiments the concentrations of ox and red in the electrolyte are equal! The dashed curve included in Fig. 4.32 corresponds to the current–voltage curve of the uncoated electrode.

The intercept of the cathodic $I_d^{-1}/\omega^{-1/2}$ plot (inset, Fig. 4.32) is zero, within the experimental accuracy. This means that the polymer film does not measurably hinder the cathodic diffusion current. The anodic process, on the other hand, does not detectably depend on the electrode rotation rate. The film effectively blocks the anodic current flow. The coated electrode constitutes an electrochemical rectifier.

Such rotated-disk experiments are suitable for obtaining information on the diffusion coefficients of the redox ions in the coating. Consider results obtained with chemically cross-linked quat. PVP of Fig. 4.15. The ratio $c_{ox}^{eq(poly)}/c_{red}^{eq(poly)}$ for the $Fe(CN)_6^{3-/4-}$ system as determined from cyclic voltammetry ($E_1 - E^{0'}$, see earlier discussion) was only 10^2, and the anodic current plateau was measurable [57]. The diffusion current across the Nernst layer in the electrolyte, i_d^S, was determined from experiments with uncoated electrodes. Thus the limiting diffusion current across the polymer film, i_d^{poly}, was obtained from Eq. (4.53):

$$i_d^{poly} = \left[\frac{1}{i_d} - \frac{1}{i_d^S} \right]^{-1} \tag{4.57}$$

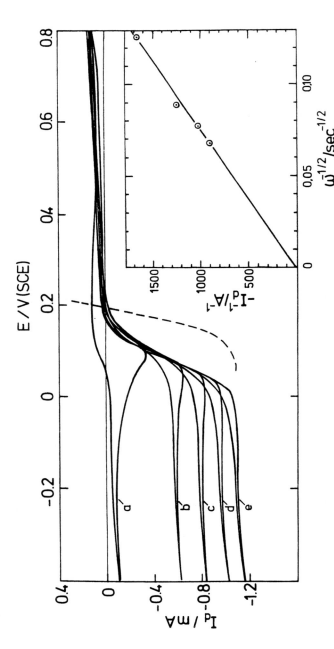

Figure 4.32. Unidirectional current flow at a glassy carbon electrode of area 0.5 cm^2, coated with a film of thickness 0.53 μm of synchrotron cross-linked poly[(4-vinyl-N-methylpyridinium-p-toluolsufonate)$_{0.83}$-co-(styrene)$_{0.17}$]. Irradiation: $D = 2190$ mJ cm^{-2}. Electrolyte: each 2.5 mM K$_3$/K$_4$[Fe(CN)$_6$)] + 0.5 M CF$_3$COONa, pH = 9. Potential sweep = 10 mV s^{-1}. Electrode rotation rate: (a) 0; (b) 600; (c) 1220; (d) 1600; and (e) 2075 min^{-1}. The dashed curve was obtained with the uncoated electrode; the anodic branch (approximately symmetrical to the cathodic branch) is cut off. I_d is the limiting cathodic current, ω the angular frequency of rotation.

According to Eq. (4.54), from the slope of a plot of i_d^{poly} versus $1/\delta$ (Fig. 4.33) the product $c_{red}^{poly} D_{red}^{poly}$ can be obtained. With the equilibrium concentration of red, $c_{red}^{eq(poly)} = 5 \times 10^{-3}$ mol dm^{-3}, the diffusion coefficient of red, with the polymer film in the anodic stationary condition, was calculated: $D_{red}^{poly} = 1 \times 10^{-9}$ cm^2 s^{-1}.

As in the experiment of Fig. 4.32, diffusion of ox through the matrix in the stationary state is so fast that its influence on the overall cathodic diffusion current is not measurable. One can therefore only obtain a low limit of D_{ox}^{poly}. From the data of Table 4.2 it is concluded that at a film thickness of 1240 nm the observed diffusion current density, i_d, is smaller than i_d^S by not more than 5 μA cm^{-2}. According to Eq. (4.57) the limiting diffusion current across the film is thus $i_d^{poly} > 1.2$ mA cm^{-2}. With $c_{ox}^{eq(poly)} = 1 \times 10^{-1}$ mol dm^{-3} one obtains from Eq. (4.54): $D_{ox}^{poly} > 1.5 \times 10^{-8}$ cm^2 s^{-1}.

This is a very remarkable result. From previous experiments (e.g., from the IR spectra), it was concluded that the Fe(CN)$_6^{3-}$ is bound more tightly to the polymer matrix than Fe(CN)$_6^{4-}$. This would point to a value $D_{ox}^{poly} < D_{red}^{poly}$, which is at variance with the preceding experimental result.

To understand this, consider the dependence of the voltammetric currents, Fig. 4.34, obtained as a function of the sweep rate on a rotated-disk electrode. At the slow sweep rate, pseudostationary currents, as in Fig. 4.32, are obtained. The

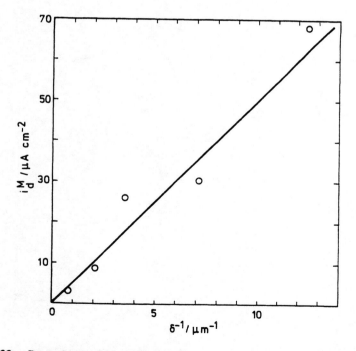

Figure 4.33. Dependence of the anodic current limited by diffusion of Fe(CN)$_6^{4-}$ across the polymer film, i_d^{poly}, on $1/\delta$, where δ is the film thickness. The i_d^{poly} values were derived from the experimental results summarized in Table 4.2.

Table 4.2 Anodic and Cathodic Diffusion Current Densities, i_d (an) and i_d (cath), as a function of film thickness δ

$\delta/\mu m$	i_d (an)/$\mu A\ cm^{-2}$	i_d (cath)/$\mu A\ cm^{-2}$
0.08	35	64
0.14	21	77
0.28	19	67
0.48	7.5	83
1.24	3	80
0 (calculated)	71.7	79.7

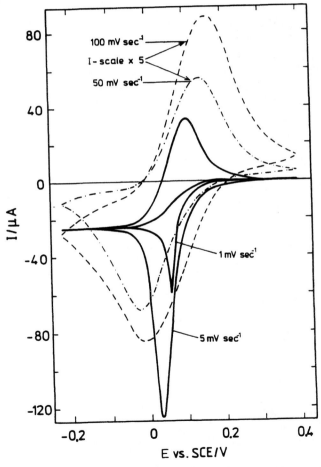

Figure 4.34. Change of the cyclic voltammogram with potential sweep rate. Film prepared from cross-linked quat. PVP; thickness: 1.5 μm. Electrolyte: each 0.1 mM $K_3/K_4[Fe(CN)_6]$ + 0.1 M KCl. Electrode of area 0.5 cm², rotated at 40 s^{-1}. Scan rates as indicated.

system has time to reach a state close to equilibrium. As the potential moves into the reduction region, a complicated nucleation-type transient is observed that is associated with a drastic increase in film solvation (cf. the IR results, Fig. 4.28). Apparently, in the anodic regime the matrix is cross-linked by the $Fe(CN)_6^{3-}$ ions. It is not unexpected that the transport of the reduced species toward the electrode is hindered by the rigid matrix. This is borne out by the small diffusion coefficient of the reduced species obtained earlier.

As the reduction of $Fe(CN)_6^{3-}$ produces a high concentration of $Fe(CN)_6^{4-}$ in the film, the matrix swells and becomes more permeable. Electroosmosis may contribute to the increased film permeability. When the film does not have sufficient time to relax to its equilibrium state, it is also permeable in the anodic regime. Therefore cyclic voltammograms assume the "normal" appearance at the higher sweep rates.

4.4 Electronically Conducting Polymer Films

4.4.1 Introductory Remarks

The typical organic polymers are good electronic insulators. Only in the second half of this century was it realized that the conductivity can be increased significantly by introducing into the polymer matrix various types of electronic states that can be reversibly occupied and emptied, thus constituting the basis for electronic conduction [86–87]. This is clearly of interest from a fundamental point of view. In addition, the conducting polymers have the potential for wide practical application. For these reasons the subject of electron conduction in polymers has been a most actively investigated research field [88–90], in which electrochemistry plays an important role (see Section 4.1). It would be well beyond the frame of this work to give a complete account of the electrochemistry of conducting polymers. Moreover, recent views discussing this subject are available [1–4]. Therefore the following discussion will focus on the fundamental aspects of electron and ion conduction that constitute the basis for understanding the electrochemical behavior of these systems.

The most appropriate general model for describing electron transfer across the redox sites in the polymer is the *electron-hopping* concept for disordered systems [91–93]. This is electron tunneling with the assistance of phonons. The electronic donor and acceptor states may each have identical energy levels (disregarding the thermal fluctuations). Typically, these are the polymers with bound redox systems, such as $Os(bipy)_3^{2+/3+}$ [93,94]. Such polymers are termed *redox polymers*. In other systems the energies of the states are spread over a range of levels. This is normally the case with the *conjugated polymers*, such as polypyrrole (PPy/PPy$^+$), where the electronic energy levels depend on the conjugation length and the site density.

The elementary step in electron transport is the transition from an occupied state (donor, red) to an unoccupied state (acceptor, ox). The situation may be represented schematically as shown in Fig. 4.35. Note that electron hopping is a bimolecular process. It is thus necessary for this process to take place to have occupied sites next to unoccupied ones; that is, the ratio between the densities of acceptor and donor

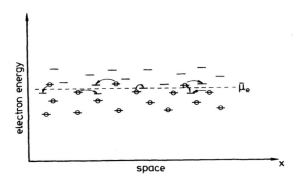

Figure 4.35. Schematic diagram of a possible electronic-energy-level distribution in a conducting polymer ($\ominus$: occupied states, $-$: unoccupied states). $\bar{\mu}_e$ is the electrochemical potential of electrons (corresponding to the Fermi level). Possible electron hops are shown by arrows.

states is very important. In solid-state work this ratio is termed *compensation* [91]; in systems of electrochemical interest the term *mixed valent state* is used [93].

The driving force for electron hoping to proceed in one preferred direction x is a gradient in the electrochemical potential of electrons. This may be formulated for redox systems exchanging one electron by

$$\frac{\partial \bar{\mu}_e}{\partial x} = \frac{\partial \bar{\mu}_{red}}{\partial x} - \frac{\partial \bar{\mu}_{ox}}{\partial x} \qquad (4.58a)$$

$$= RT \frac{\partial \ln(a_{red}/a_{ox})}{\partial x} - F \frac{\partial \phi}{\partial x} \qquad (4.58b)$$

Equation (4.58) demonstrates that both concentration gradients of the redox sites and an electric field can constitute the driving force for electron transport.

In electrochemical systems it is often of interest to oxidize/reduce the polymer, that is, to change the density of unoccupied and occupied electronic states in the bulk polymer. Consider, for example, the application of such polymers in batteries or displays. The polymer is then present in the arrangement of Fig. 4.1a. The redox sites are oxidized/reduced by applying an appropriate electrode potential E, that is, by raising or decreasing the free energy of electrons, $\bar{\mu}_e$ (Fig. 4.35). The transport of electrons to/from the sites in the polymer phase is an essential part of this oxidation–reduction process.

During oxidation/reduction of the redox sites the principle of *electroneutrality coupling* requires that a flux of ions into or out of the polymer film provides for internal charge compensation. This process is shown schematically in the reaction PPy/PPy$^+$, that is, the anodic oxidation of a polypyrrole film (Fig. 4.36), where the anions X$^-$ compensate the positive charge of the produced PPy$^+$. This ion flux will limit the rate of oxidation/reduction of the polymer when the ionic conductivity of the film is smaller than its electronic conductivity. The concentration and mobility

THIN POLYMER FILMS ON ELECTRODES

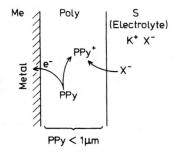

Figure 4.36. The electrochemical oxidation of a polypyrrole (PPy) film. The associated movement of anions (X$^-$) into the polymer matrix (electroneutrality coupling) is shown. Coions are disregarded.

of mobile ions in the conducting polymer is thus a subject of great importance and will also be discussed later.

4.4.2 Electronic Conduction Mechanisms

4.4.2.1 Redox Polymers

For conducting polymers it is possible that either ion transport or the electronic current limit the overall charge-transport rate. For the following discussion it is assumed that electron hopping (not electroneutrality coupling) limits the rate of charge transport. When the redox sites are solvated, electron exchange as known from redox systems in liquid solutions proceeds. The activation energy is defined by the Franck–Condon barrier [95]. An analysis of electron transport based on this type of electron exchange in electrolytes was first given by Dahms in 1968 [96]. Consider a phase in which the concentration gradients of ox and red in the x-direction constitute the driving force for electron transfer. At two positions, x_1 and $(x_1 + \lambda)$, where λ is the distance across which electron exchange proceeds, the concentrations of the considered species, c_i, differ by $\lambda(dc_i/dx)$. The situation is illustrated in Fig. 4.37. Electron exchange between the redox sites at the two positions produces a "forward" and "back" electronic current (e$^-$; Fig. 4.37). The resulting net electronic current density, i_e, is

$$i_e = Fk_{ex}\lambda^2 \left(c_{ox}\frac{dc_{red}}{dx} - c_{red}\frac{dc_{ox}}{dx} \right) \tag{4.59}$$

where k_{ex} is the homogeneous second-order rate constant for electron exchange between ox and red. Under the condition that the sites are fixed and the total concentration of sites is constant, the concentration gradients of ox and red are approximately equal. The current density is then

$$i_e = Fk_{ex}\lambda^2(c_{ox} + c_{red})\frac{dc_i}{dx} \tag{4.60}$$

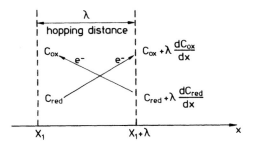

Figure 4.37. Forward and backward electron hops across the distance λ in a phase in which concentration gradients of both components of a redox system (ox, red) prevail.

Equation (4.60) resembles Fick's first law, and by comparing the coefficients one can define an *electronic diffusion coefficient*, D_e:

$$D_e = k_{ex} \lambda^2 (c_{ox} + c_{red}) \qquad (4.61)$$

This concept was later further developed [93,97] and applied successfully to electronic charge transport via redox sites (*redox hopping*) in polymers [93,98,99]. The model is now generally accepted for this process; it is represented schematically in Fig. 4.38. Experimental values of the electronic diffusion coefficients can be obtained, for example, via electrochemical impedance measurements [100], Cottrell-type potential step experiments [101], or probably best from steady–state experiments [93,102].

However, values of the D_e calculated on the basis of Eq. (4.61) can be at variance with experimental values by orders of magnitude [103]. This is understandable because in a real system electron transfer is not likely to proceed only across one value of λ, but over a distribution of distances. At each distance a different rate

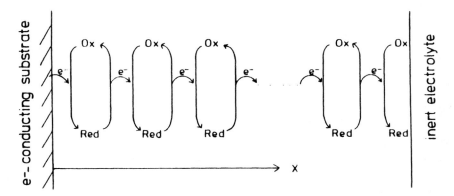

Figure 4.38. Schematic representation of the electron-hopping process, which leads to reduction of the film. The process is associated with a flux of charge-compensating ions in or out of the film.

constant k_{ex} will be effective. However, usually neither the exact distribution of redox sites in the polymer nor the dependence of k_{ex} on λ in the polymer phase are known. It has been found that k_{ex} decreases upon increasing the average site–site distance [93]. This appears to demonstrate that the effective hopping distance, that is, the thickness of the electron-transfer barrier, increases [91].

Quantitative evaluations of redox-hopping experiments involve significant uncertainties. There is the fact that a variation of site density (associated with the oxidation/reduction of the film or with preparing films with different total site concentrations) leads normally to changes of the density of osmotically active ions. This leads to differences in the degree of swelling [54]. Such structure changes tend to affect the ionic mobilities and the electron-transfer rate [104]. Special experimental arrangements have been designed to minimize this problem [93]. Another difficulty is the fact that it is not always clear if the observed currents are indeed defined by redox hopping or if the redox sites move themselves. Diffusion-type current–time and current–voltage relations are obtained in both situations. The situation is particularly serious with systems like Os(bipy)$_3^{2+/3+}$ or Ru(bipy)$_3^{2+/3+}$ in NAFION. The hydrophobic interactions between the bipyridyl ligands and the hydrophobic polymer matrix lead to small physical diffusion coefficients that may just be of the same order of magnitude as the D_e values [103,105]. These systems are therefore still the subject of intense research.

Even in cases where the redox sites are completely immobilized by chemical bonding to the matrix, they will still have a certain local mobility. One can therefore describe the electron transport by assuming not that electron exchange between the redox sites takes place over the distance λ, as was done earlier (Fig. 4.37), but that the donor and acceptor sites come in intimate contact for electron transfer [99,106]. This mechanism was proposed to be operative with iron redox sites embedded in a plasma polymer matrix [99]. Figure 4.39 illustrates the situation. The process was described in a semiquantitative way using the collision theory of reaction rates. It was assumed that the chloride-bridged activated complex (Fig. 4.39), constitutes the charge-transfer geometry. With this model, agreement with the observed electron transport rate ($D_e \approx 10^{-11}$ cm^2 s^{-1}) was obtained, whereas the analysis based on the classical Dahms–Ruff model [Eq. (4.61)] gave a result ($D_e \ll 10^{-12}$ cm^2 s^{-1}) that was at variance with the experimental findings.

The electron transport caused by an electric field was described in the following way [92,93,96]: Consider that the concentration gradients of Fig. 4.37 are zero, while an electric field changes the electronic energy level at position $x_1 + \lambda$ relative to the one at position x_1 by $\Delta G_{el} = -F \lambda d\phi/dx$. This leads to a net electronic current in the x-direction:

$$i_e = F \lambda k_{ez} c_{ox} c_{red} \left[\exp\left(\frac{\alpha F \lambda}{RT} \frac{d\phi}{dx} \right) - \exp\left(-\frac{(1-\alpha)F\lambda}{RT} \frac{d\phi}{dx} \right) \right] \tag{4.62}$$

where α is the fraction of ΔG_{el} that is effective in changing the rate of the second-order charge-transfer reaction in the x-direction [93]. Considering a typical hopping

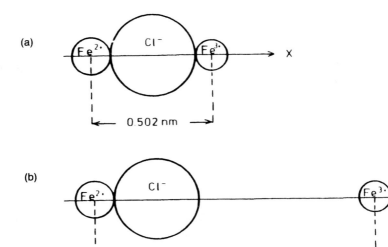

Figure 4.39. (a) Chloride-bridged activated complex of the iron(II/III) electron exchange reaction. (b) Average Fe–Fe distance in the polymer matrix of Ref. 99.

distance, $\lambda = 1$ nm, a value of $\alpha = 1/2$, and that at room temperature $RT = 2.46$ kJ/mol, it follows that the exponents of Eq. (4.62) assume the value of 1 at a field strength of $d\phi/dx = 5 \times 10^5$ V cm^{-1}. When the field strength is much lower than that, the system can be well described by the linear term of the series-developed exponential terms. Then one obtains the following expression for the electronic current density:

$$i_e = \frac{F^2 \lambda^2 k_{ex} c_{ox} c_{red}}{RT} \frac{d\phi}{dx} \tag{4.63}$$

One may formulate an "electronic mobility," u_e, defined as

$$u_e = \frac{F^2 \lambda^2 k_{ex}}{RT} \tag{4.64}$$

however, one should be aware that this parameter differs from the conventional "mobility" of ions [Eqs. (4.26) and (4.27)], because the elementary process of electron transport is a bimolecular reaction; that is,

$$i_e = u_e c_{ox} c_{red} \frac{d\phi}{dx} \tag{4.65}$$

The conductivity of the redox-polymer phase is thus:

$$\sigma_e = u_e c_{ox} c_{red} \tag{4.66}$$

which may be expressed in terms of the electronic diffusion coefficient [Eq. (4.61)]:

$$\sigma_e = \frac{F^2 D_e}{RT} \left(\frac{c_{ox} c_{red}}{c_{ox} + c_{red}} \right) \tag{4.67}$$

Clearly, neither the Einstein relation in the conventional form [Eq. (4.27)] nor the usual Nernst–Planck equation [Eq. 4.26)] are applicable for the case of redox-hopping conduction. It has been shown by Savéant [107] that the equivalent of the Nernst–Planck equation, describing redox-hopping conduction under the influence of both a concentration gradient and an electric-potential gradient, reads (for a one-electron transfer between redox sites):

$$J_e = -D_e \left[\frac{dc_{ox}}{dx} + \frac{F}{RT} \left(\frac{c_{ox} c_{red}}{c_{ox} + c_{red}} \right) \frac{d\phi}{dx} \right] \tag{4.68}$$

Important work on the subject of redox hopping in polymers has been directed by R. W. Murray, using mainly metal–polymer–metal sandwiches and interdigitized array electrodes [93]. Since the electronic conductivity of the redox polymer is proportional to the product $c_{ox} c_{red}$, it is zero when either c_{ox} or c_{red} are zero, and it has a maximum at $c_{ox} = c_{red}$. This behavior has been established experimentally, for example [93] with the ladder polymer poly(benzimidazobenzophenanthroline), Fig. 4.40.

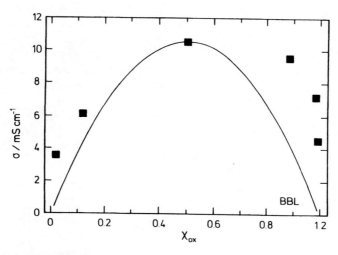

Figure 4.40. Conductivity (σ) versus mole fraction of acceptor sites of a redox conducting polymer [poly(benzimidazobenzophenanthroline)]. Black squares: experimental points; solid line: theoretical behavior calculated by normalizing Eq. (4.67) to the conductivity in the polymer in its 1:1 mixed valent state. From Ref. [93] with permission of Elsevier Science Publishers B.V.

4.4.2.2 Conjugated Polymers

The conjugated polymers can be oxidized and reduced in a way that is in principle similar to the redox polymers. During reduction anionic sites are formed that require cations for charge compensation (or anion expulsion). Upon oxidation (e.g., of polypyrrole, PPy/PPy$^+$, Fig. 4.41), cationic sites [polarons (P$^+$), bipolarons (P^{2+}) [108]] are normally formed on the polymer chains according to

$$P + X^{-(S)} = P^+X^- + e^- \tag{4.69a}$$

$$P^+ + X^{-(S)} = P^{2+}(X^-)_2 + e^- \tag{4.69b}$$

where X^- are anions (the superscript S indicates that they are in the solution phase), and P symbolizes a segment of the polymer chain of a few monomer units length. The polymer oxidation or reduction is conveniently conducted electrochemically [3,4], as shown schematically in Fig. 4.36.

The produced cationic or anionic redox sites are mobile along the polymer chain. They are the basis for the electronic conductivity of these polymers. When the undisturbed conjugated molecules extend over long distances or when the phase is crystalline (e.g., charge-transfer complexes [109]) one may find conductivities comparable to those of metals and a negative temperature coefficient. These observations point to a metallic character of such polymers [110].

Under electrochemical conditions the polymer in the electron-conducting state usually contains a high concentration of ions. It is thus swollen and amorphous. The polymer chains may be "spaghetti"-like, twisted and intermixed. In the process of conduction the electron (hole) has to jump between chains. This electron transfer has been shown [111] to correspond again to redox electron hopping as discussed earlier for redox polymers. This hopping process is believed to determine the rate of the overall electron conduction process (Fig. 4.42). Thus, with such conjugated polymers one should expect fundamentally the same behavior as for the redox polymers discussed earlier. The reason for the high conductivity of the conjugated polymers is apparently the combined effect of both the high mobility of charge

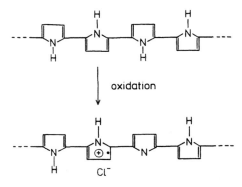

Figure 4.41. Oxidation of polypyrrole, leading to the radical cation (polaron) and a counterion (X$^-$) from the electrolyte [Eq. (4.69a)].

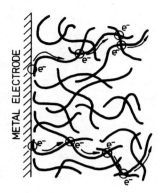

Figure 4.42. Two electron-conduction paths across a conducting polymer matrix. The polymer molecules are shown as solid lines. The hopping sites between polymer chains are marked by circles.

carriers along the polymer chain and a high rate of electron exchange (large k_{ex}) between the conjugate chains [112]. The conductivity is, in fact, normally higher than the ionic conductivity in the polymer phase. This means that the rate of oxidation/reduction of the polymer is usually determined by the rate of counterion movement required for electroneutrality coupling (Fig. 4.36) between the electronic and ionic charges. This is discussed in the following section.

4.4.3 Electroneutrality Coupling in Conducting Polymers

It has been shown by various techniques [22,113] that the ions in the common "conducting polymers" are more or less mobile. This means that, for example, the oxidized polypyrrole PPy^+ constitutes an anion exchanger. The ions present in the polymer lead to ionic conductivity, which is defined by the concentrations of the ions and their mobilities [Eq. (4.28)]. Such systems constitute "mixed conductors" [24]. The rate of oxidation/reduction may thus be limited by the rate of either electron or ion transport, depending on the value of σ_e^{poly} relative to σ_{ion}^{poly}.

In the case of redox polymers, the electronic diffusion coefficients are of the order of $D_e \approx 10^{-10} - 10^{-13}$ cm^2 s^{-1}. Assuming a typical value of 10^{-11} cm^2 s^{-1} and concentrations of redox sites, corresponding to the 1:1 mixed valent state ($c_{ox} = c_{red} = 5 \times 10^{-4}$ mol dm^{-3}, one obtains a redox hopping conductivity of $\sigma_e = 10^{-8}$ S cm^{-1}. An aqueous electrolyte, say 1 M KCl, has a conductivity of the order of 10^{-1} S cm^{-1}. In a redox polymer the typical ion concentration is of the same order of magnitude. Even if the mobility of the counterions is lower in the polymer than in the aqueous phase by several orders of magnitude, the condition $\theta_e^{poly} \ll \sigma_{ion}^{poly}$ should be fulfilled. The rate of oxidizing/reducing the film will thus be determined by the rate of the electron-hopping process. The electrochemical response of an electrode coated with a redox–polymer film may be described in a way analogous to a thin-layer cell containing a redox electrolyte along with a background electrolyte [114].

When a polymer of $\sigma_e \gg \sigma_{ion}$, such as polypyrrole in the oxidized state, is subjected to changes of the applied electrode potential, during the transient state electric fields develop in the polymer matrix, and then disappear, as electronic and ionic charge carriers migrate to new equilibrium positions [115]. A satisfactory general analysis of such dynamics has apparently not been conducted. However, from the preceding discussion it follows that the concepts derived for redox polymers under the condition that the hopping mobility of the electrons exceeds the counterion mobility should be applicable. It has been shown [116] that in this case the system's behavior is again "diffusional" in character. The coated electrode behaves like a porous-metal electrode with pores of limited depth [117,118]. Numerous experimental reports on this behavior of the conducting polymers have appeared in the literature; the first was probably by Bull, Fan, and Bard in 1982 [119].

4.4.4 The Electrochemistry of Conjugated Polymers

Consider for the following discussion that electron transport across the polymer is fast. Furthermore, the polymer shall contain a reasonably large concentration of mobile ions. One might then anticipate that under conditions of cyclic voltammetry one should obtain reversible and easily interpretable current–voltage–time curves. However, this is in general not the case. Figure 4.43 shows the results obtained with a linear polyphenylene, deposited as a thin film on a platinum electrode. The coated electrode was subjected to slow voltammetric potential scans. The film is solvated well enough that ionic charge transport required for electroneutrality coupling does not determine the rate of the redox process.

This voltammogram demonstrates two remarkable features. First, the anodic and cathodic voltammetric peaks are separated by 380 mV, which means that the electrochemical redox reaction is "irreversible" in the electrochemical sense [75]. Second, large anodic and cathodic currents of nearly constant value are observed over a potential range of the order of 1 V. Both these features have aroused considerable discussion [117,120].

The explanations appearing most probable and plausible at present are the

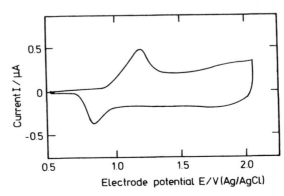

Figure 4.43. Cyclic voltammogram of a polyphenylene in CH_2Cl_2/0.1 M $TBAPF_6$; T = −60°C, sweep rate: $v = 10$ mV/s. (From Ref. 123 with permission of Elsevier Sequoia S.A.)

following: The voltammogram of Fig. 4.44 obtained with polyacetylene [121] shows that the large peak separation is essentially not a function of the potential sweep rate. This fact points to a chemical reaction following the charge transfer [75,117,122]. Upon oxidation the molecules stabilize themselves from the twisted into a partially planar structure with better conjugation; in the case of aromatic polymers that is the transition between a benzoide and the quinoide structure. This interpretation of the observed irreversibility is supported by systematic work by Heinze and co-workers with the corresponding oligomers [123,124]. It has been shown that the voltammograms obtained with various oligomers dissolved in liquid solutions react in an electrochemically reversible way [124,125]. However, when the same oligomers are deposited as films on an electrode, that is, when the free rotation is hindered by the solid matrix, the electrochemical irreversibility appears. In Fig. 4.45 the cyclic voltammograms of two oligomer films (one of them in two solvents) are shown. The electrochemical irreversibility is clearly dramatic. For the case of polythiophene the same authors [126] presented a further explanation for the irreversibility. They assumed that during oxidation crystalline domains are formed by π-interactions between polymer chain segments, which stabilize the system. This view is supported by experiments with thiophene oligomers in solution, which show that the cationic species form dimeric π complexes [127].

These oligomer results also yield a plausible explanation for the constant current region in the cyclic voltammogram of Fig. 4.43. Note that in the usual preparation procedures for conducting polymers little attention is given to the degree of poly-

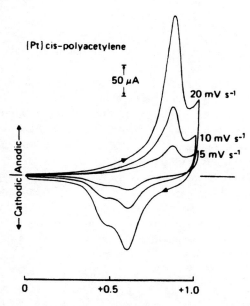

Figure 4.44. Cyclic voltammogram of a polyacetylene film on a platinum surface measured in acetonitrile containing 0.1 M Et_4NBF_4. (From Ref. 121 with permission of Elsevier Sequoia S.A.)

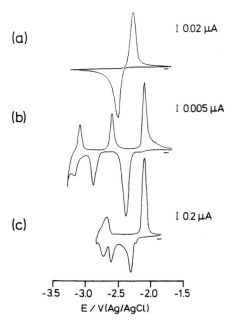

Figure 4.45. Cyclic voltammograms of thin layers on a Pt-electrode of (a) *p*-quaterphenylene (Me$_2$NH/0.1 M TBABr, $T = -75°C$, $v = 10$ mV/s). (b) *p*-sexiphenylene (Me$_2$NH/0.1 M TBABr, $T = -75°C$, $v = 10$ mV/s). (c) *p*-sexiphenylene (THF/0.1 M NaBPh$_4$, $T = -60°C$, $v = 10$ mV/s). (From Ref. 123 with permission of Elsevier Sequoia S.A.)

merization. Normally one obtains a mixture of oligomers and polymers of different chain length and possibly cross-linking. Considering the results of Fig. 4.45, it appears likely that numerous closely neighbored redox potentials are effective in an actual polymer. They are distributed over a considerable electrode potential range. The typical voltammogram is the superposition of all these redox waves. Based on this interpretation it is immediately clear that slight variations in the conditions of film preparation and of the electrolyte have considerable consequences for the resulting voltammograms (see Fig. 4.45b,c).

Truly irreversible electrochemistry in the chemical sense is a well known and annoying feature with conjugated polymers. It may lead to cross-linking by further oxidative coupling [123,124] or by nucleophilic attack of aromatic rings, typically introducing —OH or halogen groups [128]. Such reactions are often termed *overoxidation* of the polymer. They normally reduce the redox capacity of the polymer films in a most undesirable way.

Acknowledgments

The author is very grateful for valuable help received in preparing this review. The work has profited in particular from comments and contributions by the following scientists: Prof. R. P. Buck, University of

North Carolina, Chapel Hill; Prof. H. Gerischer, Fritz-Haber-Institut, Berlin; Prof. J. Heinze, Universität Freiburg; Prof. M. A. Vorotyntsev, Frumkin Institute, Moscow (presently Université P. & M. Curie, Paris VI). The original figures were prepared by Ms. Martina Purgand.

References

1. "Electrochemical Science and Technology of Polymers," Linford, R. G., Ed. Elsevier Applied Science, London, 1987, Vol. 1; 1990, Vol. 2.

2. Kaneko, M.; Wöhrle, D. In "Advances in Polymer Science," Cantow, H. J., Ed. Springer Verlag, Berlin, 1988, Vol. 84, pp. 143-228.

3. Abruna, H. D. In "Electroresponsive Molecular and Polymeric Systems," Vol. 1, Skotheim, T. A., Ed. Marcel Dekker, New York, 1988, pp. 98-160.

4. Evans, G. P. In "Advances in Electrochemical Science and Engineering," Gerischer, H., and Tobias, C. W., Eds. VCH Verlag, Weinheim, 1990, Vol. 1, pp. 1-74.

5. Heinze J. In "Topics in Current Chemistry," Steckhan, E., Ed. Springer Verlag, Berlin, 1990, Vol. 152, pp. 2-47.

6. Rubinstein, I. In "Applied Polymer Analysis and Characterization," Mitchell, J., Jr., Ed. Carl Hanser Verlag, Munich, 1992, Vol. II, pp. 233-258.

7. Murray, R. W., Ed., "Molecular Design of Electrode Surfaces", John Wiley & Sons, Inc., New York, 1992.

8. (a) Beck, F., *Electrochim. Acta* **1988**, *33*, 839; (b) Zorll, U., *Fresenius Z. Anal. Chem.* **1984**, *319*, 675; (c) Eliseeva, V. I.; Chernyi, V. N. *Prog. Org. Coat.* **1989**, *17*, 251.

9. (a) Doblhofer, K.; Eiselt, I. *Corrision Sci.* **1987**, *27*, 947, (b) Mansfeld, F. *Corrosion* **1988**, *44*, 856.

10. Taylor, S. R. *IEEE Trans. Electr. Insul.* **1989**, *24*, 787.

11. Stratmann, M. *Adv. Mater.* **1990**, 2, 191.

12. Cotton, T. M.; Kim, J. H.; Uphaus, R. A. *Microchem. J.* **1990**, 42, 44.

13. Doblhofer, K.; Figura, J.; Fuhrhop, J.-H. *Langmuir* **1992**, *8*, 1811.

14. Chidsey, C. E. D.; Loiacono, D. N. *Langmuir* **1990**, *6*, 682.

15. Wegner, G., *Ber. Bunsenges. Phys. Chem.* **1991**, *95*, 1326.

16. Wrighton, M. S. *Science* **1986**, *231*, 32.

17. Espenscheid, M. W.; Ghatak-Roy, A. R.; Moore, R. B., III; Penner, R. M.; Szentirmay, M. N.; Martin, C. R. *J. Chem. Soc., Faraday Trans. 1*, **1986**, 82, 1051.

18. Freiser, H. *J. Chem. Soc., Faraday Trans. 1*, **1986**, *82*, 1217; *Pure Appl. Chem.* **1987**, 59, 539.

19. Buck, R. P. (a) In "Theory, Design, and Biomedical Applications of Solid State Chemical Sensors," Cheung, P. W.; Fleming, D. G.; Neumann, M. R.; Ko, W. H., Eds. CRC Press, West Palm Beach, 1977, pp. 3-39. (b) In "Comprehensive Treatise of Electrochemistry," White, R. E.; Bockris, J. O'M.; Conway, B. E.; Yeager, E., Eds. Plenum, New York, 1984, Vol. 8, pp. 137-248. (c) In "Chemically Sensitive Electronic Devices: Proceedings of NATO Advanced Study Institute," Bergveld, P.; Zemel, J.; Middelhoek, S., Eds. Elsevier, Amsterdam, 1981, pp. 197-260. (d) Ibid., pp; 137-196.

20. Doblhofer, K.; Armstrong, R. D. *Electrochim. Acta* **1988**, *33*, 453.

21. Doblhofer, K. *Makromol. Chem., Macromol. Symp.* **1987**, *8*, 323; (b) Macdonald, J. R. "Impedance Spectroscopy: Emphasizing Solid materials and Systems." Wiley-Interscience, New York, 1987.

22. Burgmayer, P.; Murray, R. W. In "Handbook of Conducting Polymers," Skotheim, T. A., Ed., Marcel Dekker, New York, 1986, Vol. 1, pp. 507–523.
23. Zhong, C.; Storck, W.; Doblhofer, K. (a) *Ber. Bunsenges. Phys. Chem.* **1990**, *94*, 1149; (b) *Electrochim. Acta* **1990**, *35*, 1971.
24. Buck, R. P. (a) *J. Phys. Chem.* **1989**, *93*, 6212; (b) *J. Electroanal. Chem.* **1989**, *258*, 1; (c) *Mat. Res. Soc. Symp. Proc.* **1989**, *135*, 83; (d) Buck, R. P.; Vanýsek, P. *J. Electroanal. Chem.* **1990**, *292*, 73.
25. "Polymer Electrolyte Reviews," MacCallum, J. R.; Vincent, C. A., Eds. Elsevier Applied Science, London, 1987, Vol. 1; 1989, Vol. 2.
26. "Polymer Science and Materials," Tobolsky, A. V.; Mark, H. F., Eds. Wiley-Interscience, New York, 1971.
27. Hosesmann, R. *Polymer* **3**, *349* (1962).
28. "Polymer Handbook," 2nd ed., Brandrup, J.; Immergut, E. H., Eds. Wiley-Interscience, New York, 1975.
29. (a) Frumkin, A., *Ergebnisse der exakten Naturwiss.* **1928**, *7*, 235; (b) Mohilner, D. M. In "Electroanalytical Chemistry," Bard, A. J., Ed., Marcel Dekker, New York, 1966, Vol. 1, pp. 241–409.
30. Gucker, F. T.; Seifert, R. L. "Physical Chemistry," Norton, New York, 1966, p. 468.
31. Morf, W. E. "The Principles of Ion-Selective Electrodes and of Membrane Transport." Elsevier, Amsterdam, 1981.
32. Van Krevelen, D. W. "Properties of Polymers, Correlations with Chemical Structre." Elsevier, Amsterdam, 1972 (new edition 1990), Chapter 18.
33. Barrie, J. A. In "Diffusion in Polymers," Crank, J.; Park, G. S., Eds. Academic, London, 1968, pp. 259–313.
34. (a) Barker, R. E. *Pure Appl. Chem.* **1976**, *46*, 157; (b) Rogers, C. E. In "Polymer Permeability," Comyn, J., Ed. Elsevier Applied Science, London, 1985, pp. 11–74.
35. (a) Parlin, R. B.; Eyring, H. "Ion Transport Across Membranes," Clarke, H. T., Ed. Academic, New York, 1954; (b) Zaikov, G. E.; Iordanskii, A. L.; Markin, V. S. "Diffusion of Electrolytes in Polymers." VSP BV, Utrecht, The Netherlands, 1988, p. 11.
36. Buck, R. P. *J. Member, Sci.* **1984**, *17*, 1.
37. Williams, M. L.; Landel, R. F.; Ferry, J. D. *J. Am. Chem. Soc.* **1955**, *77*, 3701.
38. Horvai, G.; Graf, E.; Toth, K.; Pungor, E.; Buck, R. P. *Anal. Chem.* **1986**, *58*, 2735, 2741.
39. Harris, C. S.; Nitzan, A.; Ratner, M. A.; Shriver, D. F. *Solid State Ionics* **1986**, *18/19*, 151.
40. (a) Klotz, I. M. "Chemical Thermodynamics." W. A. Benjamin, New York, 1964, Chapter 21; (b) Vetter, K. "Elektrochemische Kinetik," Springer Verlag, Berlin, 1961; English translation: "Electrochemical Kinetics," Academic, New York, 1967.
41. Koryta, J.; Vanýsek, P. In "Adavances in Electrochemistry and Electrochemical Engineering," Gerischer, H.; Tobias, C. W., Eds. Wiley-Interscience, New York, 1981, pp. 113–176.
42. Born, M. *Z. Phys.* **1920**, *1*, 45.
43. MacCallum, J. R.; Vincent, C. A. In "Polymer Electrolyte Reviews," Vol. 1, MacCallum, J. R.; Vincent, C. A., Eds. Elsevier Applied Science, London, 1987, pp. 23–38.
44. Conway, B. E.; Bockris, J. O'M. In "Modern Aspects of Electrochemistry," Bockris, J. O'M.; Conway, B. E., Eds. Butterworths, London, 1954, Vol. 1, pp. 47–102.

45. "CRC Handbook of Chemistry and Physics," 67th ed., Weast, R. C., Ed. CRC Press, Boca Raton, 1986/1987, p. E-55.

46. Wranglén, G. "Korrosion und Korrosionsschutz." Springer Verlag, Berlin, 1984, Chapter 15.4 [Original in English: "An Introduction to Corrosion and Protection of Metals," Institut för Metallskydd, Stockholm 26].

47. (a) Pettit, L. D.; Bruckenstin, S. *J. Am. Chem. Soc.* **1966**, *88*, 4783; (b) Fuoss, R. M.; Accascina, F. "Electrolytic Conductance," Interscience, New York, 1959.

48. Watanabe, M.; Ogata, N. In Ref. 43, pp. 39–68.

49. Vögtle, F. "Supramolekulare Chemie." B. G. Teubner, Stuttgart, 1989; English translation: "Supramolecular Chemistry," John Wiley, Chichester, England, 1991.

50. Vanýsek, P.; Buck, R. P. *J. Electroanal. Chem.* **1991**, *297*, 19.

51. Sudhölter, E. J. R.; van der Wal, P. D.; Skowronska-Ptasinska, M.; van den Berg, A.; Bergveld, P.; Reinhoundt, D. N. *Anal. Chim. Acta* **1990**, *230*, 59.

52. Yeo, R. S.; Yeager, H. L. In "Modern Aspects of Electrochemistry," Conway, B. E.; White, R. E.; Bockris, J. O'M., Eds. Plenum, New York, 1985, Vol. 16, pp. 437–505.

53. (a) Weber, J.; Janda, P.; Kavan, L.; Jegorov, A. *J. Electroanal. Chem.* **1986**, *200*, 379; (b) Solution Technology, P. O. Box 171, Mendenhall, PA 19357.

54. Helfferich, F. "Ionenaustauscher," Band I. Verlag Chemie, Weinheim, 1959; English translation: "Ion Exchange," McGraw-Hill, New York, 1962.

55. Braun, H.; Storck, W.; Doblhofer, K. *J. Electrochem. Soc.* **1983**, *130*, 807.

56. Lindholm, B.; Sharp, M. *J. Electroanal. Chem* **1986**, *198*, 37.

57. Braun, H.; Decker, F.; Doblhofer, K.; Sotobayashi, H. *Ber. Bunsenges. Phys. Chem.* **1984**, *88*, 345.

58. De Castro, E. S.; Huber, E. W.; Villarroel, D.; Galiatsatos, C.; Mark, J. E.; Heinemann, W. R.; Murray, P. T. *Anal. Chem.* **1987**, *59*, 134.

59. Schnabel, W.; Sotobayashi, H. *Prog. Polym. Sci.* **1983**, *9*, 297.

60. Mauritz, K. A.; Hopfinger, A. J. In "Modern Aspects of Electrochemistry," Conway, B. E.; White, R. E.; Bockris, J. O'M., Eds. Plenum, New York, 1982, Vol. 14, pp. 425–508.

61. Lane, R. F.; Hubbard, A. T. *J. Phys. Chem.* **1973**, *77*, 1401.

62. Ulman, A. "An Introduction to Ultrathin Organic Films," Academic, San Diego, 1991.

63. Armstrong, R. D.; Lindholm, B.; Sharp, M. *J. Electroanal. Chem.* **1986**, *202*, 69.

64. Fabricius, G.; Kontturi, K. *Electrochim. Acta*, **1991**, *36*, 333.

65. Chu, D.; Tryk, D.; Gervasio, D.; Yeager, E. B. *J. Electroanal. Chem.* **1989**, *272*, 277.

66. Lakshminarayanaiah, N. "Membrane Electrodes," Academic, New York, 1976, Chapter 3.

67. Donnan, F. G. *Z. Elektrochem.* **1911**, *17*, 572.

68. Reichenberg, D.; Pepper, K. W.; McCauley, D. J. *J. Chem. Soc.* **1951**, *1951*, 493.

69. Donnan, F. G. *Z. Phys. Chem.* **1934**, *A168*, 369.

70. Gregor, H. P. *J. Am. Chem. Soc.* **1948**, *70*, 1293; **1951**, *73*, 643.

71. Cappadonia, M.; Doblhofer, K.; Woermann, D. *J. Colloid Interface Sci.* **1991**, *143*, 222.

72. Doblhofer, K.; Braun, H.; Lange, R. *J. Electroanal. Chem.* **1986**, *206*, 93.

73. Doblhofer, K.; Lange, R. *J. Electroanal. Chem.* **1987**, *229*, 239.
74. Lange, R.; Doblhofer, K. *J. Electroanal. Chem.* **1987**, *216*, 241.
75. Bard, A. J.; Faulkner, L. R. "Electrochemical Methods." John Wiley, New York, 1980.
76. Daum, P.; Murray, R. W. *J. Phys. Chem.* **1981**, *85*, 389.
77. Lange, R.; Doblhofer, K. *J. Electroanal. Chem.* **1987**, *237*,13; *Ber Bunsenges. Phys. Chem.* **1988**, *92*, 578.
78. Buck, R. P. *J. Electroanal. Chem.* **1973**, *46*, 1; **1986**, *210*, 1.
79. Gerischer, H.; Kolb, D. M.; Sass, J. K. *Adv. Phys.* **1978**, *27*, 437.
80. Buck, R. P. *J. Electroanal. Chem.* **1987**, *219*, 23.
81. (a) Tsou, Y.-M.; Anson, F. C. *J. Elecctrochem. Soc.* **1984**, *131*, 595; (b) Naegeli, R.; Redepenning, J.; Anson, F. C. *J. Phys. Chem.* **1986**, *90*, 6227.
82. Fitch, A. *J. Electroanal. Chem.* **1990**, *284*, 237.
83. Sharp, M. *Anal. Chim. Acta* **1972**, *59*, 137; **1972**, *61*, 99; **1972**, *62*, 385; **1975**, *65*, 405.
84. Kulesza, P. J.; Doblhofer, K. *J. Electroanal. Chem.* **1989**, *274*, 95.
85. Niwa, K.; Doblhofer, K. *Electrochim. Acta* **1986**, *31*, 549.
86. Manecke, G., Strock, W. In "Encyclopedia of Polymer Science and Engineering, 2nd ed. Kroschwitz, J. I., Ed. John Wiley, New York, 1986, Vol. 5, pp. 725-755.
87. Hamann, C., Heim, J., Burghardt, H. "Organische Leiter, Halbleiter und Photoleiter." Vieweg Verlag, Braunschweig, 1981.
88. *Makromol. Chem., Macromol. Symp.* 8 (Proceedings of the International Workshop Electrochemistry of Polymer Layers, Duisburg, 1986), Beck, F., Symposium ed. Hthig und Wepf Verlag, Heidelberg, 1987.
89. Faraday Discussions of the Chemical Society 88, "Charge Transfer in Polymeric Systems," The Royal Society of Chemistry, London, 1989.
90. *Synthetic Metals* **1993**, *55-57*, pp. 1-5123 (Proceedings of the international conference on Science and Technology of Synthetic Metals, Göteborg, 1992).
91. Overhof, H. In *Festkörperprobleme* Teusch, J., Ed. Vieweg, Braunschweig, 1976, Vol. 16, pp. 239-265.
92. Buck, R. P. *J. Phys. Chem.* **1988**, *92*, 4196, 6445.
93. Dalton, E. F.; Surridge, N. A.; Jernigan, J. C.; Wilbourn, K. O.; Facci, J. S.; Murray, R. W. *Chem. Phys.* **1990**, *141*, 143.
94. Sharp, M.; Lindholm, B.; Lind, E.-L. *J. Electroanal. Chem.* **1989**, *274*, 35.
95. Basolo, F.; Pearson, R. G. "Mechanisms of Inorganic Reactions." John Wiley, New York, 1968, Chapter 6.
96. Dahms, H. *J. Phys. Chem.* **1968**, *72*, 362.
97. Ruff, I.; Friedrich, V. J. *J. Phys. Chem.* **1971**, *76*, 3297.
98. Laviron, E. *J. Electroanal. Chem.* **1980**, *112*, 1.
99. Doblhofer, K.; Dürr, W.; Jauch, M. *Electrochim. Acta* **1982**, *27*, 677.
100. Gabrielli, C.; Haas, O.; Takenouti, H. *J. Appl. Electrochem.* **1987**, *17*, 82.
101. Buttry, D. A.; Anson, F. C. *J. Am. Chem. Soc.* **1983**, *105*, 685.

102. Chen, X.; He, P.; Faulkner, L. R. *J. Electroanal. Chem.* **1987**, *222*, 223.

103. White, H. S.; Leddy, J.; Bard, A. J. *J. Am. Chem. Soc.* **1982**, *104*, 4811.

104. Inzelt, G. *Electrochim. Acta* **1989**, *34*, 83.

105. (a) Forster, R. J; Vos, J. G. *J. Electroanal. Chem.* **1991**, *314*, 135; (b) Rubinstein, I.; Rishpon, J.; Gottesfeld, S. *J. Electrochem. Soc.* **1986**, *133*, 729.

106. Oh S.-M.; Faulkner, L. R. *J. Am. Chem. Soc.* **1989**, *111*, 5613.

107. Savéant, J. M. *J. Electroanal. Chem* **1986**, *201*, 211; **1987**, *227*, 299.

108. Brédas, J. L.; Street, G. B. *Acc. Chem. Res.* **1985**, *18*, 309.

109. Schmeißer, D.; Bätz, P.; Göpel, W.; Jaegermann, W., et al. *Ber. Bunsenges. Phys. Chem.* **1991**, *95*, 1441.

110. Ku, C. C.; Liepins, R. "Electrical Properties of Polymers." Hanser, Munich, 1987, p. 256.

111. Wegner, G.; Rühe, J. In Ref. 89, p. 33.

112. Peover, M. E. In "Electroanalytical Chemistry," Bard, A. J., Ed. Marcel Dekker, New York, 1967, Vol. 2, pp. 1–51.

113. Schlenoff, J. B.; Chien, J. C. W. *J. Am. Chem. Soc.* **1987**, *109*, 6269.

114. Woodard, F. E.; Reilley, C. N. In "Comprehensive Treatise of Electrochemistry," Yeager, E.; Bockris, J. O'M.; Conway, B. E.; Sarangapani, S., Eds. Plenum, New York, Vol. 6, pp. 353–392.

115. Buck, R. P. *J. Electroanal. Chem.* **1989**, *271*, 1.

116. Andrieux, C. P.; Savéant, J. M. *J. Phys. Chem.* **1988**, *92*, 6761.

117. Albery, W. J.; Chen, Z., et al., in Ref. 89, p. 247.

118. De Levie, R. In "Advances in Electrochemistry and Electrochemical Engineering," Delahay, P.; Tobias, C. W., Eds. John Wiley, New York, 1967, Vol. 6.

119. Bull, R. A.; Fan, F.-R. F.; Bard, A. J. *J. Electrochem. Soc.* **1982**, *129*, 1009.

120. Doblhofer, K. *J. Electroanal. Chem.* **1992**, *331*, 1015.

121. Diaz, A. F.; Clarke, T. C. *J. Electroanal. Chem.* **1980**, *111*, 115.

122. Heinze, J. *Angew. Chem.* **1984**, *96*, 823 (*Angew. Chem. Int. Ed. Engl.* **1984**, *23*, 831).

123. Meerholz, K.; Heinze, J. *Synthetic Metals* **1991**, *43*, 2871.

124. Meerholz, K.; Heinze, J. *Angew. Chem.* **1990**, *102*, 695 (*Angew. Chem. Int. Ed. Engl.* **1990**, *29*, 692.

125. Meerholz, K.; Heinze, J. *J. Am. Chem. Soc.* **1989**, *111*, 2325.

126. Heinze, J.; Meerholz, K., private communication (to be published).

127. Hill, M. G.; Han, K. R.; Miller, L. L.; Penneau, J.-F. *J. Am. Chem. Soc.* **1992**, *114*, 2728.

128. Beck, F.; Braun, P.; Oberst, M. *Ber Bunsenges. Phys. Chem.* **1987**, *91*, 967.

CHAPTER

5

Transition Metal Oxides: Versatile Materials for Electrocatalysis

Sergio Trasatti

5.1 Introduction

5.1.1 Retrospect

The "Silver Jubilee" of DSA® (dimensionally stable anodes) was celebrated by the scientific community in 1989 on the occasion of the spring meeting of the Electrochemical Society [1]. The twenty-fifth anniversary probably marked the issue of the first patent, but H. Beer's discovery took place somewhat earlier, as the author himself tells in a specific paper [2], presumably at the beginning of the 1960s.

The discovery was the consequence of an intensive technological research aimed at finding suitable substitutes for graphite anodes in chlor-alkali cells [3,4]. The continuous wear on these electrodes under anodic load led to both the production of a less pure chlorine gas and an increasing separation between anode and cathode with time [5]. The solution of this problem with the introduction of activated Ti anodes was the origin of the commercial name, DSA, by which these electrodes are universally known [6].

After some attempts using noble-metal-activated Ti electrodes, which did not entirely solve the problem of corrosion, Beer [2,6] observed much better behavior using the oxides of precious metals, in particular RuO_2 and IrO_2, prepared with a thermal procedure. The first anodes used in the electrochemical technology basically consisted of a thin layer of RuO_2 + TiO_2 mixed oxides deposited on Ti [7], and this composition has become the prototype DSA. Although additives have

---

This work has been supported by the National Research Council (CNR), Rome.

subsequently been introduced to improve the catalytic properties and the stability, such electrodes are still in service in chlor-alkali cells with excellent performances [1].

Applications have largely anticipated fundamental research in this field [8,9]. A couple of papers dealing with DSA appeared in 1970 [5,7], but they were essentially technical reports on industrial performances. The first paper with data of fundamental interest was published in 1971 by this writer [10]. In his laboratory, research on RuO_2 started in late 1967 [11]. Fundamental studies were needed after the technological success since the reasons for such excellent performance were essentially unknown. In order to control the properties and to tailor them for specific needs, it is necessary to identify the factors responsible for the behavior of these electrodes.

After the first paper, the fundamental research on these materials developed almost exponentially, spreading from the investigation of purely electrochemical aspects to the study of solid-state physical and chemical properties [12]. In parallel, the extraordinary properties of DSAs suggested their use in more and more varied applications [13,14]. Although DSAs started as anodes for Cl_2 evolution, suitable modifications have been proposed and tested for a number of other applications, as shown in Table 5.1. Moreover, the possibilities for their application in nonaqueous [15] and molten systems [16,17] have also been investigated. At the same time, new materials have been explored with the aim of finding cheaper substitutes for RuO_2 and IrO_2, and of improving the electrocatalytic and stability properties.

5.1.2 Previous Reviews

Oxide electrodes have been mentioned in many review papers and chapters, especially in those dealing with the chlor-alkali process [4,8,18-21]. However, the only existing book entirely dedicated to DSA materials was edited by this author in 1980 [12]. At almost the same time (1979), a whole chapter was devoted by this author and O'Grady to the properties and applications of RuO_2 [22]. Later, a few specific reviews concerning oxides were published, in particular on Cl_2 evolution [23], O_2 evolution [24,25], and surface properties [26,27]. However, no other comprehensive work covering the whole field of oxide properties and performance has appeared since 1980.

In the preceding context, the purpose of this chapter is to review the essential advances in the understanding of the properties of electrocatalytic oxides that have occurred during the past 10 years or so, although reference to earlier work will be made where necessary to provide a sufficient background for those not directly involved in this research field.

After discussing the technological demands for electrode materials and their impact on the development of DSAs, the chapter will illustrate their physicochemical and surface characterization to gain insight into the structural and morphological properties of these materials. The special features of the oxide-solution interface will then be described as a prerequisite to understanding the details of the mechanism of electrode reactions.

Oxides will not be described individually or in classes, although examples for

TRANSITION METAL OXIDES: VERSATILE MATERIALS FOR ELECTROCATALYSIS

Table 5.1 Recent Examples of (Proposed) Applications of Electrocatalytic Oxides Other than in Chlor-alkali Cells and Water Electrolysis[a]

Applications	References
Chlorate electrosynthesis	277, 438, 465
Bromate electrosynthesis	456, 460
Persulfate electrosynthesis	455
Sodium peroxyborate electrosynthesis	464
Chlorine dioxide electrosynthesis	572
Electrolytic MnO_2	440
CO_2 reduction	451
Molten salt electrolysis	16, 17
Ozone production	442
Wastewater treatment	439
Cathodic protection	443
Metal electrowinning	444, 463, 571
Gold electroplating	594
Chromium electroplating	466, 448–450, 595
pH sensors and actuators	139–141, 447, 613
Detectors for liquid chromatography	630
Electronics	452, 467
Resistors	462, 468–470, 624
Double-layer capacitors	204, 205
Photocatalysis	176, 454, 597, 614
Photoelectrolysis	453, 461, 466, 576, 577, 627
Photovoltaics	579
Electroless redox catalysis	458, 580, 589
Lead–acid batteries	441, 445
Li batteries	457, 459, 584, 596
Hydrogen cathodes	See Table 5.7
Oxygen cathodes	See Table 5.8
Organic electrosynthesis	See Table 5.9

[a] More examples can be found in previous reviews [13,21].

specific materials will be referred to. The subdivision will rather be based on properties and specific reactions. Sections will be devoted to gas evolution reactions, organic reactions, factors of electrocatalysis, and problems of stability.

5.1.3 Technological Demands

The electric potential difference (ΔV) applied to an electrolysis cell consists of several contributions:

$$\Delta V = \Delta E + \Sigma \eta + \Delta V_\Omega + \Delta V_t \tag{5.1}$$

where ΔE is the thermodynamic potential difference, which depends not on the electrode materials but on the electrode reactions. $\Sigma \eta$ is the sum of the anodic and cathodic overpotentials, which can be minimized by treatment of the electrode materials. ΔV_Ω is the sum of the various ohmic drops in the interelectrode gap, in the connections, and in the electrode structure. In part, ΔV_Ω can depend directly on the electrode material, but inevitably depends indirectly on it—its control is essentially an engineering problem. Finally, ΔV_t is a term that takes into account any drift of the previous potential contributions with time. This last parameter quantifies the *stability* of the electrode materials. At the beginning of performance, $\Delta V_t = 0$. Thus $\Delta V_t = \Delta V - \Delta V_0$ where ΔV_0 is the initial electric potential difference.

To be of interest for technological applications, electrode materials must satisfy a number of requirements [21,28], including high surface area, good electronic conductivity, high electrocatalytic activity, and good mechanical, chemical, and electrochemical stability. These features are necessary to minimize all the terms on the r.h.s. of Eq. (5.1) (except ΔE). Any search for new electrode materials is therefore aimed at developing electrodes for which the preceding properties can be optimized. Transition metal oxides satisfy most of these demands, and this has been the main reason for their success and for the increasing number of new applications.

5.1.4 Oxides of Interest

Many types of oxide have been investigated [28], but most interest has focused on three main classes [12]: dioxides, spinels, and perovskites, with the recent addition of pyrochlores [29]. Typical examples of the first class are RuO_2 and IrO_2 together with TiO_2 (and SnO_2) and of the second class, Co_3O_4 and $NiCo_2O_4$; in the third class the variety is larger, although doped $LaNiO_3$ and $LaCoO_3$ are the most investigated. $Pb_2Ru_2O_{7-y}$ and $Pb_2Ir_2O_{7-y}$ are the main representatives of the pyrochlores. In addition, other oxides that do not belong to these specific classes but have been studied intensively are Ta_2O_5 and ZrO_2 (as substitutes for TiO_2), Cr_2O_3, Rh_2O_3, PdO_x, and so on. A more complete picture can be obtained from the various tables in the following sections. Readers are also referred to previous reviews for work up to 1980.

5.2 Properties of Oxides for Electrodes

5.2.1 Preparation

Oxides containing precious metals are expensive materials. The amount used is in practice minimized by preparing so-called activated electrodes consisting of a metallic support on which a layer of oxide, a few micrometers, is deposited by thermal decomposition of a suitable precursor. Details of the procedure can be found in the literature [12]. For example, RuO_2 is prepared by thermal decomposition of $RuCl_3$ at temperatures between 300°C and 500°C. The precursor is routinely dissolved in isopropanol and mechanically spread onto the support, but aqueous solutions have also been used [28]. The nature of the solvent [30] and of the precursor [31] has been found to affect the properties of the resulting oxide. Table 5.2 summarizes the most common oxides and their precursors. Deposition can also be performed by other means (e.g., plasma-spray [32] and reactive sputtering [33]).

Table 5.2 Common Precursors for Some Oxide Electrodes[a]

Oxide	Precursors
RuO_2	$RuCl_3 \cdot xH_2O$
IrO_2	$IrCl_3 \cdot xH_2O$, $H_2IrCl_6 \cdot 6H_2O$
MnO_2	$Mn(NO_3)_2 \cdot 4H_2O$
Co_3O_4	$Co(NO_3)_2 \cdot 6H_2O$
NiO_x	$Ni(NO_3)_2 \cdot 6H_2O$
TiO_2	$TiCl_3$, $Ti(OR)_4$, $TiCl_4$
SnO_2	$SnCl_2 \cdot 2H_2O$, $SnCl_4$
PdO_x	$PdCl_2$
Cr_2O_3	$CrCl_3 \cdot 6H_2O$, $Cr(NO_3)_3 \cdot 9H_2O$

[a] See the original literature for mixed oxides such as perovskites and pyrochlores.

Titanium is widely used as a support since it is *the* industrial support [2,3,7]. However, other valve metals are in principle also suitable [13,34]. Nickel is also useful as a support for anodes [20,35], whereas steel and Ni are more suitable for activated cathodes since Ti is affected by hydrogen embrittlement [36].

When the aim is to increase the surface area or to prepare an especially porous structure, or when ceramic preparation does not permit the deposition on supports, powders are prepared and then compacted by just pressing, sometimes in the presence of a binder [37,38]. The resulting structure may introduce problems of electrical conduction and mechanical stability [39].

5.2.2 Structure and Surface Characterization

The properties of thermal oxides are widely dependent on many experimental variables such as the temperature of calcination and the procedure for deposition. In order to relate the physicochemical properties with the electrochemical performances, a detailed characterization of the active oxide layer is necessary before and after use. This characterization includes the determination of the crystal structure, the bulk chemical composition, the surface composition, the morphology, the surface area, the electrical conductance, and the acid–base properties. The experimental approaches include both in situ electrochemical and nonelectrochemical techniques and ex situ methods. A list is given in Table 5.3. An interdisciplinary approach is essential since oxide electrodes lie at the border between materials chemistry and physics, and electrochemistry. A summary of recent papers devoted to oxide characterization is reported in Table 5.4.

Thermogravimetric analysis (TGA) and differential thermal analysis (DTA) are essential to investigate the mechanism of thermal decomposition [40]. However, the kinetics of decomposition can be influenced by both the support and the solvent of the precursor. Thus, bulk $IrCl_3$ is seen by TGA to decompose at higher temperatures than 500°C [41]. Nevertheless, IrO_2 electrodes can be obtained at 400°C since decomposition of the chloride takes place in a thin film on a support [42]. Similarly, RuO_2 and TiO_2 show little reciprocal solubility when prepared at high temperatures

Table 5.3 Physicochemical Characterization of Oxides[a]

Thermal decomposition	TGA, DTA [40]
Lattice structure, crystallinity, crystallite size	X-ray diffraction [51]
Morphology, surface area	SEM [49], BET [472], ion absorption [48]
Electronic properties	Electrical conduction [26], reflectance [607], ellipsometry [212]
Chemical composition (bulk, surface, profile)	RBS [68], AES [474], XPS [90], SIMS [575], PAS [119], PCS [121], Raman [223], XRF [601], IR [136], ICPES [100], EPR [95], LEED [221], HREELS [92]
Hydration	Tritium exchange [26], immersion heat [631], nuclear techniques (NRA) [111]
Acid–base properties	Potentiometric titration [151]
Point of zero charge	Electrophoretic mobility [147]
Ionic surface charge	Beam probe deflection (BPD) [201]
Electrochemical surface spectrum	Cyclic voltammetry [53], impedance [212]
Electrochemical surface charge	Chronocoulometry [54]

[a] References are given as examples.

[43,44]. However, at 400°C on a Ti support, solid solutions are formed over the entire composition range, although these are metastable phases [43,45].

When examined by SEM, oxide layers very often exhibit a characteristic "cracked-mud" look caused mainly by thermal shocks in the materials [46–49]. Such a feature is, *in principle,* favorable because it gives rise to a higher macroscopic surface area. However, it can also be the source of problems if the active layer does not cover the support. If the support becomes electrochemically polarized, it passivates (being a valve metal) with formation of a nonconducting TiO_2 layer, which can dramatically increase the ΔV_Ω term in Eq. (5.1).

Thermal decomposition gives rise to oxide layers whose structure is that of a compressed powder (Fig. 5.1). X-ray analysis of these layers has shown that the crystal structure is a function of the temperature of calcination. In some cases with bulk materials, such as IrO_2 and RuO_2, diffraction peaks shift with the preparation temperature [26,50], indicating that the size of the unit cell changes with calcination [47]. This effect can be related to the large number of defects present in oxides formed at low temperatures. However, the diffraction peak position of IrO_2 is not shifted as layers are prepared [26], suggesting that the kinetics of decomposition in the large grains of powder may be different.

The crystallinity and the crystallite size of thermal oxides are also a function of the temperature of calcination. As is known from the field of catalysis, crystallites grow as the temperature of calcination is increased. This is observed for both powders and layers [26,51,52]. Moreover, the crystallite size has been found to depend on the solvent of the precursor for Co_3O_4 [51] and IrO_2 [42]. It also depends on the nature of the precursor, since the kinetics and the mechanism of decomposi-

TRANSITION METAL OXIDES: VERSATILE MATERIALS FOR ELECTROCATALYSIS   213

Table 5.4 Recent Work on the Surface Characterization of Electrocatalytic Oxides[a]

Electrode Material	References
RuO_2	15, 31, 40, 48, 54, 66, 68, 92, 111–113, 119, 138, 188, 196, 198, 206, 208, 210, 212, 216, 424, 472–478, 629
RuO_2 single crystals	221–224, 226
$RuO_2 + Ta_2O_5$	417, 479
$RuO_2 + IrO_2$	30, 53, 119, 219
$RuO_2 + ZrO_2$	417
$RuO_2 + SnO_2 + IrO_2$	96
$RuO_2 + SnO_2$	98
$RuO_2 + SnO_2 + TiO_2$	390, 480, 600, 602, 603
$Fe_xCo_{3-x}O_4$	481, 491
SnO_2 (+Sb)	482
$NiO_x + MoO_3$	481
$RuO_2 + TiO_2$	55, 56, 76, 78, 80, 94, 394, 417, 466, 483, 484, 574, 592, 599
$IrO_2 + SnO_2$	98, 612
IrO_2	40, 42, 48, 59, 64, 90, 98, 115, 119, 137, 189, 207, 571, 607, 608, 616, 629
IrO_2 single crystals	59
$IrO_2 + ZrO_2$	417
$IrO_2 + TiO_2$	49, 417, 568, 575, 606, 628
$IrO_2 + Ta_2O_5$	98, 417, 489, 570
PtO_x	40
$PtO_x + TiO_2$	417, 485
$PtO_x + ZrO_2$	417
$PtO_x + Ta_2O_5$	417
RhO_x	40
MnO_2[b]	486, 487
PdO_x	488
$Bi_{2-x}Gd_xRu_2O_7$	92
$Ag_xCo_{3-x}O_4$	490
$Ag_xLa_{1-x}CoO_3$	490
$La_{0.175}WO_3$ (111)	238
Co_3O_4	32, 51, 57, 58, 60, 61, 70, 91, 105, 144, 145, 209, 287, 492, 493, 495, 497, 583
$Co_3O_4 + RuO_2$	217, 218, 493, 495
$NiCo_2O_4$	70, 209, 287, 494, 498, 499
$Co_3O_4 + TiO_2$	495, 496
$Cr_2O_3 + TiO_2$	351, 471
$MoO_3 + TiO_2$	623
Various perovskites	500
Various metals as supports	34

[a] Previous studies have been reviewed in Ref. 12.
[b] Only pyrolytic films are considered.

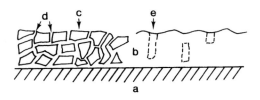

Figure 5.1. Sketch of the morphology of an oxide layer (b) deposited on a support (a). Crystallites (c), grain boundaries (d), and pores (e) are shown.

tion of different precursors are normally different. RuO_2 prepared from the nitrate has been found to be more finely dispersed than RuO_2 from the chloride [31].

The different kinetics of decomposition are also responsible for the different morphology of mixed oxides when prepared from different solvents of the precursors. In the case of an $RuO_2 + IrO_2$ mixture it has been found that the minimum particle size occurs at about 50 mol% if the precursors are dissolved in water [53], whereas no minimum is observed if isopropanol is the solvent [30]. These effects can be related to the hydrolysis of the precursors. If the solution is not sufficiently acidic, hydroxides can precipitate before the precursor is deposited on the support, leading to the formation of separate phases. For this reason, when aqueous solutions are used, they are normally acidified.

Although these oxides are customarily identified with their chemical formulas—RuO_2, IrO_2, Co_3O_4, and so on—they are in fact nonstoichiometric. RuO_2 is an oxygen-deficient material containing residual chlorine whose amount decreases as the temperature is increased [44,50,54,55]. Oxygen vacancies are balanced by the presence of Ru(III) ions, which have been identified by EPR [56]. On the contrary, Co_3O_4 is nonstoichiometric, with an excess of oxygen that is balanced by an increase in the number of Co(III) ions [51,57,58]. In other cases, dismutations are possible during the preparation. The presence of metallic Ir, formed by dismutation of Ir(III) ions in solutions, has been detected in IrO_2 films [59]. It is therefore evident that the number of variables is high, which explains the large scatter commonly observed when results obtained with different preparations are compared. Further, this makes the accurate characterization of these films essential. On the other hand, a theory of electrocatalysis for these materials can only be based on a very large spectrum of results, since a single set of data may not be sufficiently representative.

As mentioned earlier, oxide layers normally have a high surface area that decreases as the calcination temperature is increased [51]. On the other hand, too low a temperature of calcination may yield partly undecomposed, and therefore unstable, materials. For the most common oxides, such as RuO_2, TiO_2, Co_3O_4, and $NiCo_2O_4$, the optimum temperature of decomposition is around 400–500°C. At higher temperatures problems arise at the contact between the support and the active layer [60]. Besides accelerating interdiffusion [61], a high temperature promotes the oxidation of the Ti support, with formation of an insulating interlayer that bears on ΔV_Ω in Eq. (5.1) [62]. Precautions can be taken to minimize this effect, but it cannot be totally eliminated, since most precursors require an oxygenated atmosphere for decomposition.

The macroscopic morphology of an oxide layer can be modified to some extent by changing the concentration of the precursor and the preparation procedure. By using dilute solutions of $RuCl_3$ and by mechanically smoothing each layer after calcination, compact RuO_2 layers can be grown without visible cracks [50]. This procedure leads to the formation of larger crystallites that can be used to investigate the effect of the macroscopic morphology on the properties of oxide electrodes [54].

A major requirement for technological electrodes is a high electronic conductivity. Although not as important for activated layers only a few micrometers thick, it is, however, important that the conductivity not be lower than about $1 \; \Omega^{-1} \; cm^{-1}$. It is interesting that this problem does not exist with RuO_2 and IrO_2 since they are metallic conductors [63]. Despite resistivity increases because of intergrain effects due to preparation at low temperature, the conductivity remains high [47].

The resistivity of RuO_2 and IrO_2 has been found to decrease as the temperature of calcination increases (Fig. 5.2). It also depends on the procedure of preparation

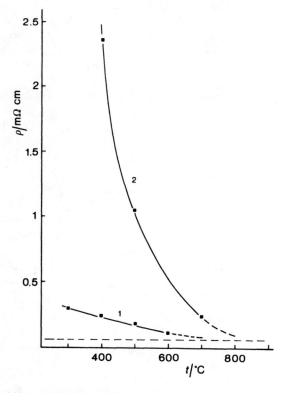

Figure 5.2. Resistivity (ρ) of RuO_2 (1) and IrO_2 (2) layers, prepared by thermal decomposition, as a function of the temperature of calcination [26,66]. (---) Resistivity of single crystals [41].

[26,64,65]. Compact layers are better conductors than cracked layers [66]. It has been suggested that the decrease in conductivity with decreasing calcination temperature observed with RuO_2 may be related to the presence of residual Cl, which expands the lattice [47]. In fact, the main factor is the intergrain resistance [65], which decreases as sintering becomes more effective at higher temperatures [26,66]. This mechanism is thus the same as the one proposed for glazy resistors (RuO_2 + SiO_2) [67].

In IrO_2, the dependence of conductivity on the temperature of calcination is more complex because of the different chemistry of decomposition. IrO_2 has been found to be nonstoichiometric with excess of oxygen, the opposite of RuO_2 [68]. Moreover, the segregation of impurities (e.g., Ir metal or nonconducting Ir_2O_3) in the grain boundaries [59] may be the reason for the observed anomalies. For instance, IrO_2 layers prepared at 350°C and calcined at higher temperatures show a decrease rather than an increase in conductivity [26]. Another anomaly is the decrease in resistivity observed with increasing thickness [64].

Other oxides, such as Co_3O_4 and $NiCo_2O_4$, are intrinsically semiconductors [57,69]. However, since the low temperature of calcination produces nonstoichiometry [51,57], they become heavily doped and therefore conductive [70], although not at the level of RuO_2. In these cases, an increase in resistance can be expected with increasing thickness [71], but the magnitude is never such as to produce important values of ΔV_Ω. In the case of perovskites, the conductivity can vary by several orders of magnitude depending on the material [72–75]. Again, the method of preparation can greatly influence the outcome.

In RuO_2 + TiO_2 mixed oxides, the conductivity is dominated by RuO_2 down to 25–30% of this oxide [43,55], then drops at lower RuO_2 contents [76] (Fig. 5.3). For this reason technological electrodes contain no more than 70–80% of valve metal oxide. At lower RuO_2 content the anatase form of TiO_2 is formed and the miscibility with RuO_2 worsens [44,77]. The prevalence of the electrical properties of RuO_2 has been attributed to the formation of an infinite chain of RuO_2 clusters in the TiO_2 matrix [78] not dissimilar to the structure proposed for RuO_2-based glazy resistors [67,79]. As the RuO_2 content decreases below 25%, the chains of clusters are interrupted and the conductivity drops [80]. A transition from semiconducting to metallic behavior has been obtained at a Ru/Ti ratio of the order of 0.1 in the case of anodically oxidized Ru-implanted Ti alloys [81].

Although conductance problems can arise if electrodes are formed from powders cemented with a binder (especially with intrinsically semiconducting materials), with layers major problems can arise at the support–active layer boundary. As already mentioned, TiO_2 films can be formed during the preparation. These can grow with time and anodic polarization [82]. The solubility of the active material in TiO_2 becomes a crucial issue. Thus, RuO_2 can dope TiO_2 quite substantially so that ohmic drops are minor even at a high temperature of calcination, at which the growth of a TiO_2 film is enhanced. IrO_2 is less soluble in TiO_2 and higher ohmic drops can be observed without specific precautions. For Co_3O_4, ohmic drops are major problems at higher temperatures than 400°C [61,83,84]. Co_3O_4 is insoluble in TiO_2 and cannot dope it effectively [85]. To avoid highly resistive contacts, in many

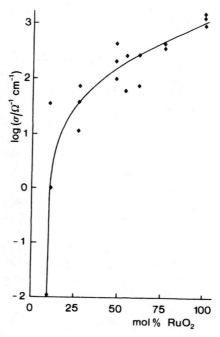

Figure 5.3. Conductivity (σ) of RuO_2 + TiO_2 layers, prepared at temperatures between 350°C and 600°C, as a function of composition. (After Ref. 55, with permission.)

instances a thin interlayer of RuO_2 has been placed between the support and the active layer [86,87]. A PdO_x interlayer has been reported to play the same role [88]. However, the possibility of interdiffusion between the adjoining layers with effects on the electrocatalytic properties of the electrodes should always be taken into account [61].

Spectroscopic techniques have been used to investigate the chemical structure of oxide electrodes. Information on both surface and in-depth profile composition have been obtained by applying IR, XPS, AES, SIMS, and RBS. In particular, RuO_2 [89], IrO_2 [90], and Co_3O_4 [60,91] have been investigated. On the surface of RuO_2 and IrO_2 the presence of the defect structure of the MO_3 oxide has been detected [89,90], but the feature is probably related to some type of adsorbed oxygen with formation of a thin, discrete solid phase [92]. In general, the resolution of the different valency states of a metal is difficult though qualitatively possible.

Three forms of oxygen are usually found on the surface of oxides, corresponding to lattice oxygen, OH, and adsorbed or chemically bound water [89–91]. The amount of the latter two forms decreases with depth, but only bound water disappears totally in the bulk [91]. The presence of surface OH groups is typical of oxides and will be discussed later. Sputtering of the surface of oxides may result in preferential removal of oxygen so that the metal/oxygen ratio at depth should be considered with caution [93].

Surface enrichment with one of the metals is normally found in mixed oxides. The surface of (Ru + Ti)O_2 is enriched with Ti [94,95] and that of (Ru + Ir)O_2 with Ir [30,53,96]. However, surface enrichment depends on the method of preparation [30,97]. Moreover, in some instances the phenomenon is indeed apparent. It has been found that thermal decomposition does not always proceed with 100% yield [98]. Thus RuO_2 + SnO_2 samples have been shown to contain excess RuO_2 even in the bulk if compared with the nominal composition. In some cases, however, surface enrichment is real, although a precise explanation cannot be provided. Possibly, a different kinetics of decomposition segregates one of the components to the surface. Nevertheless, the concentration profile in the bulk is constant [95,99]. With IrO_2 + RuO_2 the correspondence between the bulk composition and the nominal one has been confirmed by ICPES [100], which indicates that surface enrichment is real and cannot be attributed to different deposition yields.

Oxides with formally the same stoichiometric metal/oxygen ratio as the thermal ones can also be obtained electrolytically on the parent metals [101]. At constant potential or galvanostatically, only a thin film is usually obtained. Thick oxide layers can be grown by potential cycling for extended periods [102] or by complex periodic potential sequences [103]. In this way oxides with qualitatively the same voltammetric pattern as the thermal ones are obtained on Ru[104] and Ir [90]. By contrast, for anodically treated Co the structure of Co_3O_4 can be detected by x-ray diffraction only after some particular potentiodynamic treatments [105].

Despite the apparent similarity, the structure of electrolytic oxides is basically different from that of thermal oxides. The main difference is that anodic oxides are "hydrous," [106] whereas thermal oxides are "dry" [12]. Practically, there exists the same relationship as between hydroxides and oxides of the same metal. This is clear from x-ray photoelectron spectroscopy (XPS) analysis in the O(1s) binding-energy region [89,93,107–109]. In hydrous oxides, the high-binding-energy forms (i.e., —OH and —OH_2) prevail. However, as the thickness of the layer grows, the lattice oxygen form shows up [90], indicating that electrochemical treatments for extended periods lead to some form of aging or crystallization. It has been suggested [110] that the exterior (oxide–solution interface) of electrolytic oxides is more hydrous than the interior, while a thin compact film is present right on the metal surface. On the other hand, thermal oxides also show the presence of OH groups whose amount decreases with calcination temperature and depth [91]. These groups are probably either residues of the calcination process or result from the adsorption of water from the atmosphere [111].

Dry and hydrous oxides can be placed at the extreme ends of a series of compounds with the same M/O ratio but with variable proton content. If a hydrous oxide is calcined, the dry oxide is obtained. However, it may differ from other calcined samples of the same oxide simply because the precursor is different [112,113]. In principle, thermal oxides can be obtained by electrolytically growing a hydrous layer and then thermally treating it [114,115]. Alternatively, oxides can be grown directly on the parent metals by heating them in an oxygenated atmosphere [41]. However, properties turn out to be different and are usually unsatisfactory [104,116–118], since the typical morphology of the layers obtained by thermal decomposition, which is at the base of the characteristic performances, is not obtained.

Electrolytically grown Ti oxides implanted with Ru and Ir atoms have been proposed as model systems to investigate the effects of composition on the properties of DSAs. They have been characterized by XPS and photoacoustic and photocurrent spectroscopy [119–121] and have been used for kinetic studies [122,123]. However, comparison of the behavior of $RuO_2 + IrO_2$ mixed oxides with the oxides anodically grown on Ru + Ir alloys has shown that the properties are substantially different [124,125]. In particular, the XPS peak positions are not influenced by potential with thermal oxides, whereas this is not the case for anodic oxides [126]. More specifically, in the electrolytic oxides the metal ions are surrounded by an environment that differs only slightly from that of hydrated ions [110]. This is also the basis of their different behavior with respect to corrosion [25,125].

5.3 Interfacial Properties

5.3.1 The Oxide–Solution Interface

In a vacuum, oxide surfaces [127] expose two main sites: metal and oxygen ions (Fig. 5.4). Oxide surfaces are reactive toward water in the gas phase [128], especially because of the strong Lewis acidity of metal ions [129]. This property can be partly or totally quenched by high-temperature treatments, which may make an oxide surface hydrophobic [27,130]. Hydrophobicity is expected to be weaker at milder temperatures of calcination.

Low-temperature oxides are hydrophilic and react rapidly with ambient moisture. Water molecules are adsorbed on metal ions with transfer of one of the protons to a neighbouring oxygen atom. A saturated oxide surface is thus covered by a "carpet" of OH groups that mediate the interaction between the oxide surface and the envi-

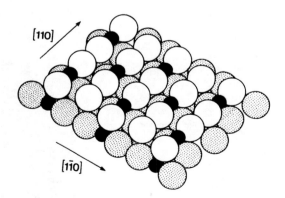

Figure 5.4. Model of the (001) face of the rutile structure [127]. (●) Metal ions. (○) Oxygen ions. Dotted circles, ions beneath the surface.

220 THE ELECTROCHEMISTRY OF NOVEL MATERIALS

ronment [27,131]. This mechanism of surface hydration is easily proved by detecting hydrogen with a nuclear analytical technique [26,111]. If just left in the ambient atmosphere, a profile of hydrogen can be detected inside an oxide layer, with a gradient that depends on the time of exposure. If placed in water, oxides become rapidly hydrated throughout the whole thickness (Fig. 5.5).

Differences are apparent between RuO_2 and IrO_2, on one hand, and Co_3O_4, on the other. The amount of water taken up by the dioxides is much higher than for the spinel (inset in Fig. 5.5). Nevertheless, the M/H ratio always remains high [111]. This indicates that water penetrates through pores and grain boundaries, thus suggesting a distinction between an internal and an external surface. The different behavior of RuO_2 and IrO_2 compared to Co_3O_4 is a consequence of the different

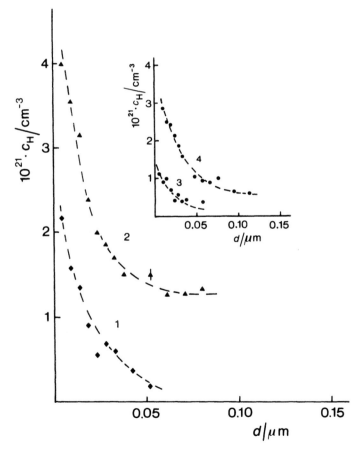

Figure 5.5. In-depth hydrogen concentration profile in an IrO_2 layer (d is the thickness) prepared at 400°C [26]. (1) After immersion in boiling water for 2 h; (2) after immersion in water at room temperature for 3 months. Inset: Hydrogen concentration profile in Co_3O_4 (3) [26] and RuO_2 (4) [111] prepared at 400°C and treated as (2).

TRANSITION METAL OXIDES: VERSATILE MATERIALS FOR ELECTROCATALYSIS

texture of the oxides. Co_3O_4 is less porous and has a lower tendency to become hydrous. Whereas the anodic oxide has the same formal composition as the thermal oxide for Ru and Ir, they differ for Co. The spinel structure of Co_3O_4, under prolonged contact with a solution or on gentle cathodic treatment, tends to decompose to simple Co oxides [132,133].

The OH groups on the surface of oxides in solution behave as weak acids or weak bases [27,135]. Different sites may be present on the same surface, depending on the coordination in the lattice [136]. If amphoteric behavior is assumed for simplicity, surface OH groups can give rise to acid dissociation:

$$—M—OH + OH^- \rightarrow —M—O^- + H_2O \qquad (5.2a)$$

or to basic behavior:

$$—M—OH + H^+ \rightarrow —M—OH_2^+ \qquad (5.2b)$$

Equations (5.2a) and (5.2b) illustrate the charging mechanism of oxide surfaces, which is different from the charging of metal surfaces. Figure 5.6 shows a sketch of the molecular structure of an oxide–solution interface where a few charged sites are present. Oxide surfaces behave as nonpolarizable interfaces whose electrical state is controlled by the solution pH and influenced by the ionic composition of the electrolyte. The linear response of the open-circuit potential of these oxides to the solution

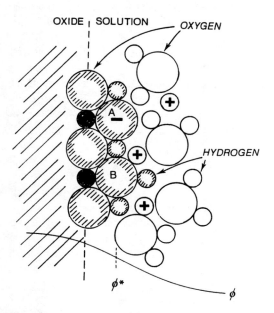

Figure 5.6. Model of an oxide–solution interface showing undissociated as well as ionized surface OH groups, and the electric potential drop arising from surface charging [27]. (●) Metal ions. Dashed area, solid phase. (⊕) Free hydrogen ions (acidic solution). (A) Ionized group according to Eq. (5.2a). (B) Ionized group according to Eq. (5.2b).

pH [137,138] has prompted the study of pH sensors based on these materials, particularly IrO_2 [139,140] and PdO_x [141]. The possibility of constructing pH-sensitive microelectrodes is attractive. However, the pH response of hydrous oxides may differ from that of the corresponding dry oxides because of the involvement of surface hydrolysis effects [142,143]. Moreover, the pH response depends on the pH range, for instance in the case of Co_3O_4, since the nature of the surface sites probably changes [144]. Thus a 120-mV slope of the E–pH relationship has been reported for Co_3O_4 in acidic solutions [145].

There exists a specific pH for each oxide at which the surface is electrically neutral. This is called the *point of zero charge* (pzc) and is usually determined by potentiometric titration or electrophoretic measurements. At pH < pzc the surface is positively charged, and at pH > pzc it is negatively charged. Specific ion adsorption shifts the pzc in a characteristic way, depending on the experimental technique for its determination [27].

There exists an extensive literature on the surface acid–base properties of oxides in solution (e.g., Refs. 135 and 146). The field has been exhaustively reviewed by the present author with particular reference to electrocatalytic oxides [27]. Only recent developments are surveyed in the following discussion.

The first determination of the pzc for RuO_2 was carried out in the author's laboratory in 1981 [147]. Investigations were later extended to IrO_2 [148], RuO_2 + IrO_2 [149], Co_3O_4 [150], and $NiCo_2O_4$ [131]. Table 5.5 summarizes the available data. The pzc of RuO_2 has been found to depend on the calcination temperature [151], but this is not the case for Co_3O_4 [150] and $NiCo_2O_4$ [131] (Fig. 5.7). This confirms the different structure of spinels and dioxides with respect to lattice hydration. RuO_2 is prepared from either $RuCl_3$ [151] and $Ru(OH)_4$, in turn obtained chemically by reduction of RuO_4^{2-} [152]. In both cases the same dependence of the pzc on the calcination temperature has been observed. This suggests that the pzc follows the evolution of the oxide from the *hydrous* to the *dry* structure as the calcination temperature is increased. These data do not support the view, put forward on the basis of conductivity and x-ray measurements [47], that Cl residues in the lattice are responsible for the dilatation of the unit cell and, therefore, for a decrease in conductivity. In fact, no Cl can be present in the lattice of RuO_2 from RuO_4^{2-}.

The importance of the pzc lies in its sensitivity to the surface structure and its independence of the oxide surface area. Measurement of the pzc can be a key approach to separate electronic and geometric effects. The present author has suggested [62,131] that a correlation between pzc and electrocatalytic activity must exist, since both properties depend on the same surface properties.

The dependence of the pzc on the surface chemical structure is readily understood if one considers that the "acidity" of a surface OH group is promoted by the strength of interaction between the oxygen atom and the surface metal ion. Although more elaborate theories have been developed [27,135], a straightforward correlation has been proposed by Butler and Ginley [153] between the pzc and the oxide "electronegativity" (x_{ox}) defined for an oxide M_aO_b as

$$x_{ox} = (x_M^a x_O^b)^{1/(a+b)} \tag{5.3}$$

Table 5.5 Point of Zero Charge of Electrocatalytic Oxides[a]

Oxide	pzc	References
RuO_2	4–7 (300–700°C)[b]	147, 151, 152, 501
	5.75 (402–420°C)	160, 176
	5.5 (layer)	502
	4.0	502
	3.88	339
(single crystal)	(potential of zero charge)	187
TiO_2	5.7	164
	5.7	167
PtO_x	5.0 (layer)	174
MnO_2	4.82	503
	3.7 (layer)	502
RuO_2 + (0–35%)IrO_2	4.5–5.5	149
Co_3O_4	7.5	150
	5.2	173
	5.0	502
	7.4 (layer)	502
Various perovskites	5.7–9.2	73
$Sr_{1-x}NbO_{3-\delta}$ ($0.05 \leq x \leq 0.3$)	5.60–6.25	339
NiO_x	8.2 (layer)	502
$NiCo_2O_4$	9.0	131
$Ru_{0.7}Rh_{0.3}O_2$	7.93	339
$Pb_2Ir_2O_{7-y}$	3.90	339
$Rb_2Ru_2O_{7-y}$	10.74	339
IrO_2	5.6 (layer)	502
	<1 (extrapolated)	148
	<3 (layer, ox.)[c]	504
	5.4 (layer, red.)[c]	504
	3.96	339
RuO_2 + 70% TiO_2	5.5 (layer)	502
IrO_2 + 70% TiO_2	4.7 (layer)	502

[a] Previous results are summarized in Ref. 27. Unless otherwise specified, data refer to suspended powder.
[b] Range of calcination temperature.
[c] The two values refer to the oxidized and the reduced surface.

where x_M and x_O are "absolute" electronegativities (Mulliken's scale). This correlation predicts that oxides with high electronegativity (the metal ion withdraws electrons from the OH group) are characterized by low (acid) pzc, whereas oxides with low electronegativity are characterized by high (alkaline) pzc (Fig. 5.8). In this context the very low pzc of IrO_2 (~ 1) [148] compared to RuO_2 (~ 6 at the same calcination temperature) [151] is readily understood in terms of the higher electronegativity of Ir compared to Ru [154].

The pzc's of electrocatalytic oxides fit the correlation proposed by Butler and Ginley qualitatively (Fig. 5.8). However, the effect of the calcination temperature cannot be easily accounted for in this correlation. In principle, nonstoichiometry should influence the pzc since the electronegativity of the oxide depends on the M/O ratio. However, a trend opposite to that experimentally observed is theoretically expected in the case of RuO_2 as the calcination temperature is decreased, since

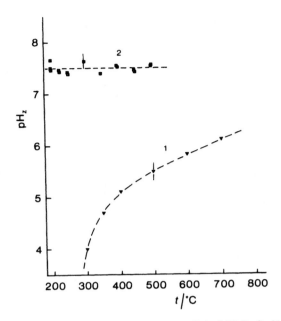

Figure 5.7. Dependence of the point of zero charge (pH_z) of (1) RuO_2 [151] and (2) Co_3O_4 [150] on the temperature of calcination.

an increase in the Ru(III) concentration should increase the pzc. The experimental results can be understood in terms of the explanation proposed for the series of MnO_2 forms. Dry β-MnO_2 shows the highest (most alkaline) pzc whereas electrolytic (hydrous) γ-MnO_2 has a more acid pzc [155].

Another important requirement in this field is the in situ estimation of the real surface area of oxides. Potentiometric titration provides surface charge–pH curves that can be converted into capacitance–pH curves by graphical differentiation if it is assumed that the oxide surface has a Nernstian response to pH. Capacitance curves for oxides invariably show a minimum at the pzc, which becomes deeper as the electrolyte concentration is lowered [156,157] (Fig. 5.9). This is similar to the minimum in the diffuse layer capacitance observed on mercury. Thus the present author has proposed [151] a Parsons–Zobel plot be used to determine the real surface area and the inner layer capacitance of oxide surfaces [158]. In practice, the Gouy–Chapman–Stern–Grahame model of the electrical double layer, which assumes that the solid–liquid interface behaves as two capacitors in series, is adopted:

$$\frac{1}{C} = \frac{1}{C_i} + \frac{1}{SC_d} \tag{5.4}$$

where C is the experimental capacitance, C_d is the calculated diffuse layer capacitance, and S is the oxide surface area. A plot of $1/C$ versus $1/C_d$ should result in a straight line whose slope is inversely proportional to S and whose intercept provides a value for C_i.

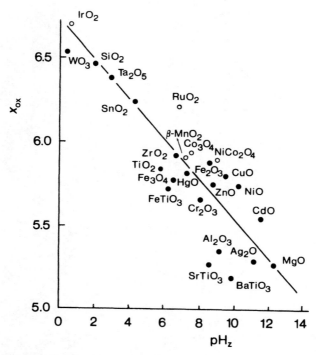

Figure 5.8. Correlation between the point of zero charge of oxides (pH_z) and their calculated electronegativity (x_{ox}). (○) Oxides investigated in the context of electrocatalysis. (Adapted from Ref. 153; reprinted by permission of the publisher, The Electrochemical Society, Inc.)

Equation (5.4) was used first with RuO_2 [151] and later with Co_3O_4 [150] and ZrO_2 [159]. Since few electrolyte concentrations are normally used, straight lines based on three data points have often been analyzed. This has led to some inconsistencies. In particular, Lyklema has suggested [160] that since disperse systems are used for these studies, the diffuse layer for spherical interfaces should be considered. Higher values of C_d (as calculated for spherical interfaces) produce higher apparent slopes for the same experimental C data, and Lyklema has concluded that the surface area is consistent with the BET [Brunauer, Emmet and Teller] value, and that the value of C_i is of the order of 300 μF cm^{-2}.

Different conclusions can be reached if the double layer is assumed planar. Then the value of C_i is of the order of 80 μF cm^{-2} [151], and the resulting active surface area is, as a rule, higher than the BET surface area. Physically, this is not unreasonable considering that BET measurements are carried out on dry powders, and part of the surface of single particles may be screened by the other particles [161]. This is not the case for powders whose particles are dispersed in solution.

This controversy has been scrutinized [162] by carrying out a simulation of the Parsons–Zobel plot for spherical as well as for planar double layers. It has been shown that plots based on too few data points can lead to an erratic linearization.

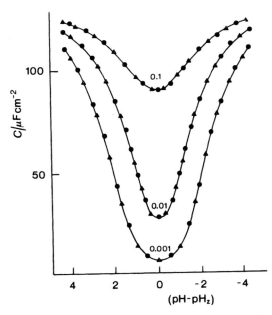

Figure 5.9. Capacitance (C) curves for an oxide–solution interface at different supporting 1:1 electrolyte concentrations (mol dm^{-3}) as a function of solution pH, calculated according to the site-binding model (●) and the simplified Grahame–Parsons model (▲), using the experimental parameters for Fe_3O_4. (After Ref. 156, with permission.)

Recent experimental verifications [159] have proved that the spherical double layer is not consistent with the available data (Fig. 5.10). Deviations from the planar double layer plot are possibly observed at very low electrolyte concentrations at which the thickness of the double layer becomes comparable with the size of the disperse particles.

The shape of the charge–pH curves around the pzc does not differ appreciably for different oxides and this has led [163] to the conclusion that the structure of the oxide–solution interface, unlike the metal–solution interface, is nonspecific. However, charge is an integral quantity and comparison should be carried out in terms of capacitance, a differential quantity. In this case, appreciable differences can be seen.

The pzc varies with the solution temperature [164,165], and its variation has been associated with the heat of proton adsorption [166,167]. Also, in this case the nature of the oxide appears to exert little effect [167], but conclusions should be drawn from more sensitive experimental quantities.

Determinations of the pzc are usually carried out with disperse systems when oxides are in a situation resembling an electrode at open circuit. Although the pzc cannot be directly relevant to the conditions of anodic load, it characterizes the oxide surface structure and composition as XPS does, with the advantage that pzc refers to the oxide–solution interface in situ. Potentiometric titration can be carried out also with polarized electrodes [27] (i.e., with oxide layers), provided efforts are

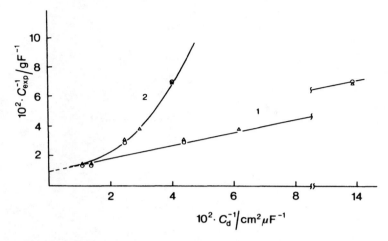

Figure 5.10. Reciprocal of the experimental capacitance (C) versus the reciprocal of the diffuse layer capacitance (C_d) calculated by means of the Gouy–Chapman theory for (1) planar interfaces and (2) spherical interfaces, for two different samples of ZrO_2 prepared at 200°C. (After Ref. 159, with permission.)

made to increase the working surface area. Experiments under polarization have shown [168–170] that the surface charge of oxides does not depend on the electrode potential (Fig. 5.11). This surprising result indicates that the state of charge of an oxide surface is governed by the pH of the solution rather than by the electrical polarization despite the high electronic conductivity. It is thought, for instance, that anodic polarization provokes withdrawal of electrons from surface sites compensated by release of protons from the surface into the solution:

$$—M^z—OH \xrightarrow{>E} —M^{z+1}—O^- + H^+ \quad (5.5)$$

In this manner, the total charge on the given surface site remains substantially the same.

The extent of ion adsorption can be determined by either potentiometric titration [160] (which gives thermodynamic surface excesses) or radiometric [171] and other analytical techniques [172] (which give total surface concentrations). In the absence of specific adsorption, the pzc can also be identified as the pH where the surface concentrations of cations and anions are equal [173,174]. Ion adsorption measurements confirm that the main mechanism of surface charging is related to the solution pH [170,175]. As a matter of fact, at constant potential, ion adsorption increases as the pH departs from the pzc (Fig. 5.12), whereas it remains constant as the potential is varied at constant pH (Fig. 5.11).

Specific adsorption can, in principle, be detected by potentiometric titration since the pzc is shifted to more alkaline values as anions are adsorbed (and vice versa for cation adsorption) [27]. Thus RuO_2 has been found to specifically adsorb Cl^- ions

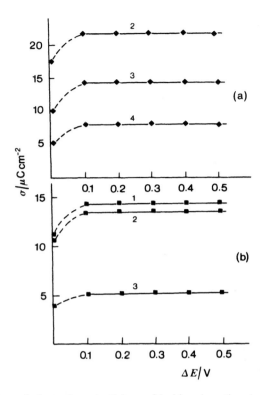

Figure 5.11. Effect of electrode potential on chloride adsorption (expressed as surface charge density, σ) on (a) Co_3O_4 and (b) $RuO_2 + TiO_2$ at different pH values: (1) 3.4; (2) 4.0; (3) 5.0; (4) 7.0. (Adapted from Ref. 169, with permission.)

[151,160] and methylviologen cations [176] the latter being used as sensitizers in the photocatalysis of water decomposition [177]. Although the double-layer model is assumed to resemble that of metals, on oxides the mechanism of specific adsorption involves surface complexation [178–180]. The presence of strongly bound water at the surface of oxides can induce a reversal in the ion adsorption sequence established with metals [135]. Small, hydrated ions can be more readily adsorbed than large, unhydrated ions. For example, F^- ions have been reported to adsorb on oxide anodes (more on IrO_2 than RuO_2) and inhibit O_2 evolution [181]. Accordingly, the possibility of hydrogen bonding with surface OH groups is the main factor inducing adsorption, so that "chemical" adsorption is more relevant than "physical" adsorption on oxides [27,174,182]. A site-binding model [183] combined with a "porous" double layer [184] has been developed [185] specifically to describe a more localized disposition of ions on oxide surfaces than on metals. Specific isotherms have been proposed for ion adsorption on oxides [27,135,186]. The effect of temperature and of solvent on ion adsorption has been investigated with PtO_x and RuO_2 [174].

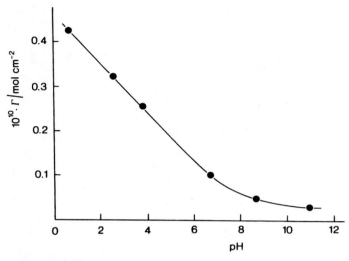

Figure 5.12. Dependence of sulfate adsorption (expressed as surface concentration, Γ) on solution pH for a $NiCo_2O_4$ electrode at 1.0 V (RHE). (After Ref. 175, with permission.)

In a manner similar to that used with metals, determination of the *potential of zero charge* of RuO_2 has been attempted by capacitance measurements as a function of potential using different single-crystal faces [187]. A minimum in the capacitance curve can indeed be observed, and this has been identified as the $E_{\sigma=0}$ of the given RuO_2 face. However, other minima and maxima are also visible in the capacitance curve. Comparison of these curves to voltammetric curves reveals that these minima and maxima appear at potentials of redox transitions. In particular, the deep minimum identified as the diffuse layer minimum corresponds to the small section of the voltammetric curve just before the region of hydrogen adsorption. Therefore the measured capacitances cannot be considered to represent double-layer charging but rather they are determined by redox surface transitions. On the other hand, if the pH is changed, the voltammetric curve of oxides shifts [61,188,189], and the capacitance curve is expected to do the same. In conclusion, the crystal face dependence of the apparent potential of zero charge determined from capacitance minima has little to do with the double-layer structure since it is the pH and not the polarization that governs the surface charging mechanism.

There is a paucity of data on the adsorption of organic compounds on electrocatalytic oxides. Data for a number of nonelectrochemical systems [27] show that adsorption on oxides is probably weakened by the surface layer of OH groups [190]. The hydrophilic heads of organic molecules are expected to be adsorbed on hydrous oxide surfaces. In fact, a correlation has been reported [191] to exist between $\Delta_{ad}G°$ and pzc, and an example is given in Fig. 5.13. Adsorption of neutral compounds on oxides thus takes place by a different process from that on metals [192].

The observed weak interaction between organic compounds and oxides [190] suggests that different intermediates may be formed in electroorganic reactions on

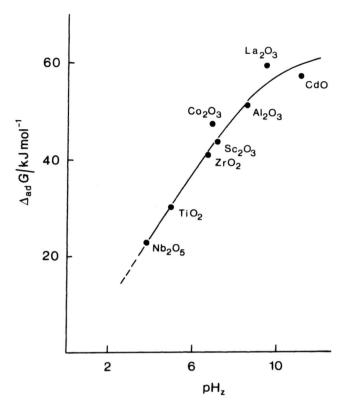

Figure 5.13. Dependence of the standard Gibbs energy of adsorption of salicylic acid on the point of zero charge of various oxides. (After Ref. 191, with permission.)

such electrodes, perhaps leading to more useful products or to higher selectivity. This has stimulated research in this direction (see later). Nevertheless, studies of adsorption mechanisms are very scanty. Adsorption of benzene has been studied at high potentials on RuO_2 [193]. Adsorption is higher in CH_3OH than in H_2O, quite the opposite of what is observed with metals. This suggests that hydrophobic interactions in the bulk of the solution are not primary factors. Adsorption has been found to be irreversible and to conform to a Freundlich isotherm. It has been pointed out that adsorption of benzene improves the corrosion resistance of RuO_2.

No adsorption of benzene has been observed at low potentials on $RuO_2 + TiO_2$ electrodes [194]. This is in line with the known hydrophobicity of the benzene molecule. It has been remarked that such a situation may be favorable in the case of benzene chlorination since no secondary processes can take place via adsorbed states. In addition adsorption of hexanol has also been studied on these electrodes as a function of potential [195]. Moreover, adsorption of ions from different solvents has been found to be proportional to the solvent donicity, which indicates the chemical nature of the interactions involved.

TRANSITION METAL OXIDES: VERSATILE MATERIALS FOR ELECTROCATALYSIS   231

Although based on a nonelectrochemical study, adsorption of organic substances on Fe_2O_3 [190] is interesting in this context because it has been conclusively observed that surfaces treated at high temperatures (and hence, more hydrophobic) normally adsorb more strongly than surfaces treated at low temperatures (and hence, hydrated).

5.3.2 *In Situ* Electrochemical Characterization

Voltammetric analysis offers the most sensitive in situ characterization of oxide electrodes. Voltammetric curves, recorded in either base or acid depending on the electrode material, are as a rule carried out in a potential range where no permanent modification of the surface takes place [e.g., between 0.4 and 1.4 V (RHE) with RuO_2 and IrO_2 in both acid and alkaline solutions; and between -0.1 and 0.5 V (SCE) in alkaline solutions with Co_3O_4 and $NiCo_2O_4$]. Some of the curves are rather featureless (Fig. 5.14b), indicating extended heterogeneity of the surface sites (e.g., RuO_2 and IrO_2 in acid) [53]. In other cases, sharp peaks are observed (Fig.

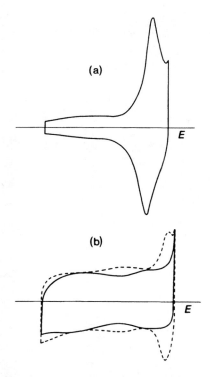

Figure 5.14. (a) Typical voltammetric curve at 20 mV s^{-1} of Co_3O_4 electrodes, calcined at 400°C, in alkaline solution. Potential range: -0.1–0.5 (SCE). (b) Voltammetric curve for RuO_2 electrodes, calcined at 400°C, in acid (———) and alkaline (-- --) solution. Potential range: 0.4–1.4 (RHE).

5.14a: Co_3O_4 in alkali) [61]. The peaks in such curves can be considered an electrochemical spectrum of the oxide surface. The position of the peak is indicative of the site's chemical nature, and the peak width indicates the heterogeneity of the site distribution. The area under the peaks is proportional to the number of sites oxidized or reduced [10]. Since peak resolution is often difficult, the voltammetric charge over the whole potential range (q^*) can be taken as a relative measure of the electrochemically active surface area [41,61,196]. This cannot be easily converted into an absolute value since the precise nature of the surface redox transitions is unknown. However, the value of q^* has proved very useful (often essential) when different electrodes or different preparation procedures are being compared.

The voltammetric charge has been found to decrease with the calcination temperature, as expected for surface area effects [42,54,197] (Fig. 5.15). q^* has been observed to differ for different precursors [31], different solvents [30,42], and different procedures. RuO_2 compact layers show much lower q^* values than cracked electrodes [54]. The voltammetric charge can in general be related to the following generalized redox reaction:

$$-MO_x(OH)_y + \delta H^+ + \delta e \leftrightarrow -MO_{x-\delta}(OH)_{y+\delta} \qquad (5.6)$$

According to reaction (5.6), protons are injected into the electrode cathodically and ejected anodically, especially in acid solutions [198–201]. In alkaline solution,

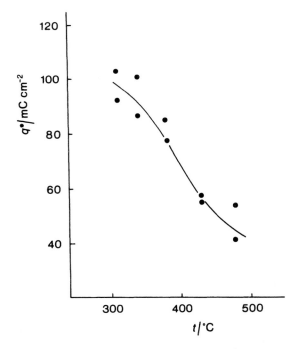

Figure 5.15. Dependence of voltammetric charge (q^*) of Ti/IrO_2 electrodes on the temperature of preparation. (From Ref. 197.)

exchange of OH^- [201] and intercalation reactions of small cations [202,203] can take place. Reaction (5.6) is the basis for the use of RuO_2 in double-layer capacitors (more appropriately termed *supercapacitors*) [204,205].

The mechanism of the voltammetric response of oxide electrodes, first suggested by the present author in his first paper on electrocatalytic oxides [10], has been confirmed recently. Reaction (5.6) produces a change in the local pH at oxide interfaces that has been directly detected [201,206]. Further, the correlation between q^* and the surface morphology has been proved by means of optical observations [207] and chemical adsorption measurements [48] (Fig. 5.16).

q^* has been observed to depend on the potential scan rate (v) [41]. The analysis of this phenomenological dependence has suggested [138,208,209] that the rate of charging and discharging is limited by proton diffusion. This has been proved also by potential step [210,211] and impedance analysis [212]. However, the evidence is that protons do not diffuse through the crystallites but around them along grain boundaries, pores, crevices, and cracks. Thus diffusion is in fact surface diffusion via a sort of Grotthus mechanism (particularly in acid solution). Extrapolation of q^* to $v = \infty$ and $v = 0$ has provided [41] a means to split the total charge into an external (outer) and an internal (inner) surface charge:

$$q_t^* = q_i^* + q_o^* \tag{5.7}$$

where $q_t^* \equiv q^*(v = 0)$ and $q_o^* = q^*(v = \infty)$.

Results [213] have shown that q_t^* and q_o^* are linearly interrelated and that the ratio is almost 2 for RuO_2 and IrO_2 (i.e., $q_i^* \approx q_o^*$, as shown in Fig. 5.17), whereas it is not far from 1 for Co_3O_4 and $NiCo_2O_4$ (i.e., $q_i^* \approx 0$). This is consistent with the analysis of other parameters, thus confirming that RuO_2 and IrO_2 are largely porous whereas spinels are not.

That protons are the diffusing species has been proved in more direct ways. Thus

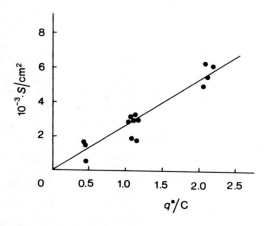

Figure 5.16. Linear correlation between the voltammetric charge (q^*) and the active surface area (S, as measured by zinc adsorption) of a series of RuO_2 electrodes. (After Ref. 48; reprinted by permission of the publisher, The Electrochemical Society, Inc.)

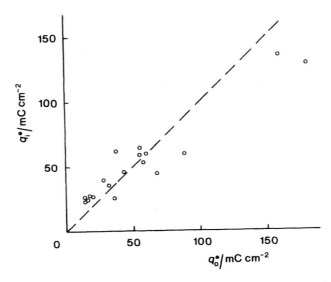

Figure 5.17. "Inner surface" charge (q_i^*) versus "outer surface" charge (q_o^*) for RuO_2 electrodes prepared at various temperatures. (From Ref. 213.)

tritium exchange experiments (in the absence of polarization) clearly show the slow establishment of an exchange equilibrium conforming to diffusional kinetics [26,138]. Again, RuO_2 and IrO_2 behave differently from Co_3O_4 for which only rapid surface exchange is observed. Thus where diffusion is operating, it is along preferential pathways not involving the bulk of crystallites. The dependence of the phenomenon on the temperature of calcination indicates that at low temperature more grain-boundary-related surface exists than at high temperature.

Voltammetric curves of fresh RuO_2 electrodes in aprotic solvents are flat and encompass a very small charge [15] (Fig. 5.18). This provides clear evidence that surface proton exchange is quenched and no charging of the surface can take place. The situation is qualitatively similar with IrO_2 layers formed anodically in aqueous solution [214]. However, in the latter case intercalation of small ions other than H^+ is possible. Freshly anodized electrodes placed in nonaqueous solvent still show high charging. If they are dried before use, the charge becomes much smaller.

If the total charge q^* is proportional to the active surface, the latter can also be estimated by plotting the voltammetric current at a given E as a function of the sweep rate if the electrode is cycled in a potential region where no peaks are present [158,215]. This has been claimed [73] to provide the "capacitance" of the oxide–solution interface ($C = dq/dE = di/dv$, where v is the sweep rate, $V s^{-1}$). However, this method cannot give absolute values either, since the precise "capacitance" of the oxide surface is unknown [27]. That this method gives the same results as q^* on a relative basis has been proved with Co_3O_4 [61].

The voltammetric charge can be taken, together with the voltammetric spectrum, as a "fingerprint" of an oxide surface. Repeated routinely with fresh electrodes and

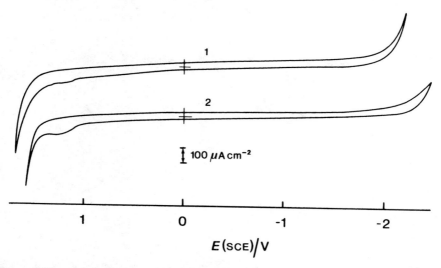

Figure 5.18. Voltammetric curves of RuO_2 prepared by spray pyrolysis, in (1) dimethylformamide and (2) dimethylsulfoxide. (After Ref. 15, reprinted by permission of the publisher, The Electrochemical Society, Inc.)

after some performances, the voltammetric curve together with q^* determinations can monitor in situ the state of the surface *in real time*. Thus q^* has been found to increase with prolonged cathodic treatment of RuO_2 [216] and IrO_2 [197]. The analysis of the data has shown [197] that hydrogen discharge promotes the wetting of the internal surface so that hydrophobic regions on fresh electrodes become hydrophilic. This implies also that a porous structure with extended grain boundaries exists. However, q^* can also increase as a consequence of roughening of the oxide surface, and it can decrease if mechanical wear (erosion) smooths down asperities on the electrode surface [86].

Voltammetric curves can be distorted by uncompensated ohmic drops. If a TiO_2 interlayer is present, the anodic and cathodic peaks of Co_3O_4 in alkali are pushed apart, whereas normally they are observed at the same potential. Thus the efficiency of electrode preparation and the evolution of the interlayer situation with time can be conveniently followed in situ [61].

Voltammetric curves are particularly useful with mixed oxides since they provide a clue to the in situ determination of surface composition. This has been shown for IrO_2 + RuO_2 [30,53,100] and Co_3O_4 + RuO_2 mixtures [26]. If typical features of the single components can be identified in the voltammetric curves, they can be used for a quantitative estimation of the surface atomic ratio. In alkali, RuO_2 shows a typical peak prior to O_2 evolution (attributable to an oxidation state of the Ru sublattice corresponding to ruthenate or perruthenate) which is absent with IrO_2, whereas IrO_2 shows a marked increase of the cathodic current with respect to RuO_2 in the low-potential range. This approach has given surface compositions in excellent agreement with XPS data (Fig. 5.19). Similarly, the surface of RuO_2 + Co_3O_4

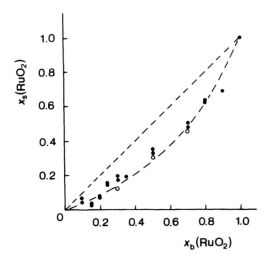

Figure 5.19. Surface (x_s) versus bulk (x_b) concentration of Ru in RuO_2 + IrO_2 layers calcined at 400°C, as determined by (○) XPS and (●) voltammetric curves [100].

electrodes turns out to be heavily enriched with Ru even at the lowest nominal RuO_2 content, which is the probable main reason for the large increase in activity exhibited by these electrodes [21,26], although the influence of RuO_2 on the conductivity of the mixed oxides has also been pointed out as an activity enhancement factor [217,218].

The position of the voltammetric peaks on the potential scale can be treated, very qualitatively, as the binding energy of XPS spectra. Chemical shifts can thus be observed. For instance (Fig. 5.20), no chemical shift is detected by XPS either as a function of composition with RuO_2 + IrO_2 electrodes [97,219] or as a function of potential with RuO_2 electrodes [126]. However, the anodic peak in alkali attributable to RuO_2 shifts with the composition of RuO_2 + IrO_2 electrodes [100]. This is possibly an indication of some interaction between the two components taking place, as suggested [97], via the ionic double layer in solution.

5.3.3 Single Crystals as Model Systems

Voltammetric curves of polycrystalline oxide layers can provide information on morphology, but only indirect insight can be gained into the atomic structure. The latter requires investigation of well-defined surfaces of single crystals. Oxide single crystals can be obtained by chemical vapor transport (CVT), although it is difficult to grow them to a sufficient size. RuO_2 and IrO_2 have been grown by CVT from the metal in an oxygen stream. The temperature of operation (>1200°C) is much higher than that for the preparation of polycrystalline materials. Details can be found elsewhere [41]. After some early attempts by this author [10], systematic work was possible with single crystals grown by Pollak and co-workers [220].

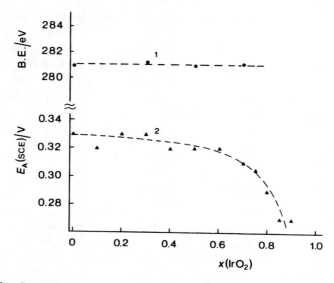

Figure 5.20. Dependence on composition of (1) the XPS line and (2) the typical voltammetric peak of Ru in $RuO_2 + IrO_2$ electrodes calcined at 400°C [100]. E_A is the potential of the most anodic peak in alkaline solution—cf. Fig. 14b.

Single crystals have been characterized by XPS [221,222], Raman spectroscopy [223] and voltammetric curves [224–226]. XPS combined with low-energy electron diffraction (LEED) has shown the precise atomic structure of different faces and has proved that faces that have lost some oxygen with use can be restored by annealing at 450°C in oxygen atmosphere. Using single crystals the mechanism of interaction between RuO_2 and Ti, leading to oxidation of the support, has been investigated [221].

Voltammetric curves have shown the effect of the crystal orientation on the adsorption of hydrogen, whose adsorbed amount correlates well with the number of metal atoms on the various faces [226]. Less informative are the voltammetric curves thus far obtained in acid solution for different faces of RuO_2, since they show only few small differences between different faces [224]. However, it has been shown by the present author [227] that voltammetric curves become featureless when extended too far into the cathodic or the anodic potential region. Thus a sound voltammetric analysis remains to be performed for single-crystal faces of RuO_2.

The redox transition occurring before oxygen evolution on the (100) face of IrO_2 has been studied [226]. The surface process is irreversible, and this was attributed to the incomplete (or inefficient) hydration of the surface of the crystals obtained at high temperature.

Other studies of single-crystal faces concern perovskites. The XPS analysis of the (111) face of $La_{0.175}WO_3$ shows [228] several forms of O, indicating hydration of the surface, while La has been found to dissolve in acid and W in alkali. It is important to point out that the hydration of single-crystal faces is promoted by the surface attack occurring during potential cycling. The reactivity of the (100) plane of the perovskite $La_{0.7}Pb_{0.3}MnO_3$ has been studied [229] for O_2 reduction and evolution in alkaline solution. The activity is relatively high but the mechanism differs from Sr-doped perovskites, suggesting the importance of Pb sites.

Single crystals are of little interest for industrial applications but they provide invaluable references as model systems. Unfortunately, this field is still underdeveloped for oxide electrodes, mostly because of the difficulty of growing crystals.

5.4 Electrocatalytic Properties

5.4.1 The Cl_2 Evolution Reaction

Since oxide electrodes were born as substitutes for graphite anodes in chlor-alkali cells, the Cl_2 evolution and reduction reactions were investigated first. DSAs show lower overpotentials and lower Tafel slopes for Cl_2 evolution than graphite [5], but they cannot be said to be exceptionally active [230] (Fig. 5.21). Their activity is not better than that of the precious metals themselves; rather it is slightly lower [231]. In fact, the main factor enhancing the activity of DSAs is their large surface area. However, decisive advantages include their chemical and electrochemical stability (see later), and the wettability, which enables bubbles of optimum size to be released [5]. Recent investigations are summarized in Table 5.6.

Chlorine evolution takes place on RuO_2-based electrodes with low Tafel slopes up to very high current densities [23,232]. The observed average Tafel slope is 40

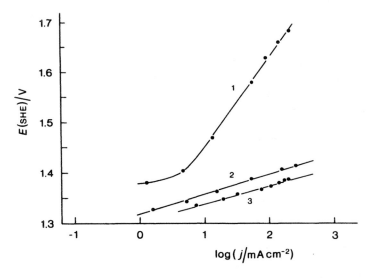

Figure 5.21. Anodic polarization curves in 1 mol dm^{-3} NaCl solution, 30°C, pH 2. (1) Graphite; (2) RuO_2; (3) Pt + 4% Ir. (After Ref. 229, with permission.)

TRANSITION METAL OXIDES: VERSATILE MATERIALS FOR ELECTROCATALYSIS 239

Table 5.6 Recent Studies of Cl_2 Evolution on Oxide Electrodes[a]

Electrode	Subject	References
$RuO_2 + TiO_2$	Acidification in pores	261–263
	Mechanism[b]	240
	Optimization of preparation	394
	Effect of porosity[b]	260
	Inhibition by anion absorption	513
	Effect of acidity[b]	249
	Effect of Cl_2 supersaturation[b]	234, 236, 237
	Selectivity, effect pH, and NaCl concentration	255, 256
	Mechanism[b]	241, 246
	Kinetics, impedance	588
	Mechanism[b]	243, 281
	Effect of composition on selectivity	587
	Absorbed intermediates	250
	Thickness effects	615
	Effect of pores on Tafel line	239
	Effect of sulfates on selectivity	434
	Mechanism	244, 507
	Exchange current and stoichiometric number	231
	Kinetics	238
	Effect of pH[b]	248
	Potential decay[b]	237
	Mechanism, intermediates	251
(doped with Sn)	Impedance study	505
(same)	Teflon-bonded, porous	578
(doped with CeO_2)	Potential decay, mass transfer	506
RuO_2	Effect of pores on Tafel line	239
	Kinetics, impedance	242
	Potential decay	237
	Redox catalysis by Ce(IV)	591
	Effect of preparation	54
	From molten electrolyte	16
(Nb/)	Thickness effects	617
(single crystal)	Mechanism	271, 272
(same)	Effect of pH	247
$RuO_2 + TiO_2 + SnO_2 + PdO_x$	Optimization	269
$SnO_2 + SbO_x + PdO_x$	Effect of Pd content	508, 514
$RuO_2 + HfO_2$	Chlorate process	277
$SnO_2 + SbO_x + RuO_2$	Effect of Ru content	265, 414, 609
	Chlorate process	277
$Bi_2Ru_2O_7$	Kinetics	509
$MnO_2 + PdO_x$	Effect of support and composition	510
$MnO_2 + RuO_2$	Same	510
$MnO_2 + PdO_x + RuO_2$	Same	510
PdO_x	Effect of preparation	511
IrO_2	Effect of pH	247
$IrO_2 + PtO_x(+PdO_x)$	Efficiency, overpotential	569

(Continued)

Table 5.6 (*Continued*)

Electrode	Subject	References
Co_3O_4	Mechanism, effect of preparation	86
	Effect of pH	247
	Effect of nonstoichiometry	83
	Effect of support	512
(doped with TiO_2)	Spectroscopic study	496
(doped with PdO_x)	Effect of composition	88
$NiCo_2O_4$	Mechanism, effect of pH[b]	245, 247
Ti_4O_7, Ti_5O_9 (bulk)	Comparison of activity	399

[a] Previous studies up to 1980 are summarized in Refs. 23 and 232. Unless otherwise specified, the support is Ti.
[b] The study of Cl_2 ionization is also included.

mV, but 30 mV has also been claimed in the low overpotential range. It has been suggested that the change in Tafel slope is related to the transition from barrierless to normal second electron transfer as the rate-determining step [233]. Lower slopes are commonly reported and are inevitably observed at the operative temperature of chlor-alkali cells (~90°C). The reasons for such slopes have been investigated particularly by Losev [234–238] and Gorodetskii [239] and attributed to slow diffusion of the product (Cl_2) away from the electrode with oversaturation of the surface layer and a shift of the reversible potential to more anodic values. This produces an initial 30-mV Tafel slope followed by a section of very low polarizability (the current steps up without increase in overpotential) before the normal polarization behavior is established. If taken point by point, the experimental data may simulate a single Tafel line of exceedingly low slope. Under similar circumstances, the 40-mV Tafel slope can be observed only in the presence of efficient mass transport.

The preceding behavior indicates both the very large activity of DSA-type electrodes and the high reversibility of the Cl_2 reaction which on RuO_2-based electrodes can be investigated under equilibrium conditions and different Cl_2 partial pressures so as to collect all relevant kinetic parameters, including the stoichiometric number [240–244].

The reaction also proceeds with a 40-mV Tafel slope on other oxides, especially Co_3O_4 [86] and $NiCo_2O_4$ [245], whose activity makes these materials promising candidates to replace, at least partly, the more expensive RuO_2 component of DSAs.

The various mechanisms thus far proposed have been reviewed recently [23,246]. A striking observation is that the reaction order with respect to Cl^- is consistently 1 with oxides electrodes [247] (although 2 is sometimes claimed), and this rules out the classic mechanisms that predict a 40-mV slope. Such an observation led Krishtalik and co-workers [233] to propose a mechanism in which Cl atoms produced in the primary discharge are further oxidized to $(Cl^+)_{ad}$ without involvement of another Cl^-. The nature of the intermediate has been established more precisely after the determination of the pH dependence of Cl_2 evolution [248]. In fact, if the activity of Cl^- is kept constant and the pH varied at constant potential, a variation of the reaction rate with pH is observed [86,247,249] and a reaction order of -1 with respect to H^+ (Fig. 5.22) is obtained.

These observations indicate that a H^+-producing step must precede the rds in the Cl_2 evolution reaction. A possible mechanism, generally valid for all oxides, is [23,246]:

$$—M—OH \leftrightarrow —M—O + H^+ + e \qquad (5.8a)$$

$$—M—O + Cl^- \rightarrow —M{<}^{Cl}_{O} + e \qquad (5.8b)$$

$$—M{<}^{Cl}_{O} + H^+ + Cl^- \rightarrow —M—OH + Cl_2 \qquad (5.8c)$$

with (5.8b) being the rds. Thus $(Cl^+)_{ad}$ may be identified with the anion of the hypochlorous acid or the acid itself [233,250,251]. An alternative mechanism, based on the acid–base properties of oxides, suggests that the concentration of active sites for Cl_2 evolution is pH dependent [247]. In both cases the mechanism includes step (5.8a), which also occurs for O_2 evolution (see later discussion), and stresses how these two reactions are interrelated at oxide electrodes. The kinetics of Cl_2 evolution turns out to be pH dependent because O_2 and Cl_2 formation take place in overlapping potential ranges, the first step being common to both mechanisms. The range of overlap may depend on the electrode material. Thus, the behavior of

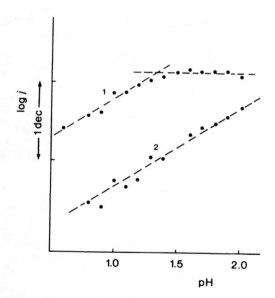

Figure 5.22. Effect of pH on the rate of Cl_2 evolution (j) at a constant potential in the Tafel region in 5 mol dm^{-3} NaCl.[247] (1) RuO_2; (2) Co_3O_4.

IrO$_2$ differs from that of RuO$_2$ in the same pH range because of the higher overpotential of O$_2$ evolution on IrO$_2$ electrodes [247].

The competition between Cl$_2$ and O$_2$ evolution is the main reason for the decrease with increasing pH of the current efficiency (selectivity) for Cl$_2$ evolution. Although efficiency is always higher than 99% under operational conditions [252], it decreases as the pH is increased [253] because the oxygen evolution reaction shifts to lower potentials (Fig. 5.23). Also, the current efficiency increases with current density since O$_2$ evolution in the presence of NaCl takes place with a higher Tafel slope than Cl$_2$ evolution [254]. Thus as the current is increased the partial current for Cl$_2$ evolution increases more rapidly than that for O$_2$ evolution [255].

The intrinsic activity of DSA electrodes for O$_2$ evolution as a side reaction is strikingly evident from the evolution of O$_2$ taking place at open circuit in Cl$_2$-saturated NaCl solutions. A short-circuit cell is established in which O$_2$ is evolved at the expenses of Cl$_2$ reduction. Extensive investigations of this particular situation have been carried out [256–259].

Chlorine evolution is a *facile* reaction and takes place with the same kinetic parameters on electrodes prepared at different calcination temperatures. Practically no dependence of the Tafel slope on the voltammetric charge is observed [23,86]. Rather, an excessive increase in surface area with formation of porosity is detrimental for two reasons: First, pores are not accessible to diffusing Cl$^-$ ions and become rapidly depleted of reactants [260]. Pores thus become the place where O$_2$ evolution occurs with an autocatalytic mechanism, since O$_2$ liberation produces acidity, which in turn depresses Cl$_2$ evolution [261–263]. The O$_2$ content of Cl$_2$ may therefore increase with porous electrodes, and the acidity formed in the pores may be dele-

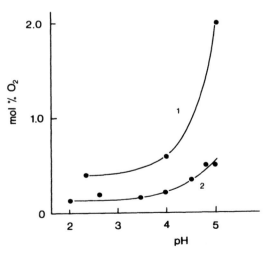

Figure 5.23. Oxygen content in the Cl$_2$ evolved from RuO$_2$ + TiO$_2$ electrodes in 5 mol dm^{-3} NaCl at different pH and different current densities. (1) 0.2 A cm^{-2}; (2) 1 A cm^{-2}. (After Ref. 409, with permission.)

terious for electrode stability. Second, with porous electrodes the real current density is relatively low and the competition between Cl_2 and O_2 evolution shifts in favor of O_2.

The difference in porosity between RuO_2 and Co_3O_4 is also evident from the results of Cl_2 evolution. Although the current is almost charge independent with "cracked" RuO_2 [54], it correlates well with q^* in the case of Co_3O_4, even at relatively low NaCl concentration (Fig. 5.24) [86]. This indicates the absence of important porosity effects with Co_3O_4.

Chlorine evolution on $TiO_2 + RuO_2$ electrodes shows that the activity is almost constant down to about 20% RuO_2; then it drops [232,264]. However, if the activity is plotted against the surface composition, a linear dependence is found, suggesting that only the Ru sites are active. For $SnO_2 + RuO_2$ mixed oxides [232,265] the minimum useful Ru content for high activity is reported to be about 10%, whereas only 1–2% RuO_2 is claimed to be sufficient to increase the activity of Co_3O_4 by an order of magnitude [21,266,267]. However, the way mixed oxides are prepared is a crucial factor. Other studies with the $Co_3O_4 + RuO_2$ system have shown [23] that the maximum activity occurs around 15 mol% RuO_2, with an increase in activity of two orders of magnitude compared to pure Co_3O_4. In the case of $PdO_x + Co_3O_4$ anodes the maximum activity has been observed at about 6 mol% PdO_x [88].

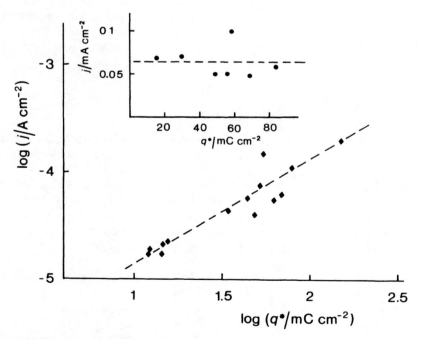

Figure 5.24. Rate of Cl_2 evolution (j) at a constant potential in the Tafel region as a function of the voltammetric charge (q^*) of Co_3O_4 electrodes; 5 mol dm^{-3} NaCl solution [86]. Inset: Data for "cracked" RuO_2 electrodes [54].

With the aim of improving the performances, coatings comprising RuO_2, TiO_2, SnO_2, and PdO_x have been proposed for both chlorate cells and seawater electrolysis [268]. Besides selectivity for Cl_2 evolution [269] and chlorate formation [270], the durability has also been optimized [268] by adopting an appropriate composition.

Chlorine evolution has also been studied on single-crystal faces of RuO_2 by means of cyclic voltammetry [271]. It has been suggested that the probable mechanism is the one proposed by Krishtalik with Cl and Cl^+ adsorbing on different sites. However, the apparent marked irreversibility of the process is at variance with the results for polycrystalline samples. In another study [247] a Tafel slope of 40 mV and a reaction order of 1 for Cl^- were observed but no pH dependence could be detected. Nevertheless, the process turned out to be reversible and measurements around the reversible potential could be carried out. The absence of a pH dependence can be explained by the high overpotential for oxygen evolution so that step (5.8a) is improbable. Alternatively, step (5.8a) might not take place because the surface is hydrophobic (dry) owing to the high temperature of preparation of single crystals.

A concentration gradient of Cl species has been found within RuO_2 layers after extensive Cl_2 evolution by XPS analysis [95,272]. This suggests that the surface is steadily covered by Cl species. However, these species cannot be the active ones in Cl_2 evolution since the low Tafel slope indicates that the reaction proceeds at low coverage of intermediates. It is more probable that these species are trapped in subsurface regions, in pores and in grain boundaries. This is consistent with the claim that Cl species originally present in RuO_2 after thermal preparation cannot be the reason why these electrodes are active in chlorine evolution.

5.4.2 The O_2 Evolution Reaction

Oxygen evolution is a "demanding" reaction and is therefore highly sensitive to the nature and the structure of electrocatalysts [25]. Oxygen evolution is an important industrial reaction, being the reaction at the anodes of water electrolyzers, in metal electrowinning processes, in cathodic protection, in electroorganic reductions [13], and so on. Oxygen evolution is also very important in the light-assisted decomposition of water from oxide suspensions [273–275]. Although materials with a low overpotential for O_2 evolution are generally necessary, in other cases materials of high overpotential are sought, as in ozone production [276] or to increase the selectivity of the Cl_2 process [277,278].

In general, materials active for chlorine evolution are also active for O_2 evolution [28] but their anodic stability is a factor. Thus RuO_2 and IrO_2 are good in acid solutions, whereas spinels and perovskites are recommended for alkaline solutions.

A summary of recent investigations is given in Table 5.7. A large spectrum of Tafel slopes is observed with oxide electrodes [24,25,232]. They range from about 30 mV for some forms of RuO_2 [279] to about 120 mV with PtO_x [280]. The Tafel slope most commonly observed is 40 mV, in both acid and alkali for RuO_2 [279,281,282] (although in alkali the overpotential is slightly higher) [283] and in

TRANSITION METAL OXIDES: VERSATILE MATERIALS FOR ELECTROCATALYSIS

Table 5.7 Recent Studies of O_2 Evolution on Oxide Electrodes[a]

Electrode	Subject	pH[b]	Reference
Ti/RuO_2	Photochemical system (on TiO_2)	a	273
	On-line mass spectrometry	b	295
	XPS study	a	89, 126
	Fractional reaction order	a	281, 301, 304
	Kinetics in perchlorate synthesis	a	282
	Effect of calcination	a	279, 515
	Kinetics as a side reaction[c]	a	258, 299
	Effect of NaCl concentration[c]	a	255
	Single crystals	a	227
	SPE water electrolysis	a	618
Ti/RuO_2 + TiO_2	Kinetics in perchlorate synthesis	a	282
	Kinetics as a side reaction[c]	a	258, 259, 299
	Impedance	a, b	516, 517
	Different compositions	b	592
	Effect of NaCl concentration[c]	a	255
	Inhibition by phosphates	a	513
	Effect of electrolyte anion	a	591
Ti/RuO_2 + ZrO_2	Effect of composition and T	b	418
Ti/RuO_2 + Ta_2O_5	Different compositions	a	479
Ti/SnO_2	Effect of dopants	a	482
PdO_x	Photochemical system (on TiO_2)	a	275
Ti/RuO_2 + IrO_2	Effect of composition	a	30, 97
Ti/TiO_2 + PtO_x	Effect of composition, spectroscopic study	b	485, 518
Ti/RuO_2 + SnO_2	Effect of composition	a	416, 609
Ti/RuO_2 + SnO_2 + TiO_2	Effect of substitution of Sn for Ti	a	390
Ti/IrO_2 + SnO_2	Effect of composition	a	612
Ti/IrO_2	Effect of TiO_2 interlayer	a	519
	Effect of heat treatment	a	520
	Photochemical system	a	274
	Anodes for Cr electrodeposition	a	449
Pt/PbO_2	Ozone formation, effect of anions	a	276
$Pb_{2+x}Ir_{2-z}O_{7-y}$ (pellets)	Activity comparison	a, b	29, 326, 339
$Pb_{2+x}Ru_{2-x}O_{7-y}$ (Teflon)	Effect of composition, effect of T	a, b	29, 326, 521, 526
Pt/$Bi_2Ru_2O_7$ (or pellets)	Effect of electrochemical pretreatment	b	526, 529
$LaNiO_3$ (pellets)	Activity comparison	b	500, 522, 525
Ti/RuO_2 + MnO_2	Effect of composition, interlayer	a	523
$A_{1-x}A_x'Bo_3$ (pellets)	Mechanism, XPS, impedance, effect of substituents	b	73, 500, 525
(A = La, Gd, Nd, Sr; A' = Sr, Ce, Th, K, Ca; B = Co, Fe, Mn, Cr, V)			
Ni/$La_{0.8}Sr_{0.2}Ni_{0.2}Co_{0.8}O_3$	Mechanism, activity	b	524
Ni/$Cu_{1.4}Mn_{1.6}O_4$	Activity comparison	b	345
Ti/SnO_2(+SbO_x) + MO_x (M = Rh, Ru, Pd, Ir, Pt)	XPS, effect of composition, activation energy	b	514
$La_{1-x}Sr_xFeO_{3-y}$ (pellets)	Mechanism, activity	b	346, 527

(Continued)

Table 5.7. (*Continued*)

Electrode	Subject	pH[b]	Reference
$La_{1-x}Sr_xCoO_3$ (pellets)	Mechanism, impedance	b	346, 528, 581, 621
Ti_4O_7 (bulk)	Comparison of activity	a, b	398, 399
$La_{0.5}Ba_{0.5}CoO_3$	Potential decay, capacitance	b	530, 531, 582
	Reaction order, mechanism	b	619
	Theoretical impedance	b	620
SrFe(or Co)$O_{3-\delta}$ (pellets)	Effect of composition	b	342
	Mechanism from MO theory		383
$SrFe_{0.9}M_{0.1}O_3$ (M = Mn, Ni, Co, Ti)	Mechanism	b	74
Ni/NiO_x	Mechanism, effect of Li doping	b	288
Ti/Co_3O_4	Kinetics electrocatalysis	b	286, 287, 419
	Photochemical system	a	274
	Effect of support	b	284
	Effect of preparation	b	60
(Teflon) Li-doped	Kinetic parameters and mechanism	b	297, 302
	Redox catalysis	b	533, 534
Ti or Ni/	Intermediates	b	535, 573
Ni/	Isotopic study	b	297, 536
Ni/	In situ activation	b	537
Au/(Teflon)	Mechanism	b	289
C/	Current efficiency	b	543
Ti or Ni/Co_3O_4+RuO_2	Effect of composition	a	218
$Ti/NiCo_2O_4$	Kinetics, electrocatalysis	b	286, 287, 419
Ni/(Teflon bonded)	Effect of preparation and T, mechanism	b	38, 293, 538–540, 585
Ni/	Influence of preparation	b	35
	Photochemical system	a	274
Ni/	Isotopic study	b	296, 297
Au/(Teflon)	Mechanism	b	289
Co_3O_4 + TiO_2 (bulk)	Effect of composition	a	420
Ti/Rh_2O_3	Photochemical system	a	274
$Ti/Fe_xCo_{3-x}O_4$	Effect of composition	a, b	491
Steel/$La_{1-x}Ba_xMnO_3$	Effect of composition and solution temperature	b	532
$La_{0.7}Pb_{0.3}MnO_3$(100) (bulk)	Mechanism	b	229
MnO_2 (Co-doped)	Identification of active sites	b	531
Ni/(Teflon)	Activity comparison, effect of T	b	532
Ti/	Anodes for Cr electrodeposition	a	446
$Ti/ZrCo_2O_4$	Activity comparison	a	419
$Au/MnCo_2O_4$ (Teflon)	Mechanism	b	289
Several	Activity comparison	a, b	421, 590
	Activation energy (exp. and calc.)	a, b	377, 378
	Factors of electrocatalysis		291
	Mechanism, reaction order		281
	Electrocatalysis at hydrous oxides		110
	Spinels: effect of trivalent ion nature		567

[a] Previous studies have been reviewed in Refs. 25 and 232. See also Ref. 24. Studies on electrolytic oxides are not included.
[b] a = acid; b = basic.
[c] During Cl_2 evolution.

alkali for Co_3O_4 [284] and IrO_2 [285] (although higher Tafel slopes [60,286], close to 60 mV and decreasing with increasing OH^- concentration [287], are also reported for Co_3O_4), and 60 mV is observed with IrO_2 in acid solution [30,97]. In some cases a higher Tafel slope (~120 mV) is observed at higher overpotentials [30,288–290]. PtO_x is as inactive as PbO_2, on which a Tafel slope of 120 mV is also observed [276,291].

Solely on the basis of the Tafel slope (provided it is not determined by film resistance) [292], a generalized mechanism has been proposed that encompasses all oxide electrode materials [25,281,289]. Thus the first step is suggested to be the primary discharge of water molecules (in acid) or of OH^- (in alkali):

$$-M-OH + H_2O \rightarrow -M\begin{subarray}{l}OH^*\\OH\end{subarray} + H^+ + e \quad (5.9a)$$

Reaction (5.9a) represents the oxidation of a surface-active site. The intermediate OH* is then converted, before further oxidation, into a more suitable species:

$$-M\begin{subarray}{l}OH^*\\OH\end{subarray} \rightarrow -M\begin{subarray}{l}OH\\OH\end{subarray} \quad (5.9b)$$

The surface complex is further oxidized:

$$-M\begin{subarray}{l}OH\\OH\end{subarray} \rightarrow -M\begin{subarray}{l}O\\OH\end{subarray} + H^+ + e \quad (5.9c)$$

and O_2 is finally liberated by decomposition of such an unstable compound:

$$2-M\begin{subarray}{l}O\\OH\end{subarray} \rightarrow 2-M-OH + O_2 \quad (5.9d)$$

According to this mechanism, step (5.9c) is rate determining on oxides exhibiting a 40-mV Tafel slope, step (5.9b) on oxides with a 60-mV Tafel slope (IrO_2 in acid), and step (5.9a) on oxides with a 120-mV Tafel slope. The main factor determining the Tafel slope on the various oxides is the strength of interaction between the oxide surface and oxygenated intermediates.

The preceding mechanism includes steps in which the oxide surface takes active part in the reaction. On the whole, the mechanism entails the formation of a higher oxide and its subsequent decomposition [126,232,279,293,294]:

$$MO_x + H_2O \rightarrow MO_{x+1} + 2H^+ + 2e \quad (5.10a)$$

$$MO_{x+1} \rightarrow MO_x + \tfrac{1}{2}O_2 \quad (5.10b)$$

Since the oxygen liberated is not necessarily from the water involved in the primary discharge, the oxide lattice can take part in this mechanism. This has been proved experimentally for RuO_2 and $NiCo_2O_4$ by preparing an isotopically marked oxide layer and detecting the isotopic nature of the O_2 evolved by means of an on-line mass spectrometer [295–298].

In some cases a lower Tafel slope (30 mV) is reported for RuO_2 [279,283,299]. It could be due to the slow removal of the product from a very active surface, but this seems unlikely in the case of O_2 evolution since the exchange current is intrinsically rather small (poor reversibility) [25]. Another possibility is that a barrierless step (5.9c) precedes a normal step (5.9c), but in this case the 30-mV slope should be followed by a 40-mV slope [300]. More probably a change in mechanism is involved. On surfaces with a high concentration of active sites, recombination of the intermediate formed in the primary discharge is possible [279]:

$$2\text{—M}\begin{matrix}\text{OH}\\\text{OH}\end{matrix} \rightarrow 2\text{—M—OH} + H_2O + \tfrac{1}{2}O_2 \tag{5.9c'}$$

This step, though rate determining, must be faster than step (5.9c); otherwise the reaction would follow the alternative route.

For low Tafel slopes (40 or 60 mV), the reaction order with respect to protons predicted by mechanism (5.9) is -1 in acid solutions. However, the observed reaction orders with respect to H^+ are normally fractional and generally close to -1.5 [301]. In alkaline solutions, fractional reaction orders are also reported (~ 1.5) [281], although higher integral values have sometimes been claimed [287,302]. Close scrutiny of the original data shows that, in these cases also, the precise value is not integral. The reported value is presumably taken as integral on account of possible experimental scatter.

Fractional reaction orders can be obtained in electrochemical kinetics if double-layer effects are operative [303]. With oxides, double-layer effects can arise, even at constant ionic strength, if the pH of the solution is changed. This is a consequence of the response of the oxide surface to the solution pH [27]. Thus the electric potential at the reacting site is a linear function of the pH:

$$\phi^* \propto -2.303\left(\frac{RT}{F}\right)\text{pH} \tag{5.11}$$

If ϕ^* is introduced into the kinetic equation describing the evolution of O_2, fractional reaction orders can arise, depending on the value of the Tafel slope [281,304]. Thus Tafel slopes of 40 mV result in a reaction order of $(1 + \alpha)$ (α = transfer coefficient), whereas no fractional reaction order is expected for a 60 mV slope. Experiments agree with predictions: For RuO_2 the reaction order is -1.5 (with respect to H^+) in acid solution [301,304] and 1.5 (with respect to OH^-) in alkaline solution [281], whereas it is -1 for IrO_2 in acid solution [30]. A fractional reaction order can be regarded as a manifestation of the acid–base properties of an oxide surface.

Unlike Cl_2 evolution (a "facile" reaction), O_2 evolution is sensitive to the surface structure of electrocatalysts. If the Tafel slope of variously prepared RuO_2 electrodes is plotted as a function of q^*, the surface charge, a smooth correlation is observed [279] (Fig. 5.25). At high values of q^* (cracked, poorly crystalline electrodes) the Tafel slope is close to 30 mV (as with the anodically grown oxide) [41,295,305], whereas it goes through 40 mV and approaches 60 mV as q^* decreases (compact, crystalline electrodes). This trend has been confirmed [227] by the data obtained with a (110) face of RuO_2 (the most common crystal face in polycrystalline material). Oxygen evolution in acid solution proceeds with a 60-mV

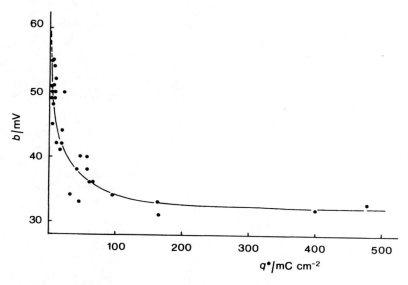

Figure 5.25. Dependence of the Tafel slope (b) for O_2 evolution on the voltammetric charge (q^*) of RuO_2 electrodes in acid solution [279].

slope and a reaction order of -1. On the whole, the results with RuO_2 indicate that the sites at the edge of small crystallites (coordinatively unsaturated) are much more active than those on the plane faces of crystallites.

The effect of the preparation procedure on the electrocatalytic activity is readily detected if the current at constant potential is plotted as a function of q^* in a log–log plot. If a linear plot of unit slope is obtained, it indicates that only geometric (surface-area) effects are operative. If deviations are observed, other effects must be invoked. For IrO_2, different electrocatalytic activities have been observed depending on the solvent used for the precursor [28,131] (Fig. 5.26).

Effects due to the presence of pores are not normally observed for O_2 evolution on oxides [62,279], unlike the case of Cl_2 liberation [54,86]. This indicates that the whole wetted surface is available for O_2 evolution, and further emphasizes the importance of pore acidification observed during Cl_2 evolution [261–263]. The absence of pore occlusion by the evolved O_2 gas indicates that a fast mechanism of transport along the pore walls (and grain boundaries) must be operative, in line with the preceding conclusions from voltammetric analyses.

It has been suggested [110] that anodic hydrous oxides must possess an open structure since they respond with the whole mass to electrical solicitations. The behavior of dry (thermal) oxides indicates that the same open structure is probably retained in pores and grain boundaries only. This indicates the existence of a link between the two extreme representatives of the large spectrum of oxide structures: electrolytic oxides (wholly hydrous) and single crystals (wholly dry).

The activity of oxide electrodes can be controlled by using mixed oxides. The

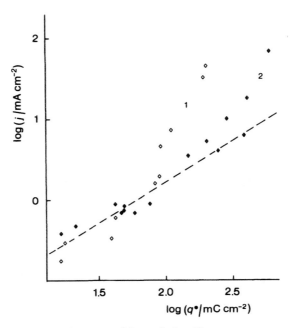

Figure 5.26. Dependence of the rate of O_2 evolution (j) at a constant potential in the Tafel region in acid solution on the voltammetric charge (q^*) of IrO_2 electrodes prepared from the same precursor dissolved in different solvent [28,131]. (1) Water; (2) isopropanol. (---) Straight line of unit slope.

final effect can be dramatically dependent on the procedure of preparation and therefore on the degree of intimate mixing achieved in the mixture. In the case of $RuO_2 + IrO_2$ the Tafel slope varies smoothly from 40 mV for pure RuO_2 to 60 mV for pure IrO_2. However, the way the Tafel slope varies for intermediate compositions is indicative of the degree of interaction between the two components [30,53,100] (Fig. 5.27). If the precursors are dissolved in isopropanol, the results are consistent with a simple additive behavior of the components; if water is used as the solvent for the precursors, the results indicate a certain degree of intimate mixing at an atomic level. In both cases, surface enrichment with Ir is observed. If the same mixed oxides are obtained by reactive sputtering [97], the Tafel slope changes with composition but with a different pattern. At the same time no surface enrichment with Ir is observed. Thus the analysis of the dependence of the electrocatalytic properties on composition can provide a clue to the understanding of the electronic structure of the mixed oxide system.

5.4.3 The H_2 Evolution Reaction

Although electrocatalytic oxides are thermodynamically predicted to be unstable in the potential range of H_2 evolution [306,307], recent patents have claimed the possible use, in particular of RuO_2 [308–310] and spinels [311], as cathodes. The

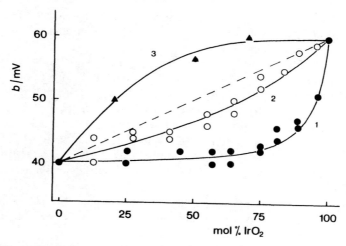

Figure 5.27. Dependence of the Tafel slope (b) for O_2 evolution in acid solution on the composition of RuO_2 + IrO_2 electrodes prepared at 400°C from the same precursors dissolved in different solvents [30]. (1) Isopropanol; (2) water. Line 3 shows the results obtained with electrodes prepared by reactive sputtering [97]. (-----) Linear variation.

first fundamental study of H_2 evolution on RuO_2 was published in 1975 by the present author [312]. However, no detailed kinetic analysis was carried out.

Detailed investigations on the mechanism of H_2 evolution have been carried out recently (Table 5.8), particularly on RuO_2, IrO_2, and $SrNbO_3$ in acid solutions. For RuO_2 and IrO_2, the cathodic treatment has been shown to promote the wetting of the electrode surface so that the charge q^* increases after extensive H_2 evolution [198,216]. Voltammetric curves have shown that after short periods of H_2 evolution the electrochemical spectrum does not differ significantly from that of a fresh electrode [216]. This points to an intrinsic stability of the oxide surface, although long-term tests are lacking. In any case no peaks specifically attributable to bulk

Table 5.8 Investigations of H_2 Evolution at Oxide Electrodes

Electrode	References
RuO_2 (electrocatalysis)	313, 314, 418, 544, 545, 547
(redox catalysis)	386, 546
(photocatalysis)	549
(photoelectrocatalysis)	548
RuO_2 single crystals	226
$Sr_xNbO_{3-\delta}$	315
IrO_2	197, 313
TiO_2	550, 551
NiO_x	430
Ti_4O_7, Ti_5O_9	398, 399
$Ni+LaNiO_3$	601

oxide reduction are visible. This may be related to the high electronic conductivity of RuO_2 and IrO_2, so that no field-assisted penetration of protons occurs, unlike the case of oxide semiconductors.

Hydrogen evolution proceeds with a single Tafel slope of 40 mV in acid solution on both IrO_2 and RuO_2 [313,314]. For $SrNbO_3$ the Tafel slope has been found to depend on composition [315]. Hystereses are observed with RuO_2 and (less) with IrO_2 [313]. For both the hysteresis tends to disappear with use, which indicates that a fully hydrated layer is finally reached. Overpotentials for both materials are comparable in acid solution, but the overpotential is higher for IrO_2 at high current density in alkaline solution where two Tafel slopes are observed [316] (the deviations from linearity in the case of RuO_2 are entirely attributable to uncompensated ohmic drop) (Fig. 5.28).

Although immediately after H_2 discharge the potential of oxides is close to the H_2 reversible potential, this is not maintained by the electrodes even in the presence of bubbling H_2, which indicates that the surface is never similar to that of a metal and that the oxide lattice is never destroyed.

The maintenance of the features of the oxide surface is confirmed by the observation of a fractional reaction order, as in O_2 evolution. Thus in acid solution the reaction order has been observed [197,313] to be about 1.5 for both IrO_2 and RuO_2. On the basis of the preceding kinetic parameters the following mechanism has been proposed for acidic solutions:

$$\text{—M—OH} + H^+ + e \rightarrow \text{—M—OH}_2 \tag{5.12a}$$

$$\text{—M—OH}_2 + e \rightarrow \text{—M—H} + OH^- \tag{5.12b}$$

$$\text{—M—H} + H_2O \rightarrow \text{—M—OH} + H_2 \tag{5.12c}$$

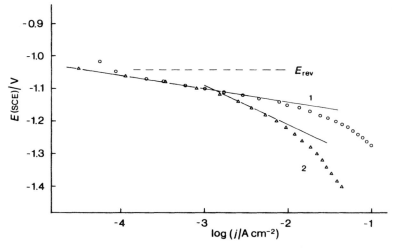

Figure 5.28. Polarization curves for H_2 evolution in alkaline solution [316]. (1) RuO_2; (2) IrO_2.

The rate-determining step is thought to be step (5.12b), while step (5.12c) is expected to be fast; otherwise permeation of hydrogen into the bulk could be induced. The mechanism is substantially the same in alkaline solution [316].

The mechanism of H_2 evolution does not appear to depend on the temperature of calcination of the electrocatalyst. Further, results show that the whole surface is active even at the highest surface area [197]. This indicates that the reactant (protons) is rapidly supplied into the pores by a mechanism other than simple diffusion in a liquid phase (cf. oxygen evolution).

One interesting aspect of the use of oxides as cathodes for H_2 evolution is their resistance to poisoning. The main cause of cathode deactivation in industrial electrolyzers is the electrodeposition of metallic impurities having high overpotential for hydrogen evolution [317]. It has been shown [314] that RuO_2 is not deactivated at short deposition times, whereas Pt is immediately deactivated. This has been explained in terms of a different mechanism for metal deposition. The oxide surface does not possess sites for strong adsorption of metallic species because of the presence of the "carpet" of OH groups [27]. Even if deposition takes place, it is kinetically retarded. A prerequisite for underpotential deposition of metals is that the support–adatom interaction be stronger than the adatom–adatom interaction [318]. This leads to the deposition of a complete monolayer before the deposition of a second layer commences. Otherwise, clusters are formed that tend to grow in the direction perpendicular to the surface rather than to spread over the surface.

Investigations of metal deposition on electrocatalytic oxides are quite scanty. Voltammetric studies of Ag deposition on RuO_2 single crystals have shown [319] that no underpotential discharge takes place. This is because $Ag-RuO_2$ interactions are weak compared to Ag–Ag interactions. Voltammetric and XPS studies of the deposition of a number of metals on thermally formed RuO_2 have led to the same conclusions [314]. The difference observed between RuO_2 and Ru metal (on which underpotential deposition does take place) confirms that the oxide maintains its hydroxylated surface structure even at cathodic potentials. This prevents oxide deactivation due to the deposition of metallic impurities.

5.4.4 The O_2 Reduction Reaction

Oxygen reduction is of importance for fuel cells and metal–air batteries [320], as well as for the so-called air cathodes in industrial electrolyzers [321]. Platinum has long been the typical catalyst for O_2 reduction [322,323]—hence the search for less expensive materials, more resistant to impurities and organic contaminants in solution.

An essential requisite for good electrocatalysis of O_2 reduction is the O–O bond breakage, which leads directly to water as the final product (four-electron reduction) [324,325]. Moreover, in many cases bifunctional catalysts are needed, since recharging requires O_2 evolution to occur on materials where O_2 reduction must also occur [326,327]. Finally, in most applications of O_2 reduction, dispersed catalysts must be used to increase the air–solid–solution contact area [328,329]. Therefore electrocatalysis of O_2 reduction is based on concepts that are not always exchangeable with those used for more typical industrial reactions.

In 1970 an article in *Nature* [330] had already drawn the attention of the scientific community to a class of "inexpensive materials" based on $LaCoO_3$, a perovskite. These oxides are also good electrocatalysts for O_2 evolution in base (cf. Sec. 4.2). They are widely known in the field of catalysis, especially as oxidation catalysts, for their properties of easily incorporating and releasing oxygen, whose mobility in the lattice is appreciable [331–333]. Besides perovskites [72], spinels [134], essentially colbatites (MCo_2O_4), have been investigated as possible electrocatalysts in alkaline solution [289,334,335], especially when supported on carbon in a dispersed form [328,329]. A further class of active catalysts includes pyrochlores of Ru or Ir (ruthenates or iridates of Pb, often partially substituted with Bi, and nonstoichiometric) [29,326,336,337]. RuO_2 has not been found to be particularly active [104,338,339]. The reduction of O_2 on this oxide proceeds to H_2O_2 at relatively high overpotentials [340]. A summary of recent studies is given in Table 5.9.

In general, O_2 reduction is difficult because the reaction is irreversible and, therefore, the overpotential is relatively high. However, in the case of $LaCoO_3$ doped with Sr it has been claimed [341] that reversibility can be attained, and the reversible potential of the O_2 reaction can be observed even though the exchange current remains low. Chemisorption of O_2 leading to the 4e reduction path is claimed not only for perovskites [342–346] but also for spinels [289,329,347] and pyrochlores [339]. However, the activity changes with overpotential [134,289,336]. As a rule, the steady-state polarization diagram consists of a section with a low Tafel slope (40–60 mV) followed by a section with a higher Tafel slope (120 mV). In the high-overpotential region O_2 reduction proceeds to H_2O_2. Nevertheless, most of these materials are also efficient H_2O_2 decomposers [348,349].

Table 5.9 Recent Studies of O_2 Reduction at Oxide Electrodes[a]

Electrode	References
$NiCo_2O_4$	284, 328, 347, 353, 538
Co_3O_4	289
$MnCo_2O_4$	289
$SrFeO_{3-\delta}$	342
$La_{1-x}Sr_xFeO_{3-\delta}$	346
$SrCoO_{3-\delta}$	342
$La_{1-x}M_xCoO_{3-\delta}$ (M = Ca, Sr)	341, 346, 610
MnO_2	542
La_xMnO_3	345, 542, 593
Ag_xMnO_3	542
$Cu_{1.4}Mn_{1.6}O_4$	346
$Pb_{2+x}Ir_{2-x}O_{7-y}$	29, 326, 339
$Pb_{2+x}Ru_{2-x}O_{7-y}$	336, 337, 339, 521, 526
$La_xM_{1-x}MnO_3$ (M = Ca, Ba, Sr)	345
$La_{0.7}Pb_{0.3}MnO_3$ (100)	229
$Bi_{2-2x}Pb_{2x}Ru_2O_{7-x}$	336, 521
$Fe_xCo_{3-x}O_4$	491
Mechanism at oxides	353

[a] Previous work is summarized in Refs. 41, 72, and 134.

By changing the metal in the octahedral sites of cobaltites it has been possible to show [134,289] that the low-overpotential reduction takes place on the M(II) ion whereas the 120-mV Tafel slope refers to the Co(III) ions in tetrahedral sites. The O_2–oxide interaction depends on surface hydration and it has been suggested [336,339] that the strongly bound water structure near the electrode surface makes oxides intrinsically less active than metallic electrodes. It has been suggested that the presence of lattice defects promotes activation: in the case of ferrites [342], the activity has been observed to be proportional to the fraction of lattice vacancies (Fig. 5.29). Accordingly, anodic prepolarization tends to decrease the activity of spinels because a more-ordered (stoichiometric) structure is promoted [350,351].

Studies have been carried out on Ni–Co spinels of variable composition. It has been found [352] that the most active electrocatalyst is the spinel $NiCo_2O_4$. As the heat treatment exceeds 400°C, the activity drops because the spinel decomposes to the simple oxides [334]. NiO is in fact poorly active for O_2 reduction; in any case, the reaction proceeds to H_2O_2 because O_2 is not suitably adsorbed on NiO.

The mechanism of O_2 reduction has been discussed in detail by Tarasevich [289,324] and Trunov [353]. It is interesting that fractional reaction orders have been observed, and this has been interpreted [134,289] by invoking participation of the surface OH groups, thus emphasizing once more that surface hydration is important in determining the specific behavior of oxide electrodes [26,62,131,339].

5.4.5 Electroorganic Reactions

Catalytic oxides have also aroused interest in the field of organic electrochemistry (anodically grown oxides, such as PbO_2 and NiO_x have been used for a long time) [354–356]. The main reason for this interest is the prospect that the different nature

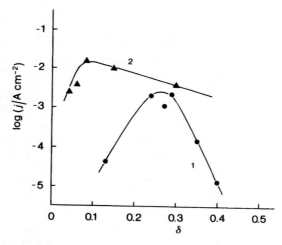

Figure 5.29. Dependence of the rate of O_2 reduction (j) at a constant potential in the Tafel region on the deviation from stoichiometry (δ) in (1) $SrFeO_{3-\delta}$ and (2) $SrCoO_{3-\delta}$. (After Ref. 342, with permission.)

of the interactions of reactants with the oxide surface might lead to different intermediates and novel reaction routes. Because of the absence of strong chemisorption with dehydrogenation, degradation of the molecule to CO_2 is less probable. On the other hand, because of the surface interaction of oxides with large organic molecules and their low overpotential for O_2 evolution, DSA-type electrodes have been proposed [357] for the destructive oxidation of contaminants in industrial effluents. Table 5.10 summarizes a number of recent studies. Although work with nonthermal oxides (in particular, NiO_x) is not reported here, a few papers dealing with methanol oxidation are included, although these do not discuss preparative reactions.

Two different mechanisms can be distinguished for organic reactions. On the one hand, direct oxidation or reduction of the molecule can take place, especially at high potentials [358,359]. It should be noted that the range of useful potentials can be very wide in organic solvents. On the other hand, because of the numerous redox transitions taking place on the surface of oxides of transition metals between H_2 and O_2 evolution in aqueous solution, indirect electrochemical oxidation can occur via

Table 5.10 Summary of Recent Electroorganic Studies with Oxide Electrodes[a]

Electrode	Support	Reaction	Year	Ref.
$NiCo_2O_4$	Ti, Ni	Oxidation of primary and secondary alcohols and of primary amines	1991	363
TiO_2	Ti	Reduction of o-nitrophenol	1991	553
		Reduction of 5-aminosalicylic acid	1991	598
SnO_2	Ti	Oxidation of organics	1991	482
		Oxidation of benzoic acid	1991	439
$Ni_xCo_{3-x}O_4$	Ti, Ni	Ethanol oxidation	1990	362
Co_3O_4	Ti, Ni	Ethanol oxidation	1990	362
$ZnCo_2O_4$	Ti, Ni	Ethanol oxidation	1990	362
RuO_2	C	Oxidation of methanol	1990	366
RuO_2	C	Oxidation of methanol	1990	364
RuO_2	Ti	Oxidation of benzyl alcohol	1990	557
RuO_2	Ti, Ta	Oxidation of formaldehyde	1989	369
PtO_x	Ti	Oxidation of ethanol	1988	373
RuO_2	Ti	Oxidation of ethanol	1988	371
RuO_2	Pt (mod.)	Oxidation of benzyl alcohol	1988	372
$Pb_{2+x}Ru_{2-x}O_{7-y}$	Pt	Oxidation of olefins	1988	526
$TiO_2 + Cr_2O_3$	Ti	Oxidation of isopropanol	1987	360
RuO_2	Ti	Oxidation of 4,4'-dichlorobiphenyl	1986	555
RuO_2	Ti	Oxidation of sulfur compounds	1986	554
RuO_2 (110)	—	Redox behavior of proteins	1985	556
RuO_2, IrO_2	Ti, Ta	Redox behavior of proteins	1984	626
$RuO_2 + 70\%TiO_2$	Ti	Oxidation of propylene glycol	1984	370
$RuO_2 + 70\%TiO_2$	Ti	Anodic Brown–Walker condensation	1983	358
RuO_2–modif.TiO_2	(powder)	Redox catalysis of phenol	1982	614
$RuO_2 + 50\%TiO_2$	Ti	Propylene chlorination	1980	625
$RuO_2 + 70\%TiO_2$	Ti	Oxidation of alcohols	1979	357
RuO_2	Ti	Oxidation of methanol	1979	365

[a] Previous studies are summarized in Refs. 355 and 356.

surface complexes at the oxide surface. The latter kind of route has been particularly investigated by Beck [360], who has proposed a Ti/(Cr$_2$O$_3$ + TiO$_2$) electrode whose activity is based on the anodic oxidation of surface Cr(III) to Cr(VI) and subsequent chemical oxidation of the organic reactant in solution by the surface Cr(VI) species. Beck has shown that oxidation of isopropanol proceeds to acetone, prior to oxygen evolution, with a 120-mV Tafel slope. However, a drawback of this electrode is the oxidative dissolution of Cr even in the absence of isopropanol, although a suitable structural arrangement of the active layer on the electrode can enhance its stability [361].

Interesting results have been reported recently for the oxidation of ethanol [362] (as an example) and of a number of other alcohols and primary and secondary amines [363] on Ti/NiCo$_2$O$_4$ electrodes. The organic reaction is clearly catalyzed by the oxidation of the electrode surface prior to O$_2$ evolution, with probable formation of Ni and Co in higher oxidation states (Fig. 5.30). The current efficiency was determined to be very high, and the selectivity is especially interesting. Practically, 100% acetic acid is formed from ethanol. Although more experimental results are needed, these electrodes do not show the dissolution problems of Cr$_2$O$_3$. However, the precise mechanism and the nature of the active sites have still to be clarified.

RuO$_2$ has not been found very active for direct CH$_3$OH oxidation [364]. Earlier experiments showed [312,365] activity in the potential region of Ru redox surface transitions. Therefore an indirect mechanism is also likely in this case [366]. The oxidation proceeds with a high Tafel slope [365]. Nevertheless, the search for suitable oxides as promoters of methanol oxidation is continuing [367,368] in view

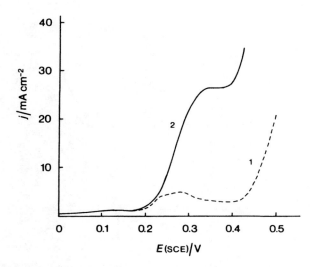

Figure 5.30. Voltammetric curves at 1 mV s^{-1} of NiCo$_2$O$_4$ electrodes in 2.5 mol dm^{-3} KOH in the (1) absence and (2) presence of 0.1 mol dm^{-3} CH$_3$OH. (After Ref. 362, with permission.)

of the higher resistance of these materials to surface poisoning. Higher alcohols have been found [365] to be oxidized to CO_2. However, the mechanism seems to depend on the potential range. Thus RuO_2 is more active than IrO_2 and Rh_2O_3 for the oxidation of HCOH to formate in alkali solution in the potential range 0.75–1.25 V(RHE). At higher potentials the reaction proceeds to CO_3^{2-}, that is, with oxidative degradation of the molecule (Fig. 5.31). A correlation with the activity for O_2 evolution has been pointed out [369].

The oxidation of propylene glycol on RuO_2 + TiO_2 electrodes gives acetic acid and CO_2 with a Tafel slope of 150 mV without any preadsorption steps [370]. Ethanol oxidation on Ti/RuO_2 electrodes has been reported to proceed with a Tafel slope of 160 mV [371]. RuO_2 particles have been incorporated into polymer-modified Pt electrodes and the system has been used for the oxidation of benzyl alcohol to benzaldehyde [372]. Also, Ti/PtO_x electrodes have been tested for ethanol oxidation [373]. In other cases no activity has been found for sorbose oxidation on Co_3O_4, MnO_2, or RuO_2, the current going entirely to O_2 [356].

Oxide electrodes offer, therefore, an alternative to more traditional materials for organic electrosynthesis. Apart from the direct oxidation at high electrode potentials, which produces less-controlled conditions, the indirect oxidation (or reduction) via oxidized (or reduced) surface active sites seems particularly worth exploring. The selectivity obtained is especially encouraging. This is a problem in organic electrosyntheses in which the first step is often a direct electron transfer to form an intermediate radical whose final fate escapes the control of the electrode.

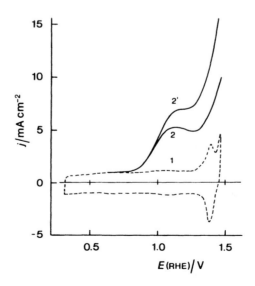

Figure 5.31. Voltammetric curves at 10 mV s^{-1} on RuO_2 electrodes in 1 mol dm^{-3} NaOH in the (1) absence and (2, 2') presence of HCOH (0.05 mol dm^{-3}). (2) Static electrode; (2') 900 rpm. (After Ref. 369, with permission of the publisher, The Electrochemical Society, Inc.)

5.5 Factors of Electrocatalysis

Geometric and electronic factors converge to determine the activity of electrocatalysts [232]. From a fundamental point of view, electrocatalytic data must be purged of geometric effects in order to single out electronic effects. The technological point of view is quite different since geometric factors are included under the general concept of electrocatalysis. Whether an electrode works better because of its extended surface area rather than because of its electronic properties is to some extent irrelevant from a purely industrial point of view. An electrode is evaluated on the basis of the balance between advantages and disadvantages, the latter including cost, stability, and pollution control.

However, from a fundamental point of view, if one wants to be able to optimize existing electrodes or design new ones, it is necessary to understand why one material works better than another, or why the activity depends on the preparation procedure. Several attempts have thus been made to develop a unifying theory for O_2 as well as Cl_2 evolution. The previous work reviewed by the present author elsewhere [25,232,374] will not be discussed in detail here.

Two basic aspects emerge from the analysis of the behavior of oxide electrodes: first, the importance of surface hydration in determining the details of the mechanisms of reaction, and second, the consistent correspondence between the potential regions where reactant oxidation and reduction occurs and the regions where redox surface transitions take place. In a general way it can be stated that the reactions occurring on oxide electrodes are catalyzed by the chemical interaction with surface-active sites which act as *electron-transfer relays*. This concept was first proposed by Tseung and Jasem [375], who have suggested that O_2 evolution accompanies surface redox transitions so that a material is more active the closer the potential of the surface redox transition is to the reversible potential for the O_2 reaction. This approach is in principle also valid for O_2 reduction, but in this case predictions may be affected by the difficulty of dissociating O_2 on oxide surfaces. In support of his proposal Tseung has shown a correlation between the thermodynamic redox potentials for surface transitions and the observed potentials of incipient O_2 evolution [291].

The same concept has been expressed by Krishtalik in a slightly different way [281]. He has proposed that one of the steps involved in oxygen (and chlorine) evolution is the oxidation of a surface-active site, which incorporates the idea of the redox potential put forward by Tseung. Alternatively, a more chemical view should take into consideration the strength of interaction between the surface and the intermediates. In this context Tamura and co-workers [376,377] have correlated the experimental activation energy for O_2 and Cl_2 evolution on different oxides with the calculated energy of activation by envisaging the formation of a surface intermediate as the modification of a surface complex. They have thus calculated, for both O_2 and Cl_2 evolution, the crystal-field-stabilization-energy variation as an intermediate is formed on the surface. Excellent linear correlations have been reported.

Along the same lines, this writer [374] has noted that, in many catalytic systems involving oxides, correlations have been observed between the catalytic activity and the metal–oxygen bond strength on the surface [378]. In practice, formation of an adsorbed oxygen intermediate is tantamount to oxidizing the active surface site from a lower to a higher valency state. Thus the relevant parameter to use in correlating the activity for O_2 evolution with the nature of the oxide electrode has been proposed [374] to be $\Delta_t H°(l \rightarrow h)$, that is, the enthalpy of transition from a lower to a higher oxide of the given metal. The same concepts have proved successful in the analysis of the experimental data for isotopic oxygen exchange from the gas phase [232].

The outcome of a similar approach is a volcano-shaped curve (Fig. 5.32), which is familiar in many fields of catalysis [379,380] and electrocatalysis [381,382]. Although the results are influenced by uncorrected geometric effects and are strictly representative only of stoichiometric oxides, such a correlation provides some rationalization of the available data, which enables the possible position of uninvestigated materials, as well as the effect of nonstoichiometry, and the effect of mixing with other oxides to be predicted.

Other approaches [24] have considered such factors as electron work functions (which is a more difficult concept to use with oxides), and exchange integrals (which is a quantum chemical way of calculating strengths of interaction). More recent approaches [383] have gone further in this quantistic direction by discussing the

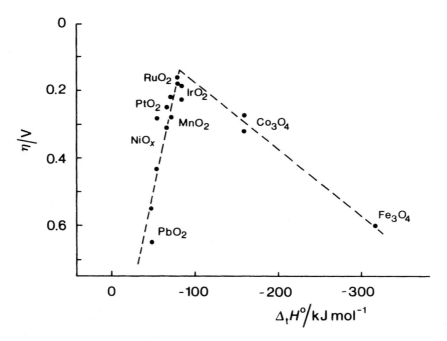

Figure 5.32. Activity (expressed as overpotential η at 0.1 mA cm^{-2}) for O_2 evolution on various oxide electrodes as a function of the enthalpy of transition from a lower to a higher oxide [232,374].

mechanism of O_2 evolution on $SrFeO_3$ within the MO theory, an approach already developed in the field of catalysis for perovskite-type oxides [384].

The same approach based on surface redox and structural transitions can be applied to Cl_2 evolution as well [385], in view of the fairly linear correlation between the overpotential for O_2 evolution and the overpotential for Cl_2 evolution at a given current density [28]. For both reactions a high overpotential is observed at the oxide surface if the adsorption strength of the intermediate is too weak for the metal ion to be oxidized to a higher valency state. For instance, this is the case for PbO_2, whose surface is oxidatively saturated before oxygen evolution commences. In such a case O_2 evolution starts only as the redox potential for the OH^-/OH couple is approached [291].

If the preceding concepts are applied to organic reactions, the activity for oxidation is expected to correlate with that for O_2 evolution since strong oxide–organic interactions are unlikely [369]. However, some stereochemical and structural correspondence between the active site and the organic molecule is probably required, and this would explain the higher activity of some oxides compared to others [362].

In the case of H_2 evolution the amount of data is still insufficient to develop a general theory. However, the creation of specific sites is required before H_2 evolution can start [191]. This involves prereduction of the surface, as has been clearly proved for the photocatalysis of H_2 evolution by RuO_2 particles dispersed in solution [386].

The mechanism of O_2 reduction does not reveal any straightforward connection with redox transitions at active sites, although this is qualitatively predicted by Tseung's model [375]. Nevertheless, the oxidation state of surface metal ions is seen to be important (e.g., for cobaltites) [134,289]. No specific theory has been developed thus far, although general views have been put forward [339]. For instance, the activity of a series of perovskites for O_2 reduction has been explained in terms of σ^* band filling. It has been suggested that a partly filled σ^* band in the oxide electrocatalyst is necessary for it to be active in O_2 reduction [72,387].

Theories of O_2 reduction are applicable to single classes of electrocatalysts such as spinels and perovskites, but a general approach encompassing all oxides, as achieved for O_2 and Cl_2 evolution, does not appear to have been developed. This is probably related to the nature of the steps in the mechanism of O_2 reduction. It is to be noted that although correlations have been attempted for O_2 reduction on metals [388], a general electrocatalytic theory (as for H_2 evolution) [389] is still lacking.

Electrocatalysis at mixed oxides can be understood in terms of electronic effects if intimate atomic interaction is assumed. The role of additives in improving the Cl_2 selectivity of DSAs is based on the idea that an oxide without a redox transition in the oxygen evolution region should retard this latter reaction when it is mixed with an active electrocatalyst for Cl_2 evolution. The reason SnO_2 is claimed [277,278] to improve the efficiency of Cl_2 anodes is that the redox transition leading to oxygen evolution on RuO_2 is pushed to more positive potentials by the presence of the inactive component. This is also the case for $RuO_2 + TiO_2$ electrodes.

Geometric effects may counteract electronic effects in mixed oxides. If TiO_2 is replaced by SnO_2, activation rather than depression of O_2 evolution is observed

(Fig. 5.33), probably because of the unfavorable morphology of the mixed oxide layer [390]. Predictions of depressing effects based only on redox transition potentials are therefore insufficiently grounded, necessitating a close investigation of the morphology and of the surface structure of the mixed oxide layer.

5.6 Problems of Electrode Stability

Stability can be quantitatively expressed by the term ΔV_t in Eq. (5.1). It can be defined as the ability of an electrode to maintain its potential constant in long-term performance. Electrode stability is one of the most important parameters and is often the decisive parameter for the choice of an electrode structure. Table 5.11 summarizes recent studies specifically directed to evaluating electrode stability.

The estimation of the real lifetime of electrodes under operational conditions is difficult since extremely long experiments would be required. A pilot plant can, of course, be used or electrodes can be mounted in the cells as part of a plant. However, on a laboratory scale the lifetime is often estimated on a relative basis by means of *accelerated tests* [82,391–394]. These possess the obvious disadvantage of not reproducing the precise operational conditions, but they offer at least the possibility of a relatively quick response. Thus electrodes can be subjected to intensive anodic or cathodic discharges and their potential measured as a function of time. Any drift of the potential is recorded and evaluated.

From the concepts discussed earlier in this chapter it is clear that the stability is related to a number of factors [25,232,395]. We can distinguish between erosion,

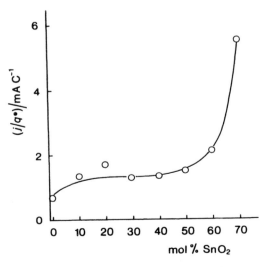

Figure 5.33. Rate of O_2 evolution (j), normalized to unit surface charge (q^*), at a constant potential in the Tafel region in acid solution as a function of the nominal SnO_2 content in $Ru_{0.3}Sn_xTi_{0.7-x}O_2$ mixed oxides [390].

TRANSITION METAL OXIDES: VERSATILE MATERIALS FOR ELECTROCATALYSIS 263

Table 5.11 Recent Studies of Corrosion and Stability of Oxide Electrodes[a]

Electrode	Subject	References
RuO_2	Wear mechanism	395
	hydrate in presence of Ce^{IV}	423, 424, 558
	Potential–pH diagram	306, 307, 605
	10% H_2SO_4 solution	559
	Effect of pH	411
	Influence of phosphates	433
	Passivation of support	82
	Effect of T of calcination	563
	Effect of various parameters	391, 392
	Corrosion by oxidants in H_2SO_4	424, 565
	Time dependence of corrosion rate	415
	RuO_4 as corrosion product	403
	Mechanism of dissolution	89, 126
	Effect of calcination	421
	Cathodic stability (poisoning)	314
$RuO_2 + TiO_2$	Wear mechanism	395
	Influence of F^-	406
	Effect of pH and current	405, 586
	Effect of pH, T, and current	409
	Different compositions	393, 413, 417
	Chlorate process, effect of NaCl concentration	562
	Influence of sulfate	434
	Selectivity of dissolution	410
	Effect of acidity	412
	High potentials, also nonaqueous solution	435
	Influence of phosphates	433
	Effect of cell shutdowns	437
	Various compositions	564
	Nonaqueous solvents	436
	Time dependence of corrosion rate	415
	Electrolysis of seawater	566
	Electrosynthesis of hypochlorite	408
	Strongly acid soultions	414
	T of calcination and composition	394
	Change in surface composition	95, 99
	Cathode stability (poisoning)	544
	Cathode stability (H_2)	428
$RuO_2 + TiO_2 + SnO_2$	Metal loss analyzed by XRF	601
$RuO_2 + TiO_2 + SnO_2 + PdO_x$	Durability in Cl_2 evolution	268
PtO_x	Wear mechanism	395
$PtO_x + TiO_2$	Different compositions	393, 417
$PtO_x + ZrO_2$	Different compositions	393, 417
$PtO_x + Ta_2O_5$	Different compositions	393, 417
$RuO_2 + ZrO_2$	Different compositions	393, 417, 418
MnO_2	In sulfate + formic acid solutions	560
	Effect of current and T in acids	561
$RuO_2 + SnO_2$	Various compositions, H_2SO_4	416, 421

(Continued)

Table 5.11 (*Continued*)

Electrode	Subject	References
$RuO_2 + MnO_2$	Different compositions	523
IrO_2	Time dependence of corrosion rate	415
	Heat treatment of Ti support	604
	Stability of overpotential	96
$Cr_2O_3 + TiO_2$	Anodic dissolution mechanism	361, 395
$RuO_2 + IrO_2$	Various compositions	96, 97
$RuO_2 + Ta_2O_5$	Different compositions	393, 417
$RuO_2 + IrO_2 + SnO_2$	Time dependence of overpotential	96
$IrO_2 + TiO_2$	Different compositions	393, 417
$IrO_2 + SnO_2$	Different compositions in H_2SO_4	612
$IrO_2 + ZrO_2$	Different compositions	393, 417
$IrO_2 + Ta_2O_5$	Different compositions	393, 417, 570
$Fe_xCo_{3-x}O_4$	Cathodic stability (O_2 reduction)	491
Co_3O_4	Stability in O_2 reduction	350, 426
	Effect of preparation	419
	Anodic dissolution, base	421
$La_{0.7}Ba_{0.3}MnO_3$	Time dependence of overpotential	532
$SrFe_{0.9}M_{0.1}O_3$	Mechanism of anodic dissolution in base	74
(M = Mn, Ti, Co, Ni)		
NiO_x	Cathodic stability (H_2)	430
$Ni_xCo_{3-x}O_4$	Different compositions	419
$NiCo_2O_4$	Time dependence of overpotential	539
$Zn_xCo_{3-x}O_4$	Different compositions	419

a Previous work has been reviewed in Refs. 35 and 232.

passivation, and wear. The first is a mechanical effect and can result in a progressive removal of the active layer. This effect can be detected by monitoring the surface state via voltammetric curves [86].

Passivation is typical of anodes and arises from the formation of an insulating layer of TiO_2 between the support and the active layer [82]. Passivation can be induced by space charge effects in the active layer (electric field penetration) [396], as well as by penetration of the solution through pores and cracks down to the Ti substrate [397]. Passivation is revealed by a sudden jump of the electrode potential to high positive values. It can also be the final result of erosion when the Ti surface becomes uncovered. In both cases the importance of the morphology of the layer and of its electrical conductivity is clear. The possibility of doping the (eventual) TiO_2 interlayer is essential. Recent investigations show the suboxides of Ti to be electronically conducting and resistant to anodic oxidation to TiO_2 [398, 399].

From a fundamental point of view the most interesting effect is wear, which includes both dissolution [400] and modification of the active layer to produce a less-active material. Examples of the latter effect are the anodic ordering of defective oxides upon anodic load and the progressive reduction of the oxide layer upon cathodic load. In both cases additives are necessary to retard the modifying reaction. Additives can be chosen on the basis of some general concepts, as illustrated later.

Anodic dissolution is the most typical anodic wear process. In general, it is not a catastrophic but a progressive event. It can end with the passivation of the support, but it can also lead to passivation owing to surface enrichment of the active oxide with an inactive component. This is, for instance, the case of RuO_2 + TiO_2 electrodes. XPS analysis [95] has shown that Ru normally dissolves preferentially so that the surface becomes enriched with TiO_2. However, if the solution is strongly acid, the chemical dissolution of TiO_2 prevails and the surface becomes enriched with Ru [99] (Fig. 5.34).

Anodic dissolution can be thermodynamically predicted [306,401]. In the case of RuO_2 if the potential limit for oxidation of the surface sites to Ru(VI) or Ru(VIII) is reached [402], Ru dissolves as ruthenate (in base) or escapes with O_2 or Cl_2 as volatile RuO_4. The latter has been demonstrated to form together with O_2 evolution [403]. For a number of precious metals it has been shown [404] that the rates of O_2 evolution and of anodic dissolution of the oxide proceed in parallel, since they possess a common reaction step [125] (Fig. 5.35). Therefore to reduce dissolution, methods to depress O_2 evolution must be devised.

In the presence of Cl_2 evolution, O_2 formation is strongly retarded [254,405] and this explains the very long lifetime of DSA in chlor-alkali cells [5,13]. The dissolution rate is only of the order of 10^{-8} g cm^{-2} h^{-1} [406]. However, dissolution is affected by the current density [391,407,408], the pH of the solution [405,409–412], and the composition of the oxide layer [391,413] (Fig. 5.36). Dissolution is lower at low pH because the O_2 evolution reaction is pushed to more anodic potentials. However, if the pH becomes negative (very concentrated acids), then TiO_2 starts to be chemically dissolved in DSA electrodes [411,414] (Fig. 5.37). The support itself (Ti metal) can be attacked by strongly acid solutions [397].

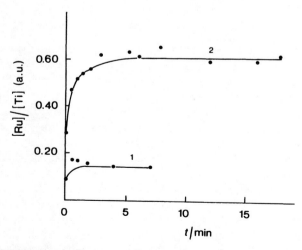

Figure 5.34. XPS in-depth profile (t is the sputtering time) of the Ru/Ti concentration ratio for RuO_2 + TiO_2 electrodes. (1) Fresh electrode; (2) after anodic polarization at 0.2 A cm^{-2} for 60 h in 2 mol dm^{-3} HCl + 3.1 mol dm^{-3} NaCl solution at 87°C. (After Ref. 99, with permission.)

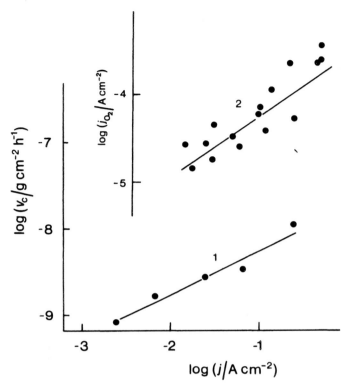

Figure 5.35. (1) Rate of corrosion (v_c) and (2) rate of O_2 evolution (j_{O2}) in 5 mol dm^{-3} NaCl as a function of the total anodic current (j) for a thermally stabilized anodic oxide on Ru + Ti. (Adapted from Ref. 125, with permission.)

IrO_2 is more resistant to anodic dissolution in acid because the transition potential to soluble intermediates is above the O_2 evolution potential [415]. Thus if IrO_2 is mixed with RuO_2 and a true solid solution is achieved, the rate of RuO_2 dissolution is pushed down exponentially as the IrO_2 content is increased [97] (Fig. 5.38). This is because the deleterious anodic transition in RuO_2 is shifted to more anodic potentials. This concept can be extended to other oxides too. In the case of pure PtO_x, SnO_2, PbO_2, and CeO_2 anodic stability is expected because no transition to soluble products is possible in the O_2 evolution region (alkaline solution for CeO_2, acid solution for SnO_2 and PbO_2) [401]. Therefore these oxides are expected to impart their inertness to a companion electrocatalytic oxide [7]. As an example, Fig. 5.39 shows the effect of SnO_2 on the lifetime of RuO_2 electrodes [416]. However, the interpretation of such results [393,417,418] should always take into account the morphology of the layer and the surface composition. In alkaline solution the stability of RuO_2 decreases appreciably [413] because the critical potential for the redox transition in the oxide is lowered at a higher rate than the O_2 evolution potential [402]. Conversely, spinels are soluble in acid solution [419,420] but are

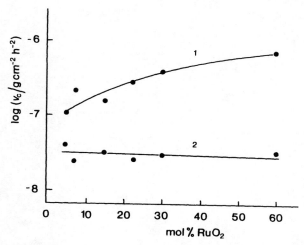

Figure 5.36. Rate of dissolution (v_c) of (1) Ru and (2) Ti at 0.2 A cm^{-2} in 5 mol dm^{-3} NaCl as a function of the composition of RuO_2 + TiO_2 electrodes. (After Ref. 413, with permission.)

quite stable in alkaline solution [421]. The rate of anodic dissolution of Co_3O_4 in acids at 1 mA cm^{-2} is of the order of 10^{-5} g cm^{-2} h^{-1} [422].

The rate of dissolution is observed to depend on the degree of hydration of the oxides [400,421,423]. Figure 5.40 shows the specific case of RuO_2 [424]. The environment of a metal ion in a hydrous oxide is more similar to an aquocation than

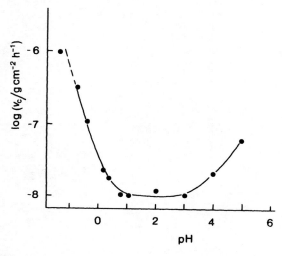

Figure 5.37. Rate of corrosion (v_c) of RuO_2 + TiO_2 electrodes in 5.13 mol dm^{-3} chloride solution at 0.2 A cm^{-2} as a function of the solution pH. (After Ref. 411, with permission.)

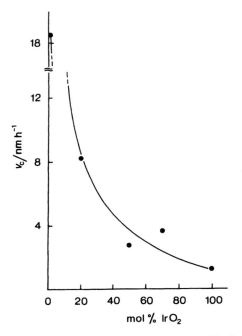

Figure 5.38. Effect of composition on the corrosion rate (v_c) of $RuO_2 + IrO_2$ electrodes at 1 A cm^{-2} in 0.5 mol dm^{-3} H$_2$SO$_4$. (After Ref. 97, with permission of the publisher, Pergamon Press PLC.)

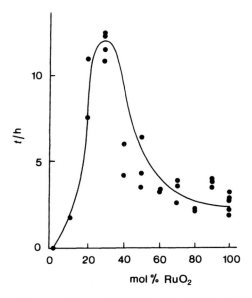

Figure 5.39. Effect of composition on the lifetime (t) of $RuO_2 + SnO_2$ electrodes at 0.5 A cm^{-2} in 0.5 mol dm^{-3} H$_2$SO$_4$. (After Ref. 416, with permission of the publisher, The Electrochemical Society, Inc.)

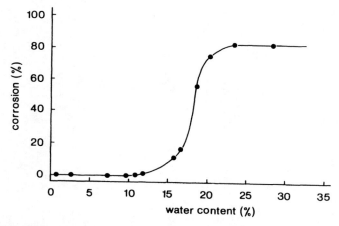

Figure 5.40. Effect of water content on the fraction of highly hydrated RuO_2 samples dissolved during O_2 liberation in 0.5 mol dm^{-3} H_2SO_4. Water content varied by heat treatment at 60–400°C. (After Ref. 424, with permission.)

in a dry oxide [110]. The work required to detach the metal ion from the lattice is therefore considerably lower [400]. This is confirmed by the much higher dissolution rate of Ru from the anodically formed oxide [232]. The rate of dissolution decreases consistently as the calcination temperature of the RuO_2 thermal oxide is increased [421]. A competition between activity and stability therefore arises. As the temperature of preparation is decreased, the activity increases but the stability decreases. The example of RuO_2 is a good illustration.

Electrocatalysts for O_2 reduction can be more stable because they operate away from the anodic dissolution region. But other factors may intervene. Co_3O_4 is stable under anodic load but the prolonged contact with alkaline solutions at open circuit causes a deterioration of the spinel structure that breaks down to a mixture of simple oxides with consequent chemical dissolution [134,425]. Consistently, a moderate cathodic treatment of Co_3O_4 can give rise to deterioration of the spinel structure with leaching of Co(II) ions [350,426]. This corresponds to what is known as *reductive dissolution* [400].

Stability under strong cathodic load entails resistance of oxides toward reduction by hydrogen. Although thermodynamically unstable, most oxides are in fact reduced at very low rate. Predictions on stability can be attempted on the basis of the reported correlation between the surface metal–oxygen bond strength and the rate of reduction by hydrogen [427] (Fig. 5.41).

Reductive deterioration (increase in overpotential for H_2 evolution) is observed at long usage times also with RuO_2 and NiO_x. Additives have been devised to retard this deterioration [428]. For RuO_2, on the basis of its effect on the reduction rate under H_2 atmosphere [429], an efficient additive seems to be ZrO_2. However, published data for RuO_2 stability tests under cathodic load are not available.

A NiO_x-based cathode has been developed for use in industry [430]. Additives are used to prolong the lifetime but their nature is unknown. For Co_3O_4, reductive

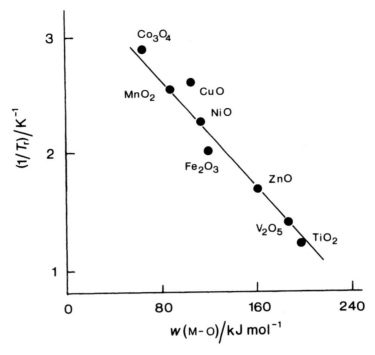

Figure 5.41. Correlation between the surface metal–oxygen binding energy in oxides (w) and the reciprocal temperature (T_r) at which a given reduction rate by H_2 gas is attained. (After Ref. 427, with permission from the Chemical Society of Japan, ref. no. CY-RT 91073.)

wear can be easily predicted on the basis of the studies on O_2 reduction [134]. Studies of reduction rates under a H_2 atmosphere, usually conducted in the field of catalysis [431,432], are useful to identify prospective cathodic stabilizers (Fig. 5.42), but their action on electrodes has not been tested. Fundamental research in this field is in its infancy. On the other hand, the claims of patents are often not supported by close examination.

Inhibitors for corrosion can be useful if they depress the O_2 evolution activity. Phosphate ions depress the RuO_2 dissolution rate [433], and F^- ions are particularly deleterious for the whole electrode structure since they can attack the support by depassivating Ti [406]. Sulfates decrease the selectivity for Cl_2 evolution [434], and hence are likely to enhance the dissolution rate of RuO_2. A similar risk exists in nonaqueous solvents: while the stability of the active layer can be increased because no O_2 evolution takes place [435], MeOH dissolves Ti since the protective TiO_2 layer is not formed [436]. A small addition of water may avoid this inconvenience. On the other hand, no corrosion of Ti has been observed [436] in aprotic solvents such as DMF and DMSO.

The consistently observed decrease of the dissolution rate (v) of oxide electrodes with time (t) is an interesting phenomenon. It has been reported [415] that a linear

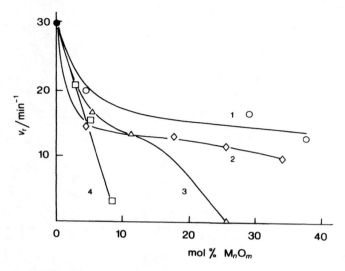

Figure 5.42. Reduction rate (v_r) of Co_3O_4 by H_2 gas as a function of type and amount of additive (M_nO_m). (●) Pure Co_3O_4. Additives: (1) ZrO_2; (2) MgO; (3) Al_2O_3; (4) La_2O_3. (After Ref. 432, with permission.)

$\log v - \log t$ plot describes the phenomenon adequately (Fig. 5.43). However, if the current is interrupted for a while, the dissolution rate jumps to a value that is higher than before the interruption but lower than the initial one (Fig. 5.44), and this occurs systematically after each interruption [407]. Since such interruptions (shutdowns) are common in industrial plants, the lifetime of DSAs depends on the average number of interruptions decreasing linearly with their number [437] (Fig. 5.45).

In the case of mixed oxides accelerated tests may often not be representative of their relative stability. It has been reported [417] that the maximum stability of IrO_2 + Ta_2O_5 mixed oxides occurs at approximately 80% Ta_2O_5. However, the surface charge shows a maximum at the same composition [393]. This suggests that the apparently higher stability is in fact related to the lower effective current density during the life test. Thus from a fundamental point of view, the meaning of these tests is ambiguous, although they may have a certain practical validity.

5.7 Conclusions and Prospects

This chapter has comprehensively reviewed the developments in the field of electrocatalytic oxide electrodes from about 1980. The picture emerging from the data discussed here shows oxides to be versatile electrocatalysts for most electrochemical reactions. The essential reason for this is the rich redox chemistry of transition metal oxides. Electrochemical reactions are activated by redox transitions on the solid surface. The thermodynamic potentials of the various reactions set the limits at which surface redox transitions can be active. Oxides with no available redox transitions near the thermodynamic potential of a given reaction are *in principle* expected to be relatively inactive. This qualitative approach can only offer broad

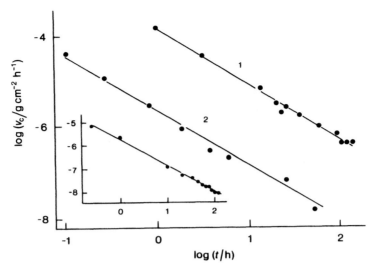

Figure 5.43. Time dependence of the corrosion rate (v_c) of (1) RuO_2 and (2) RuO_2 + 70% TiO_2 at 0.2 A cm^{-2} in 5.13 mol dm^{-3} chloride solution at 87°C. Inset: Time dependence of the corrosion rate of IrO_2 under the same conditions. (Adapted from Ref. 415, with permission.)

understanding of the observed versatility of transition metal oxides, and the related inactivity of other nontransition metal oxides. However, it is also useful as a broad criterion to predict stability.

However, the preceding concept does not offer a comprehensive understanding. The adsorption behavior of oxides is quite specific, and its understanding begins

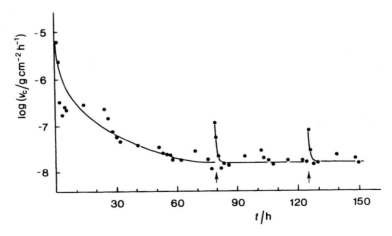

Figure 5.44. Effect of periodic current interruptions (indicated by arrows) on the time dependence of the corrosion rate (v_c) of RuO_2 + 70% TiO_2 electrodes at 0.2 A cm^{-2} in 5 mol dm^{-3} NaCl. (After Ref. 407, with permission.)

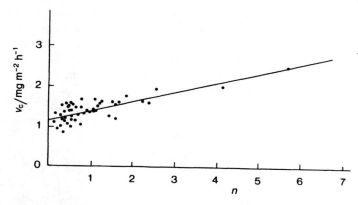

Figure 5.45. Average corrosion rate (v_c) of RuO_2 + 70% TiO_2 at 0.2 A cm^{-2} in 5 mol dm^{-3} NaCl as a function of the average monthly frequency of current interruptions (n). (After Ref. 437, with permission.)

with an appreciation of the special features of oxide surfaces. The surface hydroxylation of oxides, specifically in aqueous solution, is at the origin of the observed ion adsorbability sequence as well as of the interesting behavior with respect to metallic impurities. In other words, the interaction of the oxide surface with its aqueous environment is important, as well as the redox chemistry of the oxide.

Another specific concept to associate with oxide electrodes is that surface charging depends essentially on the pH of the solution, rather than on the electrical polarization of the electrode. This is particularly true for oxides with higher electronic conductivity. Electrocatalytic oxide electrodes therefore have little in common with oxide semiconductors. The two fields overlap precisely where interest for industrial applications in intensive electrolyzers drops, although other kinds of applications can be envisaged.

Apart from surface properties, oxide hydration proves to be of major importance in determining electrode stability in solution. As a rule, the more hydrated an oxide, the lower its anodic stability. The temperature of calcination during the preparation thus becomes a crucial variable in controlling surface area, morphology, and hence activity and stability. Very often, stability and activity vary in opposite directions with the variables of oxide preparation and treatment. Consequently, a careful exploration of the many factors contributing to oxide performances should be routine procedure, in attempts to optimize performance. Investigations of single samples or single compositions have no general validity. Experiments must have a statistical significance.

Two fields, traditionally separated as scientific disciplines, though not culturally —surface chemistry and electrochemistry—overlap substantially in the context of oxide electrodes. Acid–base properties of oxide surfaces have been shown to be linked to the electrocatalytic activity, the point of zero charge being a useful property to separate electronic from geometric factors. This allows a more significant evaluation of electrocatalysts to be carried out.

The study of oxide electrodes has contributed to the revaluation of electrochemical techniques with respect to ex situ techniques, currently regarded as more sensitive and molecularly specific, for "materials research." The surface analysis of composite materials is more significant if it is carried out in situ, and continuous monitoring of oxide surfaces is simply meaningless if current interruptions are needed to have access to the measurement.

Prospects vary, depending on the area. The possibilities for improving the performance of single-oxide systems seem exhausted. Similar studies may retain fundamental value, provided the choice of the variables to investigate is sound and sharply focused. In this context a study that still appears underdeveloped is that of single-crystal faces. Another subject is adsorption, in particular of organic substances and metal ions. More possibilities are offered by the development of mixed and composite oxides. Synergetic effects are the clue to promoting electrocatalysis and to enhancing stability. The understanding of synergetic effects is still poor, their occurrence almost unpredictable, and their investigation limited and nonsystematic.

The performance of oxide electrodes for Cl_2 evolution has been almost fully optimized. Since Cl_2 evolution is a facile reaction, further improvements can only be insignificant, therefore economically not remunerative. For these reasons there is no appreciable industrial demand for improvement, particularly in view of the decline in the market for Cl_2 gas.

However, the technological demand for improved O_2 anodes, particularly for specific new applications, continues. Thus O_2 anodes for strongly acid solutions continue to attract interest and to demand viable solutions.

Oxygen reduction is linked to the development of fuel cells and air cathodes for electrolyzers. Although spinels, perovskites, and pyrochlores appear to offer practical possibilities, most of the interest in the field is still concentrated on precious metals as immediate condidates, and transition metal macrocycles (e.g., phthalocyanines and porphyrins) as possible substitutes. Thus the development of oxide electrodes in this direction remains limited.

The use of oxides as hydrogen cathodes is an intriguing novelty that has already attracted industrial interest. Fundamental research is still sparse in this field and should be promoted. The use of oxides for hydrogen evolution is attractive in view of their resistance to poisoning; this calls for specific studies addressing metal deposition on oxides.

Organic electrochemistry, once defined by Ernest Yeager as the "dormant giant," has also taken advantage of the development of oxide electrodes. However, although the number of applications in this field is increasing, a modern breakthrough is still lacking. Oxide electrodes seem to offer a number of alternatives still to be exhaustively explored. However, reactions thus far investigated are closer to being feasibility studies than real developments. Nevertheless, prospects are encouraging.

Oxides are being discovered as versatile materials for an ever-increasing number of applications (cf. Table 5.1). Some of these uses are nonelectrochemical but have been stimulated by the interesting properties highlighted by the electrochemical applications. Oxide electrodes presently arouse interest in many branches of electrochemistry, and they are very likely to find further applications.

References

1. "Performances of Electrodes for Industrial Electrochemical Processes," Hine, F.; Tilak, B. V.; Denton, J. M.; Lisius, J. D., Eds. The Electrochemical Society, Pennington, NJ, 1989.
2. Beer, H. B. *J. Electrochem. Soc.* **1980**, *127*, 303C.
3. Horacek, S.; Puschaver, S. *Chem. Eng. Progr.* **1971**, *67*, 71.
4. Hass, K.; Schmittinger, P. *Electrochim. Acta* **1976**, *21*, 1115.
5. De Nora, O. *Chem.-Ing.-Tech.* **1970**, *43*, 182.
6. De Nora, V.; Kühn von Burgsdorff, J.-W. *Chem.-Ing.-Tech.* **1975**, *47*, 125.
7. De Nora, V.; Nidola, A. "Extended Abstracts," Electrochemical Society Meeting, Los Angeles, 1970; The Electrochemical Society, Pennington, NJ; paper 270.
8. Udupa, H. V. K.; Thangappan, R.; Yadav, B. R.; Subbiah, P. *Chem. Age India* **1972**, *23*, 545.
9. O'Leary, K. J.; Navin, T. J. In "Chlorine Bicentennial Symposium," Jeffrey, T. C.; Danna, P. A.; Holden, H. S. Eds. The Electrochemical Society, Pennington, NJ, 1974, p. 174.
10. Trasatti, S.; Buzzanca, G. *J. Electroanal. Chem.* **1971**, *29*, App. 1.
11. Tantardini, G. Thesis, University of Milan, 1968.
12. "Electrodes of Conductive Metallic Oxides," Trasatti, S., Ed. Elsevier, Amsterdam, 1980, Part A; 1981, Part B.
13. Nidola, A. In Ref. 12, Part B, Chapter 11, pp. 627–659.
14. De Nora, V. In Ref. 1, pp. 1–13.
15. Rolison, D. R.; Kuo, K.; Umana, M.; Brundage, D.; Murray, R. W. *J. Electrochem. Soc.* **1979**, *126*, 407.
16. Uchida, I.; Urushibata, H.; Toshima, S. *J. Electrochem. Soc.* **1981**, *128*, 2351.
17. Uchida, I.; Uruchibata, H.; Toshima, S. *J. Appl. Electrochem.* **1982**, *12*, 115.
18. Novak, D. M.; Tilak, V. B.; Conway, B. E. In "Modern Aspects of Electrochemistry," Bockris, J. O'M.; Conway, B. E.; White, R. E., Eds. Plenum, New York, 1982, Vol. 14, Chapter 4, pp. 195–318.
19. Hine, F.; Tilak, V. B.; Viswanathan, K. In "Modern Aspects of Electrochemistry," White, R. E.; Bockris, J. O'M.; Conway, B. E., Eds. Plenum, New York, 1986, Vol. 18, Chapter 5, pp. 249–302.
20. Hall, D. E. *J. Electrochem. Soc.* **1985**, *132*, 41C.
21. Kolotyrkin, Ya. M.; Losev, V. V.; Shub, D. M.; Roginskaya, Yu. E. *Elektrokhimiya* **1979**, *15*, 291.
22. Trasatti, S.; O'Grady, W. E. In "Advances in Electrochemistry and Electrochemical Engineering," Gerischer, H.; Tobias, C. W., Eds. Wiley-Interscience, New York, 1980, Vol. 13, Chapter 3, pp. 177–261.
23. Trasatti, S. *Electrochim. Acta* **1987**, *32*, 369.
24. Matsumoto, Y.; Sato, E. *Mater. Chem. Phys.* **1986**, *14*, 397.
25. Trasatti, S. In "Electrochemical Hydrogen Technologies," Wendt, H., Ed. Elsevier, Amsterdam, 1990, pp. 104–135.
26. Trasatti, S. *Electrochim. Acta* **1991**, *36*, 225.

27. Daghetti, A.; Lodi, G.; Trasatti, S. *Mater. Chem. Phys.* **1983**, *8*, 1.
28. Trasatti, S. *Electrochim. Acta* **1984**, *29*, 1503.
29. Manoharan, R.; Paranthaman, M.; Goodenough, J. B. *Eur. J. Solid State Inorg. Chem.* **1989**, *26*, 155.
30. Angelinetta, C.; Trasatti, S.; Atanasoska, Lj. D.; Minevski, Z. S.; Atanasoski, R. T. *Mater. Chem. Phys.* **1989**, *22*, 231.
31. Ardizzone, S.; Falciola, M.; Trasatti, S. *J. Electrochem. Soc.* **1989**, *136*, 1545.
32. Henne, R.; Schiller, G.; Schnurnberger, W.; Weber, W. *Dechema Monogr.* **1990**, *121*, 153.
33. Hackwood, S.; Schiavone, L. M.; Dautremont-Smith, W. C.; Beni, G. *J. Electrochem. Soc.* **1981**, *128*, 2569.
34. Vercesi, G. P.; Rolewicz, J.; Comninellis, C.; Hinden, J. *Thermochim. Acta* **1991**, *176*, 31.
35. Haenen, J. G. D.; Visscher, W.; Barendrecht, E. *J. Appl. Electrochem.* **1985**, *15*, 29.
36. Modestova, V. N.; Tomashov, N. D.; Yakubenko, A. R.; Kasatkina, I. V.; Skryabina, V. I.; Vashchenko, S. N.; Borisova, E. A. *Elektrokhimiya* **1984**, *20*, 39.
37. Jasem, S. M.; Tseung, A. C. C. *J. Electrochem. Soc.* **1979**, *126*, 1353.
38. Tseung, A. C. C.; Botejue, J. *Int. J. Hydrogen Energy* **1986**, *11*, 125.
39. Kudo, T.; Obayashi, H.; Yoshida, M. *J. Electrochem. Soc.* **1977**, *124*, 321.
40. Jang, G.-W.; Rajeshwar, K. *J. Electrochem. Soc.* **1987**, *134*, 1830.
41. Trasatti, S.; Lodi, G. In Ref. 12, Part A, Chapter 7, pp. 301–358.
42. Ardizzone, S.; Carugati, A.; Trasatti, S. *J. Electroanal. Chem.* **1981**, *126*, 287.
43. Gerrard, W. A.; Steele, B. C. H. *J. Appl. Electrochem.* **1978**, *8*, 417.
44. Roginskaya, Yu. E.; Bystrov, V. I.; Shub, D. M. *Zh. Neorg. Khim.* **1977**, *22*, 201.
45. Roginskaya, Yu. E.; Galyamov, B. Sh.; Lebedev, V. M.; Belova, I. D.; Venevtsev, Yu. N. *Zh. Neorg. Khim.* **1977**, *22*, 499.
46. Hine, F.; Yasuda, M.; Noda, T.; Yoshida, T.; Okuda, J. *J. Electrochem. Soc.* **1979**, *126*, 1439.
47. Pizzini, S.; Buzzanca, G.; Mari, C. M.; Rossi, L.; Torchio, S. *Mater. Res. Bull.* **1972**, *7*, 449.
48. Savinell, R. F.; Zeller III, R. L.; Adams, J. A. *J. Electrochem. Soc.* **1990**, *137*, 489.
49. Varlamova, T. V.; Belova, I. D.; Shifrina, R. R.; Galyamov, B. Sh.; Roginskaya, Yu. E.; Venevtsev, Yu. N. *Zh. Fiz. Khim.* **1990**, *64*, 385.
50. Lodi, G.; Bighi, C.; De Asmundis, C. *Mater. Chem.* **1976**, *1*, 177.
51. Garavaglia, R.; Mari, C. M.; Trasatti, S.; De Asmundis, C. *Surf. Technol.* **1983**, *19*, 197.
52. Belova, I. D.; Varlamova, T. V.; Galyamov, B. Sh.; Roginskaya, Yu. E.; Shifrina, R. R.; Prutchenko, S. G.; Kaplan, G. I.; Sevastyanov, M. A. *Mater. Chem. Phys.* **1988**, *20*, 39.
53. Angelinetta, C.; Trasatti, S.; Atanasoska, Lj. D.; Atanasoski, R. T. *J. Electroanal. Chem.* **1986**, *214*, 535.
54. Ardizzone, S.; Carugati, A.; Lodi, G.; Trasatti, S. *J. Electrochem. Soc.* **1982**, *129*, 1689.
55. Duvigneaud, P. H.; Coussment, A. *J. Solid State Chem.* **1984**, *52*, 22.
56. Kiwi, J.; Prins, R. *Chem. Phys. Lett.* **1986**, *126*, 579.

57. Belova, I. D.; Shalaginov, V. V.; Galyamov, B. Sh.; Roginskaya, Yu. E.; Shub, D. M. *Zh. Neorg. Khim.* **1978**, *23*, 286.

58. Shalaginov, V. V.; Belova, I. D.; Roginskaya, Yu. E.; Shub, D. M. *Elektrokhimiya* **1978**, *14*, 1708.

59. Lodi, G.; De Battisti, A.; Benedetti, A.; Fagherazzi, G.; Kristof, J. *J. Electroanal. Chem.* **1988**, *256*, 441.

60. Spynu, V. K.; Shalaginov, V. V.; Kozlova, N. V.; Groza, I. A.; Shub, D. M. *Elektrokhimiya* **1986**, *22*, 709.

61. Boggio, R.; Carugati, A.; Trasatti, S. *J. Appl. Electrochem.* **1987**, *17*, 828.

62. Trasatti, S. *Mater. Chem. Phys.* **1987**, *16*, 157.

63. Honig, J. M. In Ref. 12, Part A, Chapter 1, pp. 1–96.

64. Lodi, G.; De Battisti, A.; Bordin, G.; De Asmundis, C.; Benedetti, A. *J. Electroanal. Chem.* **1990**, *277*, 139.

65. Takeuchi, M.; Miwada, K.; Nagasaka, H. *Appl. Surf. Sci.* **1982**, *11/12*, 298.

66. Lodi, G.; De Asmundis, C.; Ardizzone, S.; Sivieri, E.; Trasatti, S. *Surf. Technol.* **1981**, *14*, 335.

67. Halder, N. C. *Electrocomp. Sci. Technol.* **1983**, *11*, 21.

68. Battaglin, C.; Carnera, A.; Mazzoldi, P.; Lodi, G.; Bonora, P.; Daghetti, A.; Trasatti, S. *J. Electroanal. Chem.* **1982**, *135*, 313.

69. Rao, C. N. R.; Subba Rao, G. V. *Phys. Stat. Sol.(a)* **1970**, *1*, 597.

70. Hamdani, M.; Koenig, J. F.; Chartier, P. *J. Appl. Electrochem.* **1988**, *18*, 561.

71. Konovalov, M. B.; Bystrov, V. I.; Kubasov, V. L. *Elektrokhimiya* **1976**, *12*, 1266.

72. Tamura, H.; Yoneama, H.; Matsumoto, Y. In Ref. 12, Part A, Chapter 6, pp. 261–299.

73. Bockris, J. O'M.; Otagawa, T. *J. Electrochem. Soc.* **1984**, *131*, 290.

74. Matsumoto, Y.; Kurimoto, J.; Sato, E. *Electrochim. Acta* **1980**, *25*, 539.

75. Matsumoto, Y.; Yamada, S.; Nishida, T.; Sato, E. *J. Electrochem. Soc.* **1980**, *127*, 2360.

76. Reznik, M. F.; Shub, D. M.; Lubnin, E. N. *Elektrokhimiya* **1988**, *24*, 1090.

77. Bystrov, V. I.; Avksentyev, V. V.; Sokolov, V. A. *Zh. Fiz. Khim.* **1973**, *47*, 2561.

78. Galyamov, B. Sh.; Roginskaya, Yu. E.; Lazorenko-Manevich, R. M.; Kozhevnikov, V. B.; Yanovskaya, M. I.; Kolotyrkin, Ya. M. *Mater. Chem. Phys.* **1984**, *11*, 525.

79. Biggers, J. V.; McKelvy, J. R.; Schulze, W. A. *Commun. Am. Ceram. Soc.* **1982**, C13.

80. Roginskaya, Yu. E.; Galyamov, B. Sh.; Belova, I. D.; Shifrina, R. R.; Kozhevnikov, V. B.; Bystrov, V. I. *Elektrokhimiya* **1982**, *18*, 1327.

81. Vallet, C. E. *J. Electrochem. Soc.* **1991**, *138*, 1234.

82. Loučka, T. *J. Appl. Electrochem.* **1981**, *11*, 143.

83. Shalaginov, V. V.; Shub, D. M.; Kozlova, N. V., Lomova, V. N. *Elektrokhimiya* **1983**, *19*, 537.

84. Konovalov, M. B.; Bystrov, V. I.; Kubasov, V. L. *Elektrokhimiya* **1975**, *11*, 239.

85. Del Arco, M.; Rives, V. *J. Mater. Sci.* **1986**, *21*, 2938.

86. Boggio, R.; Carugati, A.; Lodi, G.; Trasatti, S. *J. Appl. Electrochem.* **1985**, *15*, 335.

87. Morita, M.; Iwakura, C.; Tamura, H. *Electrochim. Acta* **1978**, *23*, 331.
88. Bondar, R. U.; Sorokendya, V. S.; Kalinovskii, E. A. *Elektrokhimiya* **1986**, *22*, 1653.
89. Lewerenz, H. J.; Stucki, S.; Kötz, R. *Surf. Sci.* **1983**, *126*, 893.
90. Augustynski, J.; Koudelka, M.; Sanchez, J.; Conway, B. E. *J. Electroanal. Chem.* **1984**, *160*, 233.
91. Valeri, S., Battaglin, G., Lodi, G., Trasatti, S. *Coll. Surf.* **1986**, *19*, 387.
92. Cox, P. A.; Goodenough, J. B.; Tavener, P. J.; Telles, D.; Egdell, R. G.; *J. Solid State Chem.* **1986**, *62*, 360.
93. Choudhury, T.; Saied, S. O.; Sullivan, J. L.; Abbot, A. M. *J. Phys. D* **1989**, *22*, 1185; Wandelt, K. *Surf. Sci. Rep.* **1982**, *2*, 101.
94. De Battisti, A.; Lodi, G.; Cappadonia, M.; Battaglin, G.; Kötz, R. *J. Electrochem. Soc.* **1989**, *136*, 2596.
95. Gorodetskii, V. V.; Zorin, P. N.; Pecherskii, M. M.; Busse-Machukas, V. B.; Kubasov, V. L.; Tomashpolskii, Yu. Ya. *Elektrokhimiya* **1981**, *17*, 79.
96. Hutchings, R.; Müller, K.; Kötz, R.; Stucki, S. *J. Mater. Sci.* **1984**, *19*, 3987.
97. Kötz, R.; Stucki, S. *Electrochim. Acta* **1986**, *31*, 1311.
98. Comninellis, Ch.; Vercesi, G. P.; *J. Appl. Electrochem.* **1991**, *21*, 136.
99. Gorodetskii, V. V.; Tomashpolskii, Yu. Ya.; Gorbacheva, L. B.; Sadovskaya, N. V.; Pecherskii, M. M.; Evdokimov, S. V.; Busse-Machukas, V. B.; Kubasov, V. L.; Losev, V. V. *Elektrokhimiya* **1984**, *20*, 1045.
100. Kodintsev, I. M.; Trasatti, S.; Rubel, M.; Wieckowski, A.; Kaufher, N. *Langmuir* **1992**, *8*, 283.
101. Burke, L. D. In Ref. 12, Part A, Chapter 3, pp. 141–181.
102. Burke, L. D.; Borodzinski, J. J.; O'Dwyer, K. J. *Electrochim. Acta* **1990**, *35*, 967.
103. Chialvo, A. C.; Triaca, W. E.; Arvia, A. J. *J. Electroanal. Chem.* **1987**, *225*, 227.
104. Galizzioli, D.; Tantardini, F.; Trasatti, S. *J. Appl. Electrochem.* **1974**, *4*, 57.
105. Kessler, T.; Visintín, A.; Triaca, W. E.; Arvia, A. J.; Gennero de Chialvo, M. R. *J. Appl. Electrochem.* **1991**, *21*, 516.
106. Burke, L. D.; Lyons, M. E. G. In "Modern Aspects of Electrochemistry," White, R. E.; Bockris, J. O'M.; Conway B. E., Eds. Plenum, New York, 1986, Vol. 18, Chapter 4, pp. 169–248.
107. Hall, H. Y.; Sherwood, P. M. A. *J. Chem. Soc., Faraday Trans. I* **1984**, *80*, 135.
108. Augustynski, J.; Balsenc, L. In "Modern Aspects of Electrochemistry," Conway, B. E.; Bockris, J. O'M. Eds.; Plenum, New York, 1979, Vol. 13, Chapter 4, pp. 251–360.
109. Kötz, R. In "Advances in Electrochemical Science and Engineering," Gerischer, H.; Tobias, C. W., Eds. VCH, Weinheim, 1990, Vol. 1, pp. 75–126.
110. Burke, L. D.; O'Sullivan, E. J. M. *J. Electroanal. Chem.* **1981**, *117*, 155.
111. Battaglin, G.; Carnera, A.; Della Mea, G.; Lodi, G.; Trasatti, S. *J. Chem. Soc., Faraday Trans. I* **1985**, *81*, 2995.
112. Belova, I. D.; Shifrina, R. R.; Roginskaya, Yu. E.; Popov, A. V.; Varlamova, T. V. *Elektrokhimiya* **1987**, *23*, 1208.
113. Roginskaya, Yu. E.; Belova, I. D.; Galyamov, B. Sh.; Popkov, Yu. M.; Zakharin, D. S. *Elektrokhimiya* **1987**, *23*, 1215.

114. Vuković, M. *Electrochim. Acta* **1989**, *34*, 287.

115. Vuković, M. *J. Appl. Electrochem.* **1990**, *20*, 969.

116. Burke, L. D.; Morphy, O. J.; O'Neill, J. F.; Venkatesan, S. *J. Chem. Soc., Faraday Trans. 1* **1977**, *73*, 1659.

117. Kalinovski, E. A.; Bondar, R. U.; Meshkova, N. N. *Elektrokhimiya* **1972**, *8*, 283.

118. Gennero de Chialvo, M. R.; Chialvo, A. C. *Electrochim. Acta* **1990**, *35*, 437.

119. Vallet, C. E.; Heatherley, D. E.; White, C. W. *J. Electrochem. Soc.* **1990**, *137*, 579.

120. Fateev, V. N. Guseva, M. I.; Pakhomov, V. P.; Kulikova, L. N.; Vladimirov, B. G.; Chekushkin, Yu. N.; Gordeeva, G. V. *Elektrokhimiya* **1990**, *26*, 74.

121. Vallet, C. E.; Borns, S. E.; Hendrickson, J. S.; White, C. W. *J. Electrochem. Soc.* **1988**, *135*, 387.

122. Kelly, E. J.; Heatherly, D. E.; Vallet, C. E.; White, C. W. *J. Electrochem. Soc.* **1987**, *134*, 1667.

123. Kelly, E. J.; Vallet, C. E.; White, C. W. *J. Electrochem. Soc.* **1990**, *137*, 2482.

124. Manoharan, R.; Goodenough, J. B. *Electrochim. Acta* **1991**, *36*, 19.

125. Pecherskii, M. M.; Gorodetskii, V. V.; Buné, N. Ya.; Losev, V. V. *Elektrokhimiya* **1982**, *18*, 415.

126. Kötz, R.; Lewerenz, H. J.; Stucki, S. *J. Electrochem. Soc.* **1983**, *130*, 825.

127. Henrich, V. E. *Progr. Surf. Sci.* **1983**, *14*, 175.

128. Clarke, N. S.; Hall, P. G. *Langmuir* **1991**, *7*, 678.

129. Bernholc, J.; Horsley, J. A.; Murrell, L. L.; Sherman, L. G.; Soled, S. *J. Phys. Chem.* **1987**, *91*, 1526.

130. Zettlemoyer, A. C. *J. Colloid Interface Sci.* **1968**, *28*, 343.

131. Trasatti, S. *Croat. Chem. Acta* **1990**, *63*, 313.

132. Zakharkin, G. I.; Tarasevich, M. R.; Khutornoi, A. M.; Makordei, F. V.; Sofronkov, A. N. *Zh. Fiz. Khim.* **1977**, *51*, 3068.

133. Moravskaya, O. V.; Khrushcheva, E. I.; Shumilova, N. A. *Elektrokhimiya* **1977**, *13*, 563.

134. Tarasevich, M. R.; Efremov, B. N. In Ref. 12, Part A, Chapter 5, pp. 221–259.

135. Furlong, D. N.; Yates, D. E.; Healy, T. W. In Ref. 12, Part B, Chapter 8, pp. 367–432.

136. Sorlino, M.; Busca, G. *Appl. Surf. Sci.* **1984**, *18*, 268.

137. Burke, L. D.; Mulcahy, J. K.; Whelan, D. P. *J. Electroanal. Chem.* **1984**, *163*, 117.

138. Lodi, G.; Zucchini, G.; De Battisti, A.; Sivieri, E.; Trasatti, S. *Mater. Chem.* **1978**, *3*, 179.

139. Tarlov, M. J.; Kreider, K. G.; Semancik, S.; Huang, P. *AIChE Symp. Ser.* **1989**, *85*, 36.

140. Tarlov, M. J.; Semancik, S.; Kreider, K. G. *Sensors Actuators* **1990**, *B1*, 293.

141. Bloor, L. J.; Malcolme-Lawes, D. J. *J. Electroanal. Chem.* **1990**, *278*, 161.

142. Burke, L. D.; Lyons, M. E.; O'Sullivan, E. J. M.; Whelan, D. P. *J. Electroanal. Chem.* **1981**, *122*, 403.

143. Burke, L. D.; Whelan, D. P. *J. Electroanal. Chem.* **1981**, *124*, 333; **1984**, *162*, 121.

144. Garavaglia, R.; Mari, C. M.; Trasatti, S. *Surf. Technol.* **1984**, *23*, 41.

145. Shalaginov, V. V.; Shub, D. M. *Elektrokhimiya* **1985**, *21*, 1576.

146. Blesa, M. A.; Kallay, N. *Adv. Colloid Interface Sci.* **1988**, *28*, 111.
147. Ardizzone, S.; Siviglia, P.; Trasatti, S. *J. Electroanal. Chem.* **1981**, *122*, 395.
148. Ardizzone, S.; Lettieri, D.; Trasatti, S. *J. Electroanal. Chem.* **1983**, *146*, 431.
149. Viganò, R.; Taraszewska, J.; Daghetti, A.; Trasatti, S.; *J. Electroanal. Chem.* **1985**, *182*, 203.
150. Pirovano, C.; Trasatti, S. *J. Electroanal. Chem.* **1984**, *180*, 171.
151. Siviglia, P.; Daghetti, A.; Trasatti, S. *Coll. Surf.* **1983**, *7*, 15.
152. Ardizzone, S.; Daghetti, A.; Franceschi, L.; Trasatti, S. *Coll. Surf.* **1989**, *35*, 85.
153. Butler, M. A.; Ginley, D. S. *J. Electrochem. Soc.* **1978**, *125*, 228.
154. Michaelson, H. B. *IBM J. Res. Develop.* **1978**, *22*, 72.
155. Healy, T. W.; Herring, A. P.; Fuerstenau, D. W. *J. Colloid Interface Sci.* **1966**, *21*, 435.
156. Damaskin, B. B. *Elektrokhimiya* **1989**, *25*, 1641.
157. Damaskin, B. B.; Gorichev, I. G.; Batrakov, V. V. *Elektrokhimiya* **1990**, *26*, 400.
158. Trasatti, S.; Petrii, O. A. *Pure Appl. Chem.* **1991**, *63*, 711.
159. Ardizzone, S.; Bassi, G. *J. Electroanal. Chem.* **1991**, *300*, 585.
160. Kleijn, J. M.; Lyklema, J. *J. Colloid Interface Sci.* **1987**, *120*, 511.
161. Brey, W. S.; Davis, B. H. *J. Colloid Interface Sci.* **1979**, *70*, 10.
162. Ardizzone, S.; Trasatti, S. *J. Electroanal. Chem.* **1988**, *255*, 237.
163. Fokkink, L. G. J.; De Keizer, A.; Kleijn, J. M.; Lyklema, J. *J. Electroanal. Chem.* **1986**, *208*, 401.
164. Akratopulu, K. Ch.; Kordulis, Ch.; Lycourghiotis, A. *J. Chem. Soc., Faraday Trans.* **1990**, *86*, 3437.
165. Blesa, M. A.; Figliolia, N. M.; Maroto, A. J. G.; Regazzoni, A. E. *J. Colloid Interface Sci.* **1984**, *101*, 410.
166. Lyklema, J. *Chem. Ind.* **1987**, *2*, 741.
167. Fokkink, L. G. J.; De Keizer, A.; Lyklema, J. *J. Colloid Interface Sci.* **1989**, *127*, 116.
168. Safonova, T. Ya.; Petrii, O. A.; Gudkova, E. A. *Elektrokhimiya* **1980**, *16*, 1607.
169. Kokarev, G. A.; Kolesnikov, V. A.; Gubin, A. F.; Korobanov, A. A. *Elektrokhimiya* **1982**, *18*, 407.
170. Andreev, V. N.; Heckner, K. H.; Glass, M.; Kazarinov, V. E. *Elektrokhimiya*, **1983**, *11*, 1558.
171. Wieckowski, A. In "Modern Aspects of Electrochemistry," White, R. E.; Bockris, J. O'M.; Conway, B. E., Eds. Plenum, New York, 1990, Vol. 21, Chapter 3, pp. 65–119.
172. Morrison, W. H. *J. Colloid Interface Sci.* **1984**, *100*, 121; Ardizzone, S.; Formaro, L.; Lyklema, J. *J. Electroanal. Chem.* **1982**, *133*, 147.
173. Kokarev, G. A.; Gubin, A. F.; Kolesnikov, V. A.; Skobelev, S. A. *Zh. Fiz. Khim.* **1985**, *59*, 1660.
174. Sekki, A.; Kokarev, G. A.; Kapustin, Yu. I. *Elektrokhimiya* **1985**, *21*, 1277.
175. Andreev, V. N.; Mostkova, R. I.; Kazarinov, V. E.; *Elektrokhimiya* **1984**, *20*, 1519.
176. Kleijn, J. M.; Lyklema, J. *Coll. Polym. Sci.* **1987**, *265*, 1105.
177. Furlong, D. N.; Sasse, W. H. F. *Coll. Surf.* **1983**, *7*, 29.

178. Dzombak, D. A.; Morel, F. M. M. "Surface Complexation Modeling," Wiley, New York, 1990.
179. Hayes, K. F.; Redden, G.; Ela, W.; Leckie, J. O. *J. Colloid Interface Sci.* **1991**, *142*, 448.
180. Schindler, P. W.; Stumm; W. In "Aquatic Surface Chemistry: Chemical Processes at the Particle–Water Interface," Stumm, W., Ed. Wiley, New York, 1987, Chapter 4, pp. 83–110.
181. Fukuda, K.-I.; Iwakura, C.; Tamura, H. *Electrochim. Acta* **1979**, *24*, 267.
182. Kokarev, G. A.; Kolesnikov, V. A.; Borsh, I.; Yanoshi, A.; Gubin, A. F.; Kodintsev, I. M. *Elektrokhimiya* **1985**, *21*, 1277.
183. Healy, T. W.; White, L. R. *Adv. Colloid Interface Sci.* **1978**, *9*, 303.
184. Lyklema, J. *J. Electroanal. Chem.* **1968**, *18*, 341.
185. Kleijn, J. M. *Coll. Surf.* **1990**, *51*, 371.
186. Gorichev, I. G.; Batrakov, V. V.; Damaskin, B. B. *Elektrokhimiya* **1989**, *25*, 809.
187. Tomkiewicz, M.; Huang, Y. S.; Pollak, F. H. *J. Electrochem. Soc.* **1983**, *130*, 1514.
188. Burke, L. D.; Healy, J. F. *J. Electroanal. Chem.* **1981**, *124*, 327.
189. Yuen, M. F.; Lauks, I.; Dautremont-Smith, W. C. *Solid State Ionics* **1983**, *11*, 19.
190. Kutin, A. P.; Evtyukov, N. Z.; Mishin, S. A.; Yakovlev, A. D. *Zh. Prikl. Khim.* **1985**, *58*, 1140.
191. Nechaev, E. A.; Zvonareva, G. V. *Kolloidn. Zh.* **1980**, *42*, 511.
192. Nechaev, E. A. *Kolloidn. Zh.* **1980**, *42*, 371.
193. Mirkind, L. A.; Kazarinov, V. E.; Andreev, V. N.; Albertinskii, G. L. *Elektrokhimiya* **1983**, *19*, 1144.
194. Kazarinov, V. E.; Maksimov, Kh. A.; Tedoradze, G. A.; Gorokhova, L. T. *Elektrokhimiya* **1984**, *20*, 594.
195. Kokarev, G. A.; Kapustin, Yu. I.; Kolesnikov, V. A. *Elektrokhimiya* **1985**, *21*, 1411.
196. Burke, L. D.; Murphy, O. J. *J. Electroanal. Chem.* **1979**, *96*, 19.
197. Boodts, J. F. C.; Trasatti, S. *J. Appl. Electrochem.* **1990**, *137*, 3784.
198. Rishpon, J.; Reshef, I.; Gottesfeld, S. In "Passivity of Metals and Semiconductors," Froment, M., Ed. Elsevier, Amsterdam, 1983, pp. 205–210.
199. Aurian-Blajeni, B.; Beebe, X.; Rauh, R. D.; Rose, T. L. *Electrochim. Acta* **1989**, *34*, 795.
200. Glarum, S. H.; Marshall, J. H. *J. Electrochem. Soc.* **1980**, *127*, 1467.
201. Kötz, R.; Barbero, C.; Haas, O. *J. Electroanal. Chem.* **1990**, *296*, 37.
202. McIntyre, J. D. E.; Basu, S.; Peck, W. F.; Brown, W. L.; Augustyniak, W. M. *Solid State Ionics* **1981**, *5*, 359; *Phys. Rev. B* **1982**, *25*, 7242.
203. Pickup, P. G.; Birss, V. I. *J. Electroanal. Chem.* **1988**, *240*, 171.
204. Conway, B. E. *J. Electrochem. Soc.* **1991**, *138*, 1539.
205. Sarangapani, S.; Lessner, P.; Forchione, J.; Griffith, A.; Laconti, A. B. *J. Power Sources*, **1990**, *29*, 355.
206. Jang, G.-W.; Tsai, E. W.; Rajeshwar, K. *J. Electrochem. Soc.* **1987**, *134*, 2377; *J. Electroanal. Chem.* **1989**, *263*, 383.
207. Aurian-Blajeni, B.; Kimball, A. G.; Robblee, L. S.; Kahanda, G. L. M. K. S.; Tomkiewicz, M. *J. Electrochem. Soc.* **1987**, *134*, 2637.

208. Tsai, E. W.; Rajeshwar, K. *Electrochim. Acta* **1991**, *36*, 27.
209. Hamdani, M.; Koenig, J. F.; Chartier, P. *J. Appl. Electrochem.* **1988**, *18*, 568.
210. Weston, J. E.; Steele, B. C. H. *J. Appl. Electrochem.* **1980**, *10*, 49.
211. Doblhofer, K.; Metikos, M.; Ogumi, Z.; Gerischer, H. *Ber. Bunsenges. Phys. Chem.* **1978**, *82*, 1046.
212. Rishpon, J.; Gottesfeld, S. *J. Electrochem. Soc.* **1984**, *131*, 1960.
213. Ardizzone, S.; Fregonara, G.; Trasatti, S. *Electrochim. Acta* **1990**, *35*, 263.
214. Pickup, P. G.; Birss, V. I. *J. Electroanal. Chem.* **1983**, *240*, 185; *J. Electrochem. Soc.* **1988**, *135*, 41.
215. Tilak, B. V.; Rader, C. G.; Rangarajan, S. K. *J. Electrochem. Soc.* **1977**, *124*, 1880.
216. Ardizzone, S.; Fregonara, G.; Trasatti, S. *J. Electroanal. Chem.* **1989**, *266*, 191.
217. Belova, I. D.; Galyamov, B. Sh.; Roginskaya, Yu. E. *Elektrokhimiya* **1982**, *18*, 777.
218. Burke, I. D.; McCarthy, M. M. *J. Electrochem. Soc.* **1988**, *135*, 1175.
219. Atanasoska, Lj.; Atanasoski, R.; Trasatti, S. *Vacuum* **1990**, *40*, 91.
220. Huang, Y. S.; Park, H. L.; Pollak, F. H. *Mater. Res. Bull.* **1983**, *17*, 1305.
221. Atanasoska, Lj.; Atanasoski, R. T.; Pollak, F. H.; O'Grady, W. E. *Surf. Sci.* **1990**, *230*, 95.
222. O'Grady, W. E.; Atanasoska, Lj.; Pollak, F. H.; Park, H. L. *J. Electroanal. Chem.* **1984**, *178*, 61.
223. Huang, Y. S.; Pollak, F. H. *Solid State Commun.* **1982**, *43*, 921.
224. O'Grady, W. E.; Goel, A. K.; Pollak, F. H.; Park, H. L.; Huang, Y. S. *J. Electroanal. Chem.* **1983**, *151*, 295.
225. Hepel, T.; Pollak, F. H.; O'Grady, W. E. *J. Electrochem. Soc.* **1985**, *132*, 2385.
226. Hepel, T.; Pollak, F. H.; O'Grady, W. E. *J. Electrochem. Soc.* **1984**, *131*, 2094.
227. Castelli, P.; Trasatti, S.; Pollak, F. H.; O'Grady, W. E. *J. Electroanal. Chem.* **1986**, *210*, 189.
228. Kohler, H.; Neu, W.; Dreuer, K.-D.; Schmeisser, D.; Göpel, W. *Electrochim. Acta* **1989**, *34*, 1755.
229. Matsumoto, Y.; Sato, E. *Electrochim. Acta* **1980**, *25*, 585.
230. Faita, G.; Fiori, G. *J. Appl. Electrochem.* **1972**, *2*, 31.
231. Tilak, B. V. *J. Electrochem. Soc.* **1979**, *126*, 1343.
232. Trasatti, S.; Lodi, G. In Ref. 12, Part B, Chapter 12, pp. 521–626.
233. Érenburg, R. G.; Krishtalik, L. I.; Yaroshevskaya, I. P. *Elektrokhimiya* **1975**, *11*, 1236.
234. Pecherskii, M. M.; Gorodetskii, V. V.; Evdokimov, S. V.; Losev, V. V. *Elektrokhimiya* **1981**, *17*, 1087.
235. Losev, V. V. *Elektrokhimiya* **1981**, *17*, 733.
236. Buné, N. Ya.; Chuvaeva, L. E.; Losev, V. V. *Elektrokhimiya* **1987**, *23*, 1249.
237. Losev, V. V.; Buné, N. Ya.; Chuvaeva, L. E. *Electrochim. Acta* **1989**, *34*, 929.
238. Evdokimov, S. V.; Gorodetskii, V. V.; Losev, V. V.; *Elektrokhimiya* **1985**, *21*, 1427.
239. Evdokimov, S. V.; Gorodetskii, V. V. *Elektrokhimiya* **1989**, *25*, 1139.

240. Evdokimov, S. V.; Yanovskaya, M. I.; Roginskaya, Yu. E.; Lubnin, E. N.; Gorodetskii, V. V. *Elektrokhimiya* **1987**, *23*, 1509.

241. Evdokimov, S. V.; Gorodetskii, V. V. *Elektrokhimiya* **1986**, *22*, 982.

242. Harrison, J. A.; Hermijanto, S. D. *J. Electroanal. Chem.* **1987**, *225*, 159.

243. Evdokimov, S. V.; Gorodetskii, V. V. *Elektrokhimiya* **1986**, *22*, 782.

244. Evdokimov, S. V.; Mishenina, K. A.; Gorodetskii, V. V. *Elektrokhimiya* **1988**, *24*, 1475.

245. Mostkova, R. I.; Nikolskaya, N. F.; Krishtalik, L. I. *Elektrokhimiya* **1983**, *19*, 1608.

246. Érenburg, R. G. *Elektrokhimiya* **1984**, *20*, 1602.

247. Consonni, V.; Trasatti, S.; Pollak, F. H.; O'Grady, W. E. *J. Electroanal. Chem.* **1987**, *228*, 393.

248. Érenburg, R. G.; Krishtalik, L. I.; Rogozhina, N. P. *Elektrokhimiya* **1984**, *20*, 1183.

249. Evdokimov, S. V.; Gorodetskii, V. V. *Elektrokhimiya* **1987**, *23*, 1587.

250. Shalaginov, V. V.; Reznik, M. F.; Shub, D. M. *Elektrokhimiya* **1987**, *23*, 1619.

251. Harrison, J. A.; Caldwell, D. L.; White, R. E. *Electrochim. Acta* **1983**, *28*, 1561.

252. Kuhn, A. T.; Mortimer, C. J. *J. Appl. Electrochem.* **1972**, *2*, 283.

253. Buné, N. Ya.; Serebryakova, E. V.; Losev, V. V. *Elektrokhimiya* **1979**, *15*, 745.

254. Kokoulina, D. V.; Bunakova, L. V. *Elektrokhimiya* **1986**, *22*, 689.

255. Buné, N. Ya.; Shilyaeva, G. A.; Losev, V. V. *Elektrokhimiya* **1977**, *13*, 1540.

256. Buné, N. Ya.; Perminova, E. N.; Losev, V. V. *Elektrokhimiya* **1984**, *20*, 1450; **1986**, *22*, 555.

257. Buné, N. Ya.; Perminova, E. N.; Losev, V. V. *Elektrokhimiya* **1984**, *20*, 1561.

258. Kokoulina, D. V.; Bunakova, L. V. *J. Electroanal. Chem.* **1984**, *164*, 377.

259. Kokoulina, D. V.; Bunakova, L. V. *Elektrokhimiya* **1984**, *20*, 1481.

260. Evdokimov, S. V.; Gorodetskii, V. V.; Yanovskaya, M. I.; Roginskaya, Yu. E. *Elektrokhimiya* **1987**, *23*, 1516.

261. Evdokimov, S. V.; Kasatkin, E. V. *Elektrokhimiya*, **1989**, *25*, 539.

262. Krishtalik, L. I. *Elektrokhimiya* **1979**, *15*, 462.

263. Müller, L.; Günther, H. *Elektrokhimiya* **1982**, *18*, 1670.

264. Kuhn, A. T.; Mortimer, C. J. *J. Electrochem. Soc.* **1973**, *120*, 231.

265. Bondar, R. U.; Kalinovskii, E. A.; Kunpan, I. V.; Sorokendya, V. S. *Elektrokhimiya* **1983**, *19*, 1104.

266. Shalaginov, V. V.; Markina, O. V.; Shub, D. M. *Elektrokhimiya* **1982**, *18*, 777.

267. Belova, I. D.; Galyamov, B. Sh.; Roginskaya, Yu. E. *Elektrokhimiya* **1982**, *18*, 777.

268. Krstajić, N.; Spasojević, M.; Jakšić, M. *J. Res. Inst. Catalysis, Hokkaido Univ.* **1984**, *32*, 19.

269. Spasojević, M.; Krstajić, N.; Jakšić, M. *J. Res. Inst. Catalysis, Hokkaido Univ.* **1984**, *32*, 29.

270. Spasojević, M.; Krstajić, N.; Jakšić, M. *Surf. Technol.* **1984**, *21*, 19.

271. Hepel, T.; Pollak, F. H.; O'Grady, W. E. *J. Electroanal. Chem.* **1985**, *188*, 281; *J. Electrochem. Soc.* **1986**, *133*, 69.

272. Augustynski, J. Balsenc, L.; Hinden, J. *J. Electrochem. Soc.* **1978**, *125*, 1093.

273. Mills, A.; McMurray, N. *J. Chem. Soc., Faraday Trans. 1* **1988**, *84*, 379; **1989**, *85*, 2055.

274. Harriman, A.; Pickering, I. J.; Thomas J. M.; Christensen, P. A.; *J. Chem. Soc., Faraday Trans. 1* **1988**, *84*, 2795.

275. Tampi, K. R.; Grätzel, M. *J. Mol. Catal.* **1990**, *60*, 31.

276. Kötz, E. R.; Stucki, S. *J. Electroanal. Chem.* **1987**, *228*, 407.

277. Tilak, B. V.; Tari, K.; Hoover, C. L. *J. Electrochem. Soc.* **1988**, *135*, 1386.

278. Spasojević, M.; Krstajić, N.; Jakšić, M. *J. Res. Inst. Catalysis, Hokkaido Univ.* **1984**, *32*, 29.

279. Lodi, G.; Sivieri, E.; De Battisti, A.; Trasatti, S. *J. Appl. Electrochem.* **1978**, *8*, 135.

280. Iwakura, C.; Fukuda, K.; Tamura, E. *J. Electrochem. Soc. Jpn.* **1977**, *45*, 135.

281. Krishtalik, L. I. *Electrochim. Acta* **1981**, *26*, 329.

282. Kokoulina, D. V.; Bunakova, L. V.; Khomyakova, T. I.; Sirotkina, E. B. *Elektrokhimiya* **1986**, *22*, 24.

283. Kokoulina, D. V.; Bunakova, L. V.; Eleva, M. Z. *Elektrokhimiya* **1985**, *21*, 1121.

284. Iwakura, C.; Honij, A.; Tamura, H. *Electrochim. Acta* **1981**, *26*, 1319.

285. Iwakura, C.; Tada, H.; Tamura, H. *J. Electrochem. Soc. Jpn.* **1977**, *45*, 202.

286. Singh, R. N.; Hamdani, M.; Koenig, J.-F.; Poillerat, G.; Gautier, J. L.; Chartier, P. *J. Appl. Electrochem.* **1990**, *20*, 442.

287. Singh, R. N.; Koenig, J.-F.; Poillerat, G.; Chartier, P. *J. Electrochem. Soc.* **1990**, *137*, 1408.

288. Botejue Nadesan, J. C.; Tseung, A. C. C. *J. Electrochem. Soc.* **1985**, *132*, 2957.

289. Efremov, B. N.; Tarasevich, M. R. *Elektrokhimiya* **1981**, *17*, 1672.

290. Burke, L. D.; McCharty, M. *Electrochim. Acta* **1984**, *29*, 211.

291. Rasiyah, P.; Tseung, A. C. C. *J. Electrochem. Soc.* **1984**, *131*, 803.

292. Cahan, B. D.; Chen, C. T. *J. Electrochem. Soc.* **1982**, *129*, 700.

293. Rasiyah, P.; Tseung, A. C. C.; Hibbert, D. B. *J. Electrochem. Soc.* **1982**, *129*, 1724.

294. Burke, L. D.; Murphy, O. J.; O'Neill, J. F.; Venkatesan, S. *J. Chem. Soc., Faraday Trans. 1* **1977**, *73*, 1659.

295. Wohlfarth-Mehrens, M.; Heitbaum, J. *J. Electroanal. Chem.* **1987**, *237*, 251.

296. Hibbert, D. B. *J. Chem. Soc. Chem. Commun.* **1980**, 202.

297. Hibbert, D. B.; Churchill, C. R. *J. Chem. Soc., Faraday Trans. 1* **1984**, *80*, 1965.

298. Willson, J.; Wolter, D.; Heitbaum, J. *J. Electroanal. Chem.* **1985**, *195*, 299.

299. Kokoulina, D. V.; Bunakova, L. V. *Elektrokhimiya* **1985**, *21*, 1629.

300. Krishtalik, L. I. *Elektrokhimiya* **1968**, *4*, 240.

301. Carugati, A.; Lodi, G.; Trasatti, S. *Mater. Chem.* **1981**, *6*, 255.

302. Rasiyah, P.; Tseung, A. C. C. *J. Electrochem. Soc.* **1983**, *130*, 365.

303. Conway, B. E.; Salomon, M. *Electrochim. Acta* **1964**, *9*, 1599.

304. Angelinetta, C.; Falciola, M.; Trasatti, S. *J. Electroanal. Chem.* **1986**, *205*, 347.

305. Vuković, M.; Angerstein-Kozlowska, H.; Conway, B. E. *J. Appl. Electrochem.* **1982**, *12*, 193.

306. Barral, G.; Diard, J. P.; Montella, C. *Electrochim. Acta* **1986**, *31*, 277.
307. Loučka, T. *J. Appl. Electrochem.* **1990**, *20*, 522.
308. Cairns, J. F.; Denton, D. A.; Izard, P. A. *Eur. Pat. Appl.* **1984**, EP 129 374; *Chem. Abstr.* **1985**, *102*, 102442.
309. Nidola, A. *PCT Int. Appl.* **1986**, WO86 03 790; *Chem. Abstr.* **1986**, *105*, 122974.
310. Clerc-Renand, J.; Leroux, F.; Ravier, D. *Eur. Pat. Appl.* **1987**, EP 240 413; *Chem. Abstr.* **1988**, *108*, 45864.
311. Nicolas, E. *Eur. Pat. Appl.* **1981**, 23 368; *Chem. Abstr.* **1981**, *94*, 199838.
312. Galizzioli, D.; Tantardini, F.; Trasatti, S. *J. Appl. Electrochem.* **1975**, *5*, 203.
313. Boodts, J. C. F.; Fregonara, G.; Trasatti, S. In Ref. 1, pp. 135–45.
314. Kötz, E. R.; Stucki, S. *J. Appl. Electrochem.* **1987**, *17*, 1190.
315. Manoharan, R.; Goodenough, J. B. *J. Electrochem. Soc.* **1990**, *137*, 910.
316. Trasatti, S. In "Modern Chlor-Alkali Technology," Wellington, T. C., Ed. Elsevier, Amsterdam, 1991, Vol. 5, pp. 281–294.
317. Nidola, A.; Schira, R. *J. Electrochem. Soc.* **1986**, *133*, 1653.
318. Trasatti, S. *Z. Phys. Chem. N. F.* **1975**, *98*, 75.
319. Hepel, T.; Pollak, F. H.; O'Grady, W. E. *J. Electroanal. Chem.* **1987**, *236*, 295; *J. Electrochem. Soc.* **1988**, *135*, 562.
320. Yeager, E. *J. Mol. Catal.* **1986**, *38*, 5.
321. Yeager, E.; Bindra, P. *Chem.-Ing.-Tech.* **1980**, *52*, 384.
322. Tarasevich, M. R. *Elektrokhimiya* **1973**, *9*, 599.
323. Damjanovic, A. In "Modern Aspects of Electrochemistry," Bockris, J. O'M.; Conway, B. E. Eds.; Butterworths, London, 1969, Vol. 5, Chapter 5, pp. 369–483.
324. Tarasevich, M. R. *Elektrokhimiya* **1981**, *17*, 1208.
325. Yeager, E. *Electrochim. Acta* **1984**, *29*, 1527.
326. Kannan, A. M.; Shukla, A. K.; Sathyanarayana, S. *J. Electroanal. Chem.* **1990**, *281*, 339.
327. Trunov, A. M.; Kotseruba, A. I.; Yakovleva, N. M.; Polishchuk, V. E. *Elektrokhimiya* **1978**, *14*, 1165.
328. Kaisheva, A.; Gamburtsev, S.; Iliev, I. *Elektrokhimiya* **1981**, *17*, 1362.
329. Vilinskaya, V. S.; Bulavina, N. G.; Shepelev, V. Ya.; Burshtein, R. Kh. *Elektrokhimiya* **1979**, *15*, 932.
330. Meadowcroft, D. B. *Nature* **1970**, *226*, 847.
331. Fierro, J. L. G. *Catal. Today,* **1990**, *8*, 153.
332. Teraoka, Y.; Zhang, H.-M.; Furukawa, S.; Yamazoe, N. *Chem. Lett.* **1985**, 1743.
333. Yamazoe, N.; Teraoka, Y. *Catal. Today* **1990**, *8*, 175.
334. Bagotsky, V. S.; Shumilova, N. A.; Khrushcheva, E. I. *Electrochim. Acta* **1976**, *21*, 919.
335. Trunov, A. M.; Domnikov, A. A.; Reznikov, G. L.; Yuppets, F. R. *Elektrokhimiya* **1979**, *15*, 783.
336. Egdell, R. G.; Goodenough, J. B.; Hamnett, A.; Naish, C. C. *J. Chem. Soc., Faraday Trans. 1,* **1983**, *79*, 893.

337. Almashev, B. K.; Kim, S. N.; Pushkareva, G. A.; Kokunova, V. N.; Fasman, A. B.; Sinitsyn, N. M. *Elektrokhimiya* **1989**, *25*, 1663.

338. Horkans, J.; Shafer, W. W. *J. Electrochem. Soc.* **1977**, *124*, 1202.

339. Goodenough, J. B.; Manoharan, R.; Paranthaman, M. *J. Am. Chem. Soc.* **1990**, *112*, 2076.

340. O'Grady, W.; Iwakura, C.; Huang, J.; Yeager, E. In "Electrocatalysis," Breiter, M., Ed. The Electrochemical Society, Pennington, NJ, 1982, pp. 286-302.

341. Tseung, A. C. C.; *J. Electrochem. Soc.* **1978**, *125*, 1660.

342. Takeda, Y.; Kanno, R.; Kondo, T.; Yamamoto, O.; Taguchi, H.; Shimada, M.; Koizumi, M. *J. Appl. Electrochem.* **1982**, *12*, 275.

343. Karlsson, G. *Electrochim. Acta* **1985**, *30*, 1555.

344. Manoharan, R.; Shukla, A. K. *Electrochim. Acta* **1985**, *30*, 205.

345. Raj, I. A.; Rao, K. V.; Venkatesan, V. K. *Bull. Electrochem.* **1986**, *2*, 157.

346. Bronoel, G.; Grenier, J. C.; Reby, J. *Electrochim. Acta* **1980**, *25*, 1015.

347. Trunov, A. M.; Verenikina, N. N. *Elektrokhimiya* **1981**, *17*, 135.

348. Takahashi, H.; Bindra, P.; Yeager, E. "Extended Abstracts," Electrochemical Society Meeting, Los Angeles, CA, 1979; The Electrochemical Society, Pennington, NJ; paper 195.

349. Voloshin, A. G.; Kramarenko, N. I.; Kolesnikova, I. P.; Korolenko, S. D. *Elektrokhimiya* **1977**, *13*, 1724.

350. Efremov, B. N.; Zakharkin, G. I.; Zhukov, S. R.; Tarasevich, M. R. *Elektrokhimiya* **1978**, *14*, 937.

351. Efremov, B. N.; Zakharkin, G. I.; Tarasevich, M. R.; Zhukov, S. R. *Zh. Fiz. Khim.* **1978**, *52*, 1671.

352. King, W. J.; Tseung, A. C. C. *Electrochim. Acta* **1974**, *19*, 485.

353. Trunov, A. M. *Elektrokhimiya* **1986**, *22*, 1093.

354. Couper, A. M.; Pletcher, D.; Walsh, F. C. *Chem. Rev.* **1990**, *90*, 837.

355. Conway, B. E. In Ref. 12, Part B, Chapter 9, pp. 433-519.

356. Fioshin, M. Ya.; Avrutskaya, I. A. *Usp. Khim.* **1975**, *44*, 2067.

357. Kazarinov, V. E.; Tedoradze, G. A.; Gorokhova, L. T.; Medvedev, V. A.; Babkin, V. A. *Elektrokhimiya* **1979**, *15*, 1894.

358. Mirkind, L. A.; Albertinskii, G. L.; Kornienko, A. G. *Elektrokhimiya*, **1983**, *19*, 122.

359. Iwakura, C.; Goto, F.; Tamura, H. *J. Electrochem. Soc. Jpn.* **1980**, *48*, 21.

360. Beck, F.; Schulz, H. *J. Electroanal. Chem.* **1987**, *229*, 339.

361. Beck, F.; Schulz, H. *Ber. Bunsenges. Phys. Chem.* **1984**, *88*, 155; *Electrochim. Acta* **1984**, *29*, 1569.

362. Cox, P.; Pletcher, D. *J. Appl. Electrochem.* **1990**, *20*, 549.

363. Cox, P.; Pletcher, D. *J. Appl. Electrochem.* **1991**, *21*, 11.

364. Kennedy, B. J.; Smith, A. W. *J. Electroanal. Chem.* **1990**, *293*, 103.

365. Burke, L. D.; Murphy, O. J. *J. Electroanal. Chem.* **1979**, *101*, 351.

366. Kennedy, B. J.; Smith, A. W.; Wagner, F. E. *Aust. J. Chem.* **1990**, *43*, 913.

367. Ohmori, T.; Nodasaka, Y.; Enyo, M. *J. Electroanal. Chem.* **1990**, *281*, 331.
368. Biswas, P. C.; Ohmori, T.; Enyo, M. *J. Electroanal. Chem.* **1991**, *305*, 205.
369. O'Sullivan, E. J. M.; White, J. R. *J. Electrochem. Soc.* **1989**, *136*, 2576.
370. Zyablitseva, M. P.; Safonova, T. Ya.; Petrii, O. A. *Elektrokhimiya* **1984**, *20*, 131.
371. Asokan, K.; Krishnan, V. *Bull. Electrochem.* **1988**, *4*, 827.
372. Cosnier, S.; Deronzier, A.; Moutet, J.-C. *Inorg. Chem.* **1988**, *27*, 2390.
373. Asokan, K.; Krishnan, V. *Bull. Electrochem.* **1988**, *4*, 369.
374. Trasatti, S. *J. Electroanal. Chem.* **1980**, *111*, 125.
375. Tseung, A. C. C.; Jasem, S. *Electrochim. Acta* **1977**, *22*, 31.
376. Arikado, T.; Iwakura, C.; Tamura, H. *Electrochim. Acta* **1978**, *23*, 9.
377. Inai, M.; Iwakura, C.; Tamura, H. *J. Electrochem. Soc. Jpn.* **1980**, *48*, 173, 229.
378. Pankratiev, Yu. D. *React. Kinet. Catal. Lett.* **1982**, *20*, 255.
379. Ichikawa, S. *Chem. Eng. Sci.* **1990**, *45*, 529.
380. Vijh, A. *J. Catalysis* **1974**, *33*, 385.
381. Trasatti, S. *J. Electroanal. Chem.* **1972**, *39*, 163.
382. Appleby, A. J. In "Modern Aspects of Electrochemistry," Conway, B. E.; Bockris, J. O'M., Eds. Plenum, New York, 1974, Chapter 5, pp. 369–478.
383. Mehandru, S. P.; Anderson, A. B. *J. Electrochem. Soc.* **1989**, *136*, 158.
384. Kojima, I.; Adachi, H.; Yasumori, I. *Surf. Sci.* **1983**, *130*, 50.
385. Inai, M.; Iwakura, C.; Tamura, H. *Electrochim. Acta* **1979**, *24*, 993.
386. Mills, A.; Williams, G. *J. Chem. Soc. Chem. Commun.* **1989**, 321.
387. Matsumoto, Y.; Yoneyama, H.; Tamura, H. *J. Electroanal. Chem.* **1977**, *83*, 237.
388. Trasatti, S. In "Electrocatalysis," O'Grady, W. E.; Ross, P. N.; Will, F. G., Eds. The Electrochemical Society, Pennington, NJ, 1982, pp. 73–91.
389. Parsons, R. *Trans. Faraday Soc.* **1958**, *54*, 1053.
390. Boodts, J. C. F.; Trasatti, S. *J. Electrochem. Soc.* **1990**, *137*, 3784.
391. Iwakura, C.; Inai, M.; Manabe, M.; Tamura, H. *J. Electrochem. Soc. Jpn.* **1980**, *48*, 91.
392. Barral, G.; Guitton, J.; Montella, C.; Vergara, F. *Surf. Technol.* **1980**, *10*, 25.
393. Comninellis, Ch.; Vercesi, G. P. *J. Appl. Electrochem.* **1991**, *21*, 335.
394. Spasojević, M. D.; Krstajić, V. N.; Jakšić, M. M. *J. Res. Inst. Catalysis, Okkaido Univ.* **1983**, *31*, 77; *J. Mol. Catal.* **1987**, *40*, 311.
395. Beck, F. *Corr. Sci.* **1989**, *29*, 379; *Electrochim. Acta* **1989**, *34*, 811.
396. Hine, M.; Yasuda, M.; Iida, T.; Ogata, Y.; Hara, K. *Electrochim. Acta* **1984**, *29*, 1447.
397. Makarychev, Yu. B.; Spasskaya, E. K.; Khodkevich, S. D.; Yakimenko, L. M. *Elektrokhimiya*, **1976**, *12*, 994.
398. Pollock, R. J.; Houlihan, J. F.; Bain, A. N.; Coryea, B. S. *Mater. Res. Bull.* **1984**, *19*, 17.
399. Hayfield, P. C. S.; Clarke, R. L. In Ref. 1, pp. 87–110.

400. Gorichev, I. G.; Kipriyanov, N. A. *Usp. Khim.* **1984**, *53*, 1790.

401. Pourbaix, M. "Atlas of Electrochemical Equilibria in Aqueous Solutions"; NACE, Houston, TX, 1974.

402. Bondar, R. U.; Kalinovskii, A. E. *Elektrokhimiya* **1978**, *14*, 730.

403. Kötz, R.; Stucki, S.; Scherson, D.; Kolb, D. M. *J. Electroanal. Chem.* **1984**, *172*, 211.

404. Kolotyrkin, Ya. M.; Losev, V. V.; Chemodanov, A. N. *Mater. Chem. Phys.* **1988**, *19*, 1.

405. Novikov, E. A.; Zhinkin, N. V.; Éberil, V. I.; Busse-Macukas, V. *Elektrokhimiya* **1990**, *26*, 245.

406. Pecherskii, M. M.; Gorodetskii, V. V.; Buné, N. Ya. *Elekrokhimiya* **1988**, *24*, 853.

407. Gorodetskii, V. V.; Pecherskii, M. M.; Yanke, V. B.; Shub, D. M.; Losev, V. V. *Elektrokhimiya* **1979**, *15*, 559.

408. Klementeva, V. S.; Uzbekov, A. A.; Kubasov, V. L.; Lambrev, V. G. *Elektrokhimiya* **1985**, *58*, 681.

409. Zhinkin, N. V.; Novikov, E. A.; Fedotova, N. S.; Éberil, V. I.; Busse-Machukas, V. B. *Elektrokhimiya* **1989**, *25*, 1094.

410. Uzbekov, A. A.; Klementeva, V. S. *Elektrokhimiya* **1985**, *21*, 758.

411. Pecherskii, M. M.; Gorodetskii, V. V.; Buné, N. Ya.; Losev, V. V. *Elektrokhimiya* **1986**, *22*, 656.

412. Gorodetskii, V. V.; Pecherskii, M. M.; Yanke, V. B.; Buné, N. Ya.; Busse-Machukas, V. B.; Kubasov, V. L.; Losev, V. V. *Elektrokhimiya* **1981**, *17*, 513.

413. Klementeva, V. S.; Uzbekov, A. A. *Elektrokhimiya* **1985**, *21*, 796.

414. Kishi, T. Sugimoto, Y.; Nagai, T. *Surf. Technol.* **1985**, *26*, 245.

415. Evdokimov, S. V.; Mishenina, K. A. *Elektrokhimiya* **1989**, *25*, 1605.

416. Iwakura, C.; Sakamoto, K. *J. Electrochem. Soc.* **1985**, *132*, 2420.

417. Rolewicz, J.; Comninellis, C.; Plattner, E.; Hinden, J. *Chim.* **1988**, *42*, 75.

418. Burke, L. D.; McCarthy, M. *Electrochim. Acta* **1984**, *29*, 211.

419. Rethinaraj, J. P.; Chockalingam, S. C.; Kulandaisamy, S.; Visvanathan, S. *Bull. Electrochem.* **1986**, *2*, 629.

420. Trusov, G. N.; Gochalieva, E. P.; Kryuchkova, É. Ya.; Fandeeva, M. F. *Elektrokhimiya* **1982**, *18*, 220.

421. Tamura, H.; Iwakura, C. *Int. J. Hydrogen Energy* **1982**, *7*, 857.

422. Shub, D. M.; Chemodanov, A. N.; Shalashnikov, V. V. *Elektrokhimiya* **1978**, *14*, 595.

423. Mills, A.; Giddings, S.; Patel, I. *J. Chem. Soc., Faraday Trans. 1* **1987**, *83*, 2317.

424. Mills, A.; Giddings, S.; Patel, I.; Lawrence, C. *J. Chem. Soc., Faraday Trans. 1* **1987**, *83*, 2331.

425. Efremov, B. N.; Tarasevich, M. R.; Zakharkin, G. I.; Zhukov, S. R. *Zh. Fiz. Khim.* **1978**, *52*, 1971.

426. Efremov, B. N.; Tarasevich, M. R.; Zakharkin, G. I.; Zhukov, S. R. *Elektrokhimiya* **1978**, *14*, 1504.

427. Klissurski, D.; Dimitrova, R. *J. Chem. Soc. Jpn.* **1990**, *63*, 590.

428. Nidola, A.; Schira, R. "Extended Abstracts," Electrochemical Society Meeting, Philadelphia, 1987; The Electrochemical Society, Pennington, NJ; paper 375.

429. Long, Y.-C.; Zang, Z.-D.; Dwight, K.; Wold, A. *Mat. Res. Bull.* **1988**, *23*, 631.

430. Yoshida, M.; Noaki, Y. In Ref. 1, pp. 15–37.

431. Ivanova, A. S.; Dzisko, V. A.; Moroz, É. M.; Noskova, S. P. *Kinet. Katal.* **1986**, *27*, 428.

432. Ivanova, A. S.; Dzisko, V. A.; Moroz, É. M.; Noskova, S. P. *Kinet. Katal.* **1985**, *26*, 1193.

433. Pecherskii, M. M.; Evdokimov, S. V.; Chuvaeva, L. E.; Buné, N. Ya.; Gorodetskii, V. V. *Elektrokhimiya* **1988**, *24*, 850.

434. Buné, N. Ya.; Portnova, M. Yu.; Gorodetskii, V. V.; Pecherskii, M. M.; Yanke, V. B.; Filatov, V. P.; Garkuska, A. A.; Losev, V. V. *Zh. Prikl. Khim.* **1981**, *54*, 2027.

435. Mirkind, L. A.; Gorodetskii, V. V.; Albertinskii, G. L.; Gorodetskaya, I. L. *Elektrokhimiya* **1983**, *19*, 1183.

436. Burke, L. D.; Healy, J. F.; Murphy, O. J. *J. Appl. Electrochem.* **1983**, *13*, 459.

437. Éberil, V. I.; Zhinkin, N. V.; Archakov, V. P.; Izosenkov, R. I.; Busse-Machukas, V. B.; Mazanko, A. F. *Elektrokhimiya* **1986**, *22*, 459.

438. Viswanathan, K.; Tilak, B. V. *J. Electrochem. Soc.* **1984**, *131*, 1551.

439. Stucki, S.; Kötz, R.; Carcer, B.; Suter, W. *J. Appl. Electrochem.* **1991**, *21*, 99.

440. Matsuki, K.; Sugawara, M. In "Progress in Batteries and Solar Cells"; JEC Press, Cleveland, 1989, Vol. 8, pp. 28–35.

441. Inai, M.; Iwakura, C.; Tamura, H. *J. Appl. Electrochem.* **1979**, *9*, 745.

442. Stucki, S.; Baumann, H.; Christen, H. J.; Kötz, R. *J. Appl. Electrochem.* **1987**, *17*, 773.

443. Walaszkowski, J. *Bull. Electrochem.* **1987**, *3*, 535.

444. Rethinaraj, J. P.; Chockalingam, S. C.; Kulandaisamy, S.; Viswanathan, S. *Bull. Electrochem.* **1986**, *2*, 635.

445. Inai, M.; Iwakura, C.; Tamura, H. *J. Electrochem. Soc. Jpn.* **1979**, *47*, 668; **1980**, *48*, 384.

446. Danilov, F. I.; Velichenko, A. B.; Loboda, S. M.; Shalaginov, V. V.; Shub, D. M. *Elektrokhimiya* **1988**, *24*, 855.

447. Katsube, T.; Lauks, I. R.; van der Spiegel, J.; Zemel, J. N. *Jpn. J. Appl. Phys.* **1983**, *22*, 469.

448. Danilov, F. I.; Velichenko, A. B.; Loboda, S. M.; Shalaginov, V. V.; Shub, D. M. *Elektrokhimiya* **1989**, *25*, 257.

449. Ramachandran, P.; Venkateswaran, K. V. *Bull. Electrochem.* **1988**, *4*, 901.

450. Vargalyuk, V. F.; Loshkarev, Yu. M.; Stets, N. V. *Elektrokhimiya* **1989**, *25*, 992.

451. Bandi, A. *J. Electrochem. Soc.* **1990**, *137*, 2157.

452. Krusin-Elbaum, L.; Wittmer, M. *J. Electrochem. Soc.* **1988**, *135*, 2610.

453. Noufi, R. *J. Electrochem. Soc.* **1983**, *130*, 2126.

454. Brandys, M.; Sassoon, R. E.; Rabani, J. *J. Phys. Chem.* **1987**, *91*, 953.

455. Fukuda, K.-I.; Iwakura, C.; Tamura, H. *Electrochim. Acta* **1978**, *23*, 613; **1979**, *24*, 363.

456. Pavlović, O. Z.; Krstajić, N. V.; Spasojević, M. D. *Surf. Coat. Technol.* **1988**, *37*, 177.

457. Dalard, P.; Deroo, D.; Foscallo, D.; Mouliom, C. *Solid State Ionics* **1985**, *15*, 91.

458. Kiwi, J. *J. Chem. Soc., Faraday Trans. 2* **1982**, *78*, 339.

459. Ohzuku, T.; Sawai, K.; Hirai, T. *J. Electrochem. Soc.* **1990**, *137*, 3004.

460. Pushpavanam, S.; Mohan, S.; Silaimani, S. M.; Viswanathan, R.; Narasimham, K. C. *Bull. Electrochem.* **1990**, *6*, 185.

461. Pleskov, Yu. V.; Kolbasov, G. Ya.; Kraitsberg, A. M.; Taranenko, N. I.; Lipyavka, V. G. *Elektrokhimiya* **1990**, *26*, 366.

462. Vasudevan, S.; Vest, R. W. *Ceram. Trans.* **1990**, *11*, 417.

463. Krasilova, T. Ya.; Matlis, M. Ya. *Zh. Prikl. Khim.* **1980**, *53*, 1291.

464. Kondrikov, N. B.; Kiselev, E. Yu.; Kuchma, I. V.; Ilín, I. E.; Berdyugina, V. P.; Shub, D. M. *Elektrokhimiya* **1990**, *26*, 580.

465. Lyubushkin, V. I.; Lyubushkina, E. T. *Zh. Prikl. Khim.* **1980**, *53*, 2224.

466. Gutiérrez, C.; Salvador, P. *J. Electroanal. Chem.* **1985**, *187*, 139.

467. Nieh, C. W.; Kolawa, E.; So, F. C. T.; Nicolet, M.-A. *Mater. Lett.* **1988**, *6*, 177.

468. Abe, O.; Taketa, Y. *J. Phys. D* **1989**, *22*, 1777.

469. Szpytma, A.; Kusy, A. *Thin Solid Films* **1984**, *121*, 263.

470. Halder, N. C.; Snyder, R. *J. Electrocomp. Sci. Technol.* **1984**, *11*, 123.

471. Cámara, O. R. *J. Electroanal. Chem.* **1990**, *284*, 155.

472. Burke, L. D.; Murphy, O. J. *J. Electroanal. Chem.* **1990**, *284*, 155.

473. Morea, G.; Sabbatini, L.; Zambonin, P. G.; Tangari, N.; Tortorella, V. *J. Chem. Soc., Faraday Trans. 1* **1989**, *85*, 3861.

474. Green, M. L.; Gross, M. E.; Papa, L. E.; Schnoes, K. J.; Brasen, D. *J. Electrochem. Soc.* **1985**, *132*, 2677.

475. Malitesta, C.; Morea, G.; Sabbatini, L.; Zambonin, P. G. *Ann. Chim.* **1988**, *78*, 473.

476. Wagner, N.; Brümmer, O. *Phys. Stat. Sol. (a)* **1985**, *89*, K123.

477. Anderson, D. P.; Warren, L. F. *J. Electrochem. Soc.* **1984**, *131*, 347.

478. Prosychev, I. I.; Shaplygin, I. S. *Zh. Neorg. Khim.* **1981**, *26*, 3137.

479. De Battisti, A.; Brina, R.; Gavelli, G.; Benedetti, A.; Fagherazzi, G. *J. Electroanal. Chem.* **1985**, *200*, 93.

480. Bandi, A.; Mihelis, A.; Vartires, I.; Ciortan, E.; Rosu, I. *J. Electrochem. Soc.* **1987**, *134*, 1982.

481. Habicher, B.; Fahidy, Th. Z.; *Metall (Berlin)* **1987**, *41*, 1014.

482. Kötz, R.; Stucki, S.; Carcer, B. *J. Appl. Electrochem.* **1991**, *21*, 14.

483. Wagner, N.; Horx, M.; Jacobs, M. *Cryst. Res. Technol.* **1990**, *25*, 349.

484. Wagner, N.; Kühnemund, L. *Cryst. Res. Technol.* **1989**, *24*, 1009.

485. Comninellis, Ch.; Plattner, E. *Oberfläche-Surf.* **1982**, *23*, 315.

486. Albella, J. M.; Fernández-Navarrete, L.; Martínex-Duart, J. M. *J. Appl. Electrochem.* **1981**, *11*, 273.

487. Burke, L. D.; Ahern, M. J. *J. Electrochem. Soc.* **1985**, *132*, 2662.

488. Sorokendya, V. S.; Bondar, R. U. *Elektrokhimiya* **1989**, *25*, 1553.

489. Vercesi, G. P.; Salamin, J.-Y.; Comninellis, Ch. *Electrochim. Acta* **1991**, *36*, 991.

490. Schiller, G.; Bolwin, K.; Schurnberger, W. *Thin Solid Films* **1989**, *174*, 85.

491. Kishi, T.; Takahashi, S.; Nagai, T. *Surf. Coat. Technol.* **1986**, *27*, 351.

492. Kolotyrkin, Ya. M.; Belova, I. D.; Roginskaya, Yu. E.; Kozhevnikov, V. B.; Zakharkin, D. S.; Venentsev, Yu. N. *Mater. Chem. Phys.* **1984**, *11*, 29.

493. Shalaginov, V. V.; Kasatkin, É. V. *Elektrokhimiya* **1988**, *24*, 1686.

494. Markina, E. L.; Tarasevich, M. R.; Efremov, B. N. *Elektrokhimiya* **1988**, *24*, 93.

495. Shalaginov, V. V.; Kozlova, N. V.; Lomova, V. N.; Shub, D. M. *Zh. Neorg. Khim.* **1983**, *28*, 2196.

496. Shub, D. M.; Reznik, M. F.; Shalaginov, V. V.; Lubnin, E. N.; Kozlova, N. V.; Lomova, V. N. *Elektrokhimiya* **1983**, *19*, 502.

497. Kozlova, N. V.; Shalaginov, V. V.; Lomova, V. N.; Shub, D. M. *Zh. Neorg. Khim.* **1983**, *28*, 2455.

498. Haenen, J.; Visscher, W.; Barendrecht, E. *J. Electroanal. Chem.* **1986**, *208*, 273, 279, 323.

499. Belova, I.; Roginskaya, Yu. E.; Shifrina, R. R.; Gagarin, S. G.; Plekhanov, Yu. V.; Venevtsev, Yu. N. *Solid State Commun.* **1983**, *47*, 577.

500. Bockris, J. O'M.; Otagawa, T.; Young, V. *J. Electroanal. Chem.* **1983**, *150*, 633.

501. Battaglin, G.; Carnera, A.; Lodi, G.; Giorgi, E., Daghetti, A.; Trasatti, S. *J. Chem. Soc., Faraday Trans. 1* **1984**, *80*, 913.

502. Kokarev, G. A.; Kolesnikov, V. A.; Gubin, A. F.; Korobanov, A. A. *Elektrokhimiya* **1982**, *18*, 466.

503. Tamura, H.; Oda, T.; Nagayama, M.; Furuichi, R. *J. Electrochem. Soc.* **1989**, *136*, 2782.

504. Vitinsh, A.; Safonova, T. Ya.; Petrii, O. A. *Elektrokhimiya* **1990**, *26*, 614.

505. Harrison, J. A.; Caldwell, D. L.; White, R. E. *Electrochim. Acta* **1984**, *29*, 203.

506. Kühnemund, L.; Heidrich, H.-J.; Sabela, R.; Stepan, L.; Müller, L. *Z. Phys. Chem.* **1990**, *271*, 901.

507. Heidrich, H.-J.; Múller, L.; Schatte, G. *Elektrokhimiya* **1990**, *26*, 905.

508. Bondar, R. U.; Sorokendya, V. S.; Kalinovskii, E. A. *Elektrokhimiya* **1984**, *20*, 1369.

509. Carcia, P. F.; Flippen, R. B.; Bierstedt, P. E. *J. Electrochem. Soc.* **1980**, *127*, 596.

510. Morita, M.; Iwakura, C.; Tamura, H. *J. Electrochem. Soc. Jpn.* **1980**, *48*, 12.

511. Bondar, R. U.; Kalinovskii, E. A. *Elektrokhimiya* **1980**, *16*, 1492.

512. Shalaginov, V. V.; Shub, D. M.; Kozlova, N. V.; Lubnin, E. N.; Kulkova, N. V. *Elektrokhimiya* **1983**, *56*, 1302.

513. Buné, N. Ya.; Portnova, M. Yu.; Pilatov, V. P.; Losev, V. V. *Elektrokhimiya* **1984**, *20*, 1291.

514. Iwakura, C.; Inai, M.; Uemura, T.; Tamura, H. *Electrochim. Acta* **1981**, *26*, 579.

515. Melsheimer, J.; Ziegler, D. *Thin Solid Films* **1988**, *163*, ,301.

516. Denton, D. A.; Harrison, J. A.; Knowles, R. I. *Electrochim. Acta* **1981**, *26*, 1197.

517. Harrison, J. A.; Caldwell, D. L.; White, R. E. *Electrochim. Acta* **1984**, *29*, 1139.

518. Telepnya, Yu. V.; Shub, D. M.; Zhdan, P. A.; Kasatkin, É. V. *Elektrokhimiya* **1990**, *26*, 93.

519. Matsumoto, Y.; Tazawa, T.; Muroi, N.; Sato, E.-I. *J. Electrochem. Soc.* **1986**, *133*, 2257.

520. Vuković, M. *J. Appl. Electrochem.* **1987**, *17*, 737.

521. Horowitz, H. S.; Longo, J. M.; Horowitz, H. H. *J. Electrochem. Soc.* **1983**, *130*, 1851.
522. Otagawa, T.; Bockris, J. O'M. *J. Electrochem. Soc.* **1982**, *129*, 2391.
523. Rethinaraj, J. P.; Chockalingam, S. C.; Kulandaisamy, S.; Visvanathan, S. *Bull. Electrochem.* **1988**, *4*, 969.
524. Vermeiren, Ph.; Leysen, R.; King, H. W.; Murphy, G. G.; Vandenborre, H. *Adv. Hydrogen Energy* **1986**, *5*, 431.
525. Bockris, J. O'M.; Otagawa, T. *J. Phys. Chem.* **1983**, *87*, 2960.
526. van Veen, J. A. R.; van der Eijk, J. M.; De Ruiter, R.; Huizinga, S. *Electrochim. Acta* **1988**, *33*, 51.
527. Wattiaux, A.; Grenier, J. C.; Pouchard, M.; Hagenmuller, P. *J. Electrochem. Soc.* **1987**, *134*, 1714.
528. Vondrák, J.; Doležal, L. *Electrochim. Acta* **1984**, *29*, 477.
529. Iwakura, C.; Edamoto, T.; Tamura, H. *Bull. Chem. Soc. Jpn.* **1986**, *59*, 145.
530. Willems, H.; Kobussen, A. G. C.; de Wit, J. H. W. *J. Electroanal. Chem.* **1985**, *194*, 317.
531. Kobussen, A. G. C.; Willems, H.; Broers, G. H. J. *J. Electroanal. Chem.* **1982**, *142*, 67.
532. Raj, I. A.; Chandrasekaran, R.; Venkatesan, V. K. *Bull. Electrochem.* **1987**, *3*, 443.
533. Brunschwig, B. S.; Chou, M. H.; Creutz, C.; Ghosh, P.; Sutin, N. *J. Am. Chem. Soc.* **1983**, *105*, 4832.
534. Parmon, V. N.; Elizarova, G. L.; Kim, T. V. *React. Kinet. Catal. Lett.* **1982**, *21*, 195.
535. Conway, B. E.; Liu, T. C. *Ber. Bunsenges. Phys. Chem.* **1987**, *91*, 461.
536. Hibbert, D. B. *J. Chem. Soc. Chem. Commun.* **1980**, 202.
537. Brossard, L. *J. Appl. Electrochem.* **1991**, *21*, 612.
538. Angelo, A. C. D.; Gonzalez, E. R.; Avaca, L. A. *Int. J. Hydrogen Energy* **1991**, *16*, 1.
539. Davidson, C. R.; Kissel, G.; Srinivasan, S. *J. Electroanal. Chem.* **1982**, *132*, 129.
540. Rasiyah, P.; Tseung, A. C. C. *J. Electrochem. Soc.* **1983**, *130*, 2384.
541. Kishi, T.; Shiota, K. *Surf. Coat. Technol.* **1988**, *34*, 287.
542. Raj, I. A.; Vasu, K. I. *Int. J. Hydrogen Energy* **1990**, *15*, 751.
543. Stand, N.; Sokol, H.; Ross, P. N. *J. Electrochem. Soc.* **1989**, *136*, 3570.
544. Nidola, A.; Schira, R. "Extended Abstracts," Electrochemical Society Meeting, Philadelphia, 1987; The Electrochemical Society, Pennington, NJ; paper 376.
545. Ho, S.-I.; Whelan, D. P.; Rajeshwar, K.; Weiss, A.; Murley, M.; Reid, R. *J. Electrochem. Soc.* **1988**, *135*, 1452.
546. Keller, P.; Moradpour, A.; Amouyal, E. *J. Chem. Soc., Faraday Trans. 1* **1982**, *78*, 3331.
547. Kleijn, M.; van Leeuwen, H. P. *J. Electroanal. Chem.* **1988**, *247*, 253.
548. Sakata, T.; Hashimoto, K.; Kawai, T. *J. Phys. Chem.* **1984**, *88*, 5214.
549. Kleijn, J. M.; Boschloo, G. K. *J. Electroanal. Chem.* **1991**, *300*, 595.
550. Brainina, Kh. Z.; Khodos, M. Ya.; Belysheva, G. M.; Krivosheev, N. V. *Elektrokhimiya* **1984**, *20*, 1380.

551. Torresi, R. M.; Cámara, O. R.; De Pauli, C. P.; Giordano, M. C. *Electrochim. Acta* **1987**, *32*, 1291.

552. Matsuki, K.; Kamada, H. *Electrochim. Acta* **1986**, *31*, 13.

553. Ravichandran, C.; Kennedy, C. J.; Chellammal, S.; Thangavelu, S.; Anantharaman, P. N. *J. Appl. Electrochem.* **1991**, *21*, 60.

554. Nechaev, E. A.; Silina, T. F.; Pavlichenko, V. A.; Tupikova, I. V. *Elektrokhimiya* **1986**, *22*, 1557.

555. Laule, G.; Hawk, R.; Miller, D. *J. Electroanal. Chem.* **1986**, *213*, 329.

556. Harmer, M. A.; Hill, H. A. O. *J. Electroanal. Chem.* **1985**, *189*, 229.

557. Asokan, K.; Krishnan, V. *Bull. Electrochem.* **1990**, *6*, 449.

558. Mills, A.; Davies, H. *J. Chem. Soc., Faraday Trans.* **1991**, *87*, 473.

559. Korczyński, A.; Doniec, A.; Swiderski, J. *Corr. Sci.* **1981**, *21*, 329.

560. Danilov, F. I.; Velichenko, A. B.; Lazorina, S. M.; Kalinovskaya, S. E.; *Elektrokhimiya* **1987**, *23*, 1634.

561. Kalinovskii, E. A.; Shustov, V. A.; Chaikovskaya, V. M.; Rossinskii, Yu. K. *Zh. Prikl. Khim.* **1985**, *58*, 799.

562. Uzbekov, A. A.; Klementeva, V. S.; Kubasov, V. L.; Lambrev, V. G. *Zh. Prikl. Khim.* **1985**, *58*, 686.

563. Krstajić, N.; Spasojević, M.; Atanasoski, R. *Hem. Ind.* **1982**, *36*, 121.

564. Rethinaraj, J. P.; Kulandaisamy, S.; Chockalingam, S. C.; Visvanathan, S.; Rajagopalan, K. S. "Proceedings of the Symposium on Advances in Electrochemicals," Karaikudi, 16 April 1984; CECRI, Karaikudi, pp. 2.1–2.16.

565. Mills, A.; Davies, H. *J. Chem. Soc., Faraday Trans.* **1990**, *86*, 955.

566. Klementeva, V. S.; Uzbekov, A. A.; Kubasov, V. L.; Lambrev, V. G. *Zh. Prikl. Khim.* **1984**, *57*, 2623.

567. Lyubushkin, V. I.; Smirnov, V. A.; Lyubushkina, L. M. *Elektrokhimiya* **1981**, *17*, 828.

568. De Battisti, A.; Barbieri, A.; Giatti, A.; Battaglin, G.; Daolio, S.; Boscoletto, A. B. *J. Mater. Chem.* **1991**, *1*, 91.

569. Muranaga, T.; Kanaya, Y.; Yokota, N. In Ref. 1, pp. 77–85.

570. Comninellis, C.; Plattner, E. In Ref. 1, pp. 229–243.

571. Ferron, C. G.; Duby, P. F. In Ref. 1, pp. 259–277.

572. Hardee, K. L.; Gordon, A. Z.; Pyle, C. B.; Sen, R. K. *U.S. 4 426 263* **1984**; *Chem. Abstr.* **1984**, *100*, 93550.

573. Conway, B. E.; Lui, T. C. *Mater. Chem. Phys.* **1989**, *22*, 163.

574. Roginskaya, Yu. E.; Belova, I. D.; Galyamov, B. Sh.; Chibirova, F. Kh.; Shifrina, R. R. *Mater. Chem. Phys.* **1989**, *22*, 203.

575. Daolio, S.; Facchin, B.; Pagura, C.; De Battisti, A.; Battaglin, G. *Surf. Interface Anal.* **1990**, *16*, 457.

576. Gissler, W.; McEvoy, A. J.; Grätzel, M. *J. Electrochem. Soc.* **1982**, *129*, 1733.

577. Gissler, W.; McEvoy, A. J. *J. Electroanal. Chem.* **1982**, *142*, 375.

578. Langer, S. H.; Pietsch, S. J. *J. Electrochem. Soc.* **1979**, *126*, 1189.

579. McEvoy, A. J.; Gissler, W. *J. Appl. Phys.* **1982**, *53*, 1251.

580. Kiwi, J.; Grätzel, M. *Chimia* **1979**, *33*, 289.

581. Willems, H.; Moers, M.; Broers, G. H. J.; De Wit, J. H. W. *J. Electroanal. Chem.* **1985**, *194*, 305.

582. Kobussen, A. G. C. *J. Electroanal. Chem.* **1981**, *126*, 199.

583. Kessler, T.; Visintin, A.; de Chialvo, M. R.; Triaca, W. E.; Arvia, A. J. *J. Electroanal. Chem.* **1989**, *261*, 315.

584. Thackeray, M. M.; Baker, S. D.; Adendorff, K. T. *Solid State Ionics* **1985**, *17*, 175.

585. Tseung, A. C. C.; Jasem, S. M. *J. Appl. Electrochem.* **1981**, *11*, 209.

586. Losev, V. V. Z. *Phys. Chem. (Leipzig)* **1987**, *268*, 129.

587. Buné, N. Ya.; Zaripova, É. N.; Reznik, M. F.; Losev, V. V. *Elektrokhimiya* **1986**, *22*, 396.

588. Denton, D. A.; Harrison, J. A.; Knowles, R. I. *Electrochim. Acta* **1980**, *25*, 1147.

589. Mills, A.; Cook, A. *J. Chem. Soc., Faraday Trans. 1* **1988**, *84*, 1691.

590. Yeo, R. S.; Orehotsky, J.; Visscher, W.; Srinivasan, S. *J. Electrochem. Soc.* **1981**, *128*, 1900.

591. Kokoulina, D. V.; Bunakova, L. V.; Eleva, M. Z. *Elektrokhimiya* **1985**, *21*, 1121.

592. Burke, L. D.; Murphy, O. J. *J. Electroanal. Chem.* **1980**, *109*, 199.

593. Raj, I. A.; Giridhar, V. V.; Vasu, K. I. *Bull. Electrochem.* **1990**, *6*, 407.

594. Efimov, E. A.; Gerish, T. V.; Sitnikova, T. G. *Zashch. Met.* **1982**, *18*, 295.

595. Danilov, F. I.; Velichenko, A. B.; Lobova, S. M. *Elektrokhimiya* **1990**, *26*, 1185.

596. Armand, M.; Dalard, F.; Deroo, D.; Mouliom, C. *Solid State Ionics* **1985**, *15*, 205.

597. Resch, U.; Fox, M. A. *J. Phys. Chem.* **1991**, *95*, 6316.

598. Vasudevan, D.; Chellammal, S.; Anantharaman, P. N. *J. Appl. Electrochem.* **1991**, *21*, 839.

599. Kowalska, E.; Urbański, P. *Nukleonika* **1989**, *34*, 27.

600. Pushpavanam, S.; Narasimham, K. C.; Vasu, K. I. *Bull. Electrochem.* **1988**, *4*, 979.

601. Bramstedt, W. R.; Hardee, K. L.; Johnson, B. R.; Harrington, D. E. *Talanta* **1981**, *28*, 737.

602. Bandi, A.; Vartires, I.; Mihelis, A.; Hainăroşie, C. *J. Electroanal. Chem.* **1983**, *157*, 241.

603. Wagner, N.; Kühnemund, L. *Cryst. Res. Technol.* **1988**, *23*, 1017.

604. Fukuda, K.; Iwakura, C.; Tamura, H. *Electrochim. Acta* **1980**, *25*, 1523.

605. Xue, T.; Osseo-Asare, K. *J. Less-Common Met.* **1989**, *152*, 103.

606. De Battisti, A.; Barbieri, A.; Giatti, A.; Battaglin, G.; Daolio, S.; Boscoletto, A. B. *J. Mater. Chem.* **1991**, *1*, 191.

607. Sato, Y.; Yanagida, M.; Yamanaka, H.; Tanigawa, H. *J. Electrochem. Soc.* **1989**, *136*, 863.

608. Sanjinés, R.; Aruchamy, A.; Lévy, F. *J. Electrochem. Soc.* **1989**, *136*, 1740.

609. Iwakura, C.; Taniguchi, Y. Tamura, H. *Chem. Lett.* **1981**, 689.

610. Shimizu, Y.; Uemura, K.; Matsuda, H.; Miura, N.; Yamazoe, N. *J. Electrochem. Soc.* **1990**, *137*, 3430.

611. Anani, A.; Mao, Z.; Srinivasan, S.; Appleby, A. J. *J. Appl. Electrochem.* **1991**, *21*, 683.
612. Balko, E. N.; Nguyen, P. H. *J. Appl. Electrochem.* **1991**, *21*, 678.
613. Olthius, W.; Van Kerkhof, J. C.; Bergveld, P.; Bos, M.; Van der Linden, W. E. *Sens. Actuators B* **1991**, *4*, 151.
614. Burke, L. D.; Healy, J. F.; Dhubhghaill, O. N. *Surf. Technol.* **1982**, *16*, 341.
615. Shub, D. M.; Lubnin, E. N.; Buné, N. Ya.; Reznik, M. F.; Losev, V. V. *Elektrokhimiya* **1986**, *22*, 659.
616. Kang, K. S.; Shay, J. L. *J. Electrochem. Soc.* **1983**, *130*, 766.
617. Velikodnyi, L. N.; Shepelin, V. A.; Kasatkin, É. V. *Elektrokhimiya* **1984**, *20*, 910; **1983**, *19*, 373.
618. Sedlak, J. M.; Lawrance, R. J.; Enos, J. F. *Int. J. Hydrogen Energy* **1981**, *6*, 159.
619. Kobussen, A. G. C.; Mester, C. M. A. M. *J. Electroanal. Chem.* **1980**, *115*, 131.
620. Kobussen, A. G. C.; Broers, G. H. *J. J. Electroanal. Chem.* **1981**, *126*, 221.
621. Matsumoto, Y.; Manabe, H.; Sato, E. *J. Electrochem. Soc.* **1980**, *127*, 811.
622. Rethinaraj, J. P.; Visvanathan, S. *Mater. Chem. Phys.* **1991**, *27*, 337.
623. Anbananthan, N.; Rao, K. N.; Venkatesan, V. K. *Bull. Electrochem.* **1990**, *6*, 480; *Mater. Chem. Phys.* **1991**, *27*, 351.
624. Nishizawa, T.; Senna, M.; Kuno, H. *J. Mater. Sci.* **1983**, *18*, 1346.
625. Barmashenko, V. I.; Mazanko, A. F.; Skorokhod, G. A.; Tsypenyuk, R. B. *Elektrokhimiya*, **1980**, *16*, 1487.
626. Harmer, M. A.; Hill, H. A. O. *J. Electroanal. Chem.* **1984**, *170*, 369.
627. Kraitsberg, A. M.; Pleskov, Yu. V. *Elektrokhimiya* **1987**, *23*, 1113.
628. Battaglin, G.; De Battisti, A.; Barbieri, A.; Giatti, A.; Marchi, A. *Surf. Sci.* **1991**, *251/252*, 73.
629. Lodi, G.; Zucchini, G. L.; De Battisti, A.; Giatti, A.; Battaglin, G.; Della Mea, G. *Surf. Sci.* **1991**, *251/252*, 836.
630. Leech, D.; Wang, J.; Smythe, M. R. *Electroanalysis* **1991**, *3*, 37.
631. Healy, T. W.; Fuerstenau, D. W. *J. Colloid Sci.* **1965**, *20*, 376.

CHAPTER

6

Electrochemistry of UO_2 Nuclear Fuel

D. W. Shoesmith, S. Sunder, and W. H. Hocking

6.1 Introduction

Uranium dioxide (UO_2) in the form of ceramic pellets encased by a Zircaloy cladding is the standard fuel for the water-cooled nuclear reactors that are presently used throughout much of the world for electric power generation [1]. Natural uranium (0.7 at.% ^{235}U) can be burned in the heavy-water-moderated CANDU* reactor, whereas UO_2 that has been enriched in ^{235}U or mixed with a small proportion of $^{239}PuO_2$ is required for light-water-moderated and advanced gas-cooled reactors. Thorium dioxide (ThO_2) seeded with a few percent of fissile nuclides is a potential alternative fuel for the CANDU reactor and high-temperature gas-cooled reactors [1,2]. The most common fuel for liquid-metal-cooled fast-breeder reactors (LMFBR) has been ceramic $U_{1-y}Pu_yO_2$, where y is typically ~0.2. Advanced LMFBR fuels of the (U,Pu)C and (U,Pu)N type have been developed, but less widely used. Finally, metallic uranium played a significant role in the development of nuclear power, and U_ySi_x has been favored for research reactors.

In this review our attention will be focused on the electrochemistry of UO_2, since this is the only material that has been studied electrochemically in any detail; where pertinent, reference to other fuels will be made. The major reason for interest in the redox chemistry of UO_2 is the very large change in its aqueous solubility with changes in redox potential. In the fully reduced form, UO_2 is extremely insoluble, but under oxidizing conditions the solubility of uranium (as U^{VI}) is four to 10 orders of magnitude higher. Thus stable geological deposits of uranium oxide inevitably

---

* CANada Deuterium Uranium, registered trademark of AECL.

have a composition close to UO_2 [3] and their mobilization by transport of dissolved uranium in groundwaters requires oxidizing conditions. Advantage is taken of this difference in solubility in the hydrothermal extraction of uranium ores and in the reprocessing of nuclear fuels [4–9] for which strongly oxidizing processes are used. The release of radionuclides from used fuel in a geological disposal vault will also be controlled by the redox potential of the groundwater and will be much more rapid under oxidizing conditions [10–13].

The redox conditions of importance in these situations cover a wide range over which the degree of oxidation of the surface of UO_2, and its rate of dissolution will change markedly (Fig. 6.1) [10,11,14]. By showing that the dissolution rates of UO_2 exposed to various oxidizing agents can be predicted from dissolution currents measured electrochemically on UO_2 electrodes, Nicol et al. [4,5] have demonstrated the usefulness of electrochemical techniques in elucidating the complex redox chemistry of UO_2. Indeed, electrochemical models predicting the dissolution behavior of UO_2 under hydrothermal leaching conditions and used nuclear fuel (UO_2) under waste disposal conditions have been published [4,5,14,15]. From Fig. 6.1 the similarity in redox conditions between geochemical and used fuel disposal environments is obvious, and high-grade uranium ore deposits have been studied as natural analogs for used-fuel disposal [10,16,17]. Although the general chemistry is similar, the differences between high-density sintered fuel pellets, altered by inreactor irradiation and thermal treatments, and natural uraninite deposits yield distinct differences in the dissolution/corrosion behavior.

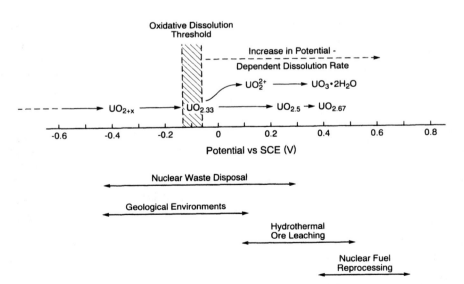

Figure 6.1. Change in the surface composition of UO_2 and the onset of oxidative dissolution as a function of the potential of a UO_2 electrode (adapted from Fig. 1 of Ref. 14). The redox conditions expected in selected natural and industrial environments are indicated.

6.2 Fuel Composition

The ceramic oxide fuels have been traditionally fabricated by sintering pressed compacts of very-fine-grained oxide powder(s) at ~1700°C under a reducing atmosphere [18,19]. Improved sintering at lower temperatures can be achieved under mildly oxidizing conditions, through enhanced diffusion rates, but must then be followed by a reduction step. These sintered ceramics typically have 92–99% of the theoretical density with grain sizes of 2–15 μm and mainly noninterconnected submicron pores. A slightly hyperstoichiometric composition is the norm for the uranium dioxide fuels (UO_{2+x}, where $0.01 \geq x \geq 0.0001$), whereas the mixed-oxide fuels are invariably hypostoichiometric [$(U,Pu)O_{2-x}$, where $0.06 \geq x \geq 0.02$] with vacancies in the oxygen sublattice [2,18,20]. Subtle variations in composition are known to occur on different scales within these ceramics, dependent upon the exact details of the fabrication process. The carbide and nitride fuels are also usually produced by sintering fine-grained powders, under an argon or nitrogen atmosphere, respectively. Deviations from the nominal stoichiometry and variations in composition are again expected [2].

As nuclear fuel is burned in a reactor, a plethora of fission products and actinides with widely varying solubilities in the host matrix is created [21]. Collision cascades initiated by fission and α-recoil events generate profuse lattice defects, although healing effects limit their final number density [22–24]. The actinides and rare earths readily substitute for uranium (or plutonium) in the uraninite lattice and hence remain atomically dispersed in the oxide fuels. Fission products that have limited solubility in UO_2 (e.g., Xe, Kr, I, Br, Cs, Rb, Zr, Mo, Te, Tc, Ru, Rh, Pd, Cd, and Sn) are initially trapped at defect sites but tend to segregate gradually (to grain boundaries) by thermal and radiation-enhanced diffusion [10,21–30]. A small increase in oxygen potential as a function of burnup is also expected with oxide fuels because the effective valence of the fission products is less than 4 [31,32]; however, this may be partly offset by diffusion of the excess oxygen into the Zircaloy cladding [33,34]. Comparable changes in the carbide and nitride fuels during irradiation have been documented. [2].

6.3 Structural Properties

A comprehensive review of the structural properties of UO_2 has been written by Vollath [35]. Here we confine ourselves to those UO_z phases ($2 \leq z \leq 3$) that appear to be important under electrochemical conditions and/or affect the properties of UO_2 in a manner that changes its electrochemical behavior.

The actinide dioxides ThO_2, UO_2, and PuO_2 (as well as their solid solutions) adopt the cubic fluorite structure (Fig. 6.2A). Every metal atom is surrounded by eight equivalent nearest-neighbor oxygen atoms, each of which in turn is surrounded by a tetrahedron of four equivalent metal atoms. An important feature of this lattice is the large, cubically coordinated interstitial sites, which can accommodate additional oxygen atoms, with relatively small distortions of the structure, up

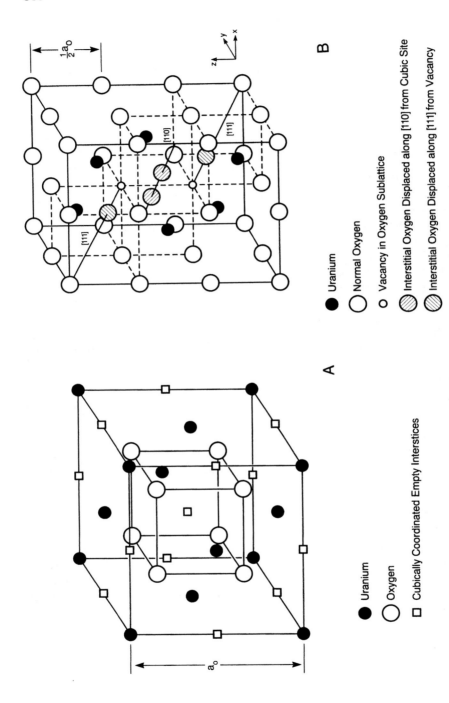

Figure 6.2. (A) Cubic fluorite structure adopted by the actinide dioxides UO_2, ThO_2, and PuO_2 (as well as their solid solutions). (B) Formation of defect clusters identified in UO_{2+x} by neutron diffraction [36,37].

to a composition near $UO_{2.33}$. Increases in oxygen content lead to derivative phases in which the oxygen interstitials and adjacent lattice ions are displaced from the ideal uraninite structure sites. Neutron diffraction measurements on $UO_{2.12}$ at 1100 K have shown that small defect clusters are formed without disruption of the overall symmetry [36,37] (Fig. 6.2B). The existence of similar clusters at lower temperatures and smaller departures from stoichiometry has been inferred but remains to be proven. Depending on heat treatment and redox conditions, phases with the nominal composition U_4O_9, which have ordered defect clusters, can occur together with UO_{2+x} [3]. These subtle alterations of the uraninite structure apparently have a modest impact on bulk electrical conductivity [38–40].

The phase α-U_3O_7 ($UO_{2.33}$) has a tetragonally distorted fluorite structure and can therefore be considered an extension of the UO_{2+x} series [41]. It is effectively the end of the fluorite-structure range as further oxidation induces a complete transformation to a more open layerlike configuration with significantly lower density [3,42]. Three distinct phases with the composition U_2O_5 have been reported (only at high pressure [43]) and appear to have structures intermediate between the fluorite structure and the layer structure encountered in U_3O_8. In the composition range from U_2O_5 to U_3O_8, up to 12 distinct phases have been reported, although α-U_3O_8 is the phase generally formed by oxidation of UO_2 [3]. The highest oxide of uranium of interest here is UO_3, which forms hydrated derivatives, with $UO_3 \cdot 2H_2O$ being stable near room temperature [44,45]. All the intermediate uranium oxides are electrically conductive, whereas UO_3 is an insulator. The phase $UO_2(OH)_2$ is of interest kinetically, since it appears to limit oxidative-dissolution rates in alkaline solution. This compound possesses a layered structure of uranyl groups bonded to hydroxyl groups and is the expected oxidation product of U_3O_8 in aqueous solutions [3]. The uranium trioxide–water system has been discussed in detail [3,45,46].

6.4 Electrical Properties

Electrochemical studies of the nuclear fuels are predicated on their ability to conduct significant electric current. The carbides and nitrides of uranium and plutonium, like the pure elements, behave as metals in this respect: The electrical resistivity is very low and has a positive temperature coefficient [47–49]. A broad and partly filled conduction band composed of overlapping, hybridized metal 7s, 6d, 5f, and C/N 2p states provides high mobility for the charge carriers [2,50,51]. Uranium dioxide [and $(U,Pu)O_2$] is best described as a Mott-Hubbard insulator—an insulator characterized by a partly filled cationic shell, which has a sufficiently narrow energy bandwidth that the mobility of the electrons is severely restricted by their mutual Coulomb interaction [52–56]. Electronic conductivity can nonetheless occur by an activated process described as small polaron hopping [57–59]: the nominally localized electrons move from one cation to the next in a series of thermally assisted jumps. Finally, ThO_2 is a classic insulator with a large gap (~ 5 eV) between the filled valence and empty conduction bands [59–62].

An energy-level diagram for UO_2 has been schematically depicted in Fig. 6.3:

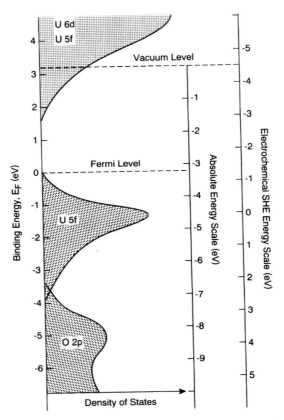

Figure 6.3. Schematic energy-level diagram for UO_2 derived from spectroscopic and electrochemical data (for $\Delta\Phi_H = 0$). The filled valence band has predominantly O 2p character, whereas the empty conduction band consists mainly of U 5f, 6d, and 7s states.

the band shapes and energies relative to the Fermi level were derived from optical, X-ray photoelectron and bremsstrahlung isochromat spectra [61–70]—supported by theoretical calculations [60,71]. The absolute (vacuum) energy level was established using the measured work function of UO_2 [72,73], and the conventional electrochemical scale was estimated by assuming a work function of 4.5 eV for the standard hydrogen electrode (SHE) [74]. A narrow U 5f band, which contains two electrons per uranium atom for stoichiometric UO_2, falls in the gap between the filled valence band and the empty conduction band. It has a maximum density of states near -1.4 eV (relative to E_F), but extends in the wings from the top of the valence band to near the Fermi level. The valence band has predominantly O 2p character, although there is also evidence for a modest contribution from U 6d and 5f orbitals [20,60,61,71,75–77]. Conversely, the conduction band consists mainly of overlapping manifolds of U 7s, 6d, and 5f states—the energy difference between the occupied and empty 5f levels (~ 5 eV) represents the Coulomb correlation energy, U_{eff} [61–65]. Because the U 5f Hubbard subbands are significantly narrower than U_{eff}, electron correlation effects must be important for emission or excitation processes that produce a $5f^1$ (or $5f^0$) final state [61,62]. Nonetheless, Fig. 6.3 does provide a coherent basis for interpreting the electrical and electrochemical properties of (doped) UO_2, which invariably involve these configurations.

Electronic conductivity in perfectly stoichiometric UO_2 would require promotion of electrons from the occupied U 5f level to the conduction band—a strongly activated process ($E_A \simeq 1.1$ eV) with extremely low probability at room temperature [52–55]. In practice, ceramic UO_2 fuel normally has a slight excess of oxygen, nominally present as interstitial O^{2-} ions. Overall charge balance is maintained by ionization of an appropriate fraction of U^{IV} ions to the U^V and/or U^{VI} valence state. This creates holes in the narrow occupied U 5f band, which can migrate by a hopping process with a low activation energy (~0.2 eV) [38–40,53–55,78–87]. The maximum number density of charge carriers in UO_{2+x} should thus be given by

$$N_h = 2xN_U \tag{6.1}$$

where N_U is the number density of uranium atoms (2.44×10^{22} cm^{-3}) [38,53–55]. There is some evidence, however, of hole localization or trapping at interstitial sites, especially for higher degrees of hyperstoichiometry [53,84]. Formally, the activation energy for electrical conductivity (≤ 0.3 eV) can be regarded as the sum of two terms, one for hole ionization and one for hole mobility, both of which could be composition dependent [58,84].

Replacement of a fraction of the U^{IV} ions in UO_2 with lower-valent species (e.g., yttrium and the rare earths) requires further ionization of the remaining uranium ions to maintain overall charge balance. This creates mobile holes in the U 5f band in the absence of interstitial O^{2-} ions. Hole localization, which decreases the mobility of holes in UO_{2+x}, appears to be less important, and the conductivity correspondingly higher than in UO_{2+x} [53]. Conversely, incorporation of higher-valent species (e.g., niobium) quenches the low-temperature (extrinsic) electrical conductivity by hole annihilation [52],

$$h^+ + Nb^{4+} \rightarrow Nb^{5+} \tag{6.2}$$

Because used fuel contains significant quantities of lower-valent (fission-product) dopants (the amount depending on the degree of burnup), its electrical conductivity should be enhanced compared with that of the unirradiated fresh fuel. A further increase in the conductivity of used fuel could arise from the anticipated increase in oxygen potential with burnup. Segregation of fission products to grain boundaries will also affect conductivity in a manner presently undetermined. These conductivity differences between unirradiated and used fuel are expected to influence the sensitivity to oxidation and dissolution and would be reflected in the results of electrochemical experiments.

6.5 Electrochemical Properties

The electrochemical properties of UO_{2+x} (as well as UO_2-based SIMFUEL and used fuel) should thus exhibit some parallels to those of p-type semiconductors;

however, the characteristics of the charge carriers must also be considered. A normal p-type semiconductor would show an abrupt transition at its flatband potential, whereas significantly more gradual changes can be expected for the uraninite materials, because they have a much larger number density of charge carriers with substantially lower mobility. The flatband potential E_{fb}, for UO_{2+x} in aqueous solution has been estimated using the expression

$$eE_{fb} = \Phi + U_R + e\,\Delta\Phi_H \tag{6.3}$$

where Φ is the work function of UO_{2+x}, U_R is the zero point of the electrochemical energy scale (-4.5 ± 0.2 eV), and $\Delta\Phi_H$ is a correction term, called the Helmholtz voltage, which accounts for the effects of adsorbed ions [74,88-91]. When H^+/OH^- adsorption predominates, $\Delta\Phi_H$ has a linear dependence on pH with a slope of ~60 mV/pH-unit [88-92]. In dilute aqueous solutions at pH = 9.5, a value of $E_{fb} = -1.7 \pm 0.4$ V (vs. SCE) has been calculated [93]. This is an unusually negative flatband potential, which has permitted conventional electrochemical studies to be performed on UO_{2+x} at potentials significantly more negative than would be feasible with a typical p-type semiconductor [74,93].

The energy-level schemes for a selection of semiconductors are compared with that of UO_2 (at pH = 0) in Fig. 6.4 [74,89,90,94-103], which also depicts their relationship to the standard potentials for a variety of redox couples [104]. Clearly, UO_2 will be polarized positive to the flatband potential in the presence of these redox reactions. Only small polarizations will be required for the surface of the electrode to become degenerate and capable of promoting electrochemical reactions, as the Fermi level is pulled into the 5f band.

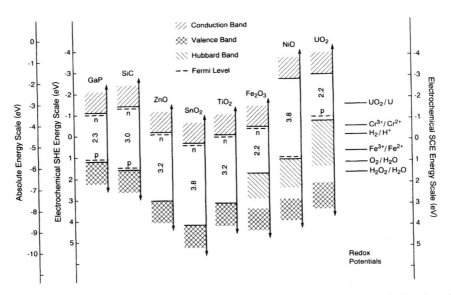

Figure 6.4. Energy-level schemes for a selection of semiconductors compared with that of UO_2 at pH = 0 [74,89,90,94-103]. The number between the energy levels is the band gap. The standard potentials for a number of redox couples are also shown [104].

6.6 Thermodynamic Properties

The relative stabilities of important uranium phases in contact with water are summarized in a potential-pH diagram in Fig. 6.5; these lines were calculated using thermodynamic data from a recent NEA review [105]. Because all the uranium oxides fall well within the stability field for water, UO_2-based nuclear fuels exhibit a rich aqueous electrochemistry [106]. Conversely, U, UC, and UN are quite unstable with respect to oxidation by water. Among the solid compounds of plutonium, PuO_2 is the thermodynamically preferred species within the water stability field [107,108].

The solubility of uranium in water as a function of pH under reducing (UO_2) and oxidizing ($UO_3 \cdot 2H_2O$) conditions (Fig. 6.6A) and the predominant soluble uranium species as a function of potential and pH (Fig. 6.6B) have also been calculated [105]. There is some uncertainty in these data, especially at low pH. The uranyl ion, UO_2^{2+}, is the soluble form of U^{VI}, whereas U^{IV} dissolves as the U^{4+} ion. However, both ions, but especially U^{4+}, are extensively hydrolyzed even in rather acidic

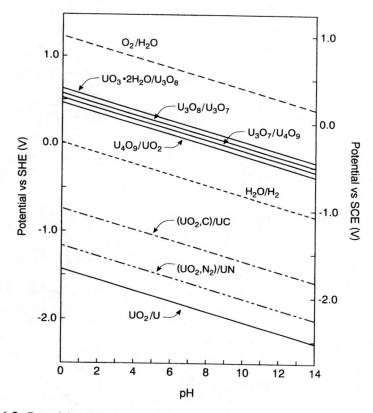

Figure 6.5. Potential-pH diagram showing the relative stabilities of important uranium phases [involving carbon (C), nitrogen (N), and oxygen (O)] in contact with water (calculated from data in Ref. 105).

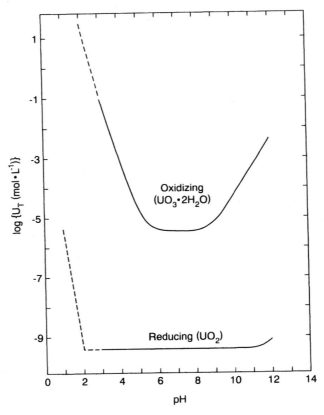

Figure 6.6A. Solubilities of uranium dioxide (UO_2) and schoepite ($UO_3 \cdot 2H_2O$) as a function of pH at 25°C (calculated from data in Ref. 105).

aqueous solutions, forming species of the type $U_x(OH)_y^{(4x-y)+}$ (for pH > 1) and $(UO_2)_x(OH)_y^{(2x-y)+}$ (for pH > 4) [105–108]. The stability of uranium in aqueous solution can be enhanced by complexation with a wide range of inorganic and organic anions [105–110]. Equilibrium constants for anions of importance in deep groundwaters (i.e., of significance to nuclear waste disposal) are collected in Table 6.1. The uranyl ion (UO_2^{2+}) is preferentially complexed by carbonate and phosphate over a wide pH range, from near neutral to alkaline, but in the latter case precipitates as well as soluble species can be readily formed. Sulfate and fluoride complexation of UO_2^{2+} are less favorable and unimportant except at low pH and high concentrations. The opposite trend is observed for U^{4+}, the stability of hydrolyzed species strongly discriminating against complexation by carbonate (and other anions) over the near neutral to alkaline regime. Conversely, at low pH, complexation of U^{4+} by sulfate and especially fluoride is quite favorable. The overall impact of the groundwater anions is thus to increase the difference between the solubility of U^{VI} and U^{IV} over the neutral to weakly alkaline pH regime important geochemically and at a nuclear waste disposal site [10,11,107–110].

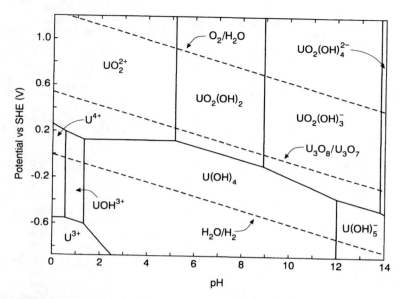

Figure 6.6B. Predominant soluble uranium species in water as a function of potential, and pH at a total uranium concentration of 10^{-6} mol · L^{-1} (calculated from data in Ref. 105). Although the solubility limits for U^{IV} species have been greatly exceeded, the diagram is still representative of the distribution of these species for concentrations at/below saturation.

6.7 Surface Composition Under Electrochemical Conditions

Figure 6.7 is a highly simplified energy-level diagram that attempts to illustrate the expected behavior of UO_2 exposed to aerated aqueous solutions at pH = 9.5 For such conditions, a steady-state corrosion potential (E_{CORR}) of about +70 mV versus SCE is obtained (0.1 mol · L^{-1} NaClO$_4$) [111], a value sufficiently positive of the equilibrium potentials for various oxide transformations (Fig. 6.5) that complex surface chemistry is to be expected due to hole injection into the U 5f level under natural corrosion conditions.

A series of oxidation and reduction peaks is obtained in the voltammetric response of a UO_2 electrode in neutral to slightly alkaline solutions (Fig. 6.8) [112]. The vertical shaded region marks the potential range over which oxidation of the surface through the sequence of phases—U_4O_9, U_3O_7, U_3O_8, and $UO_3 \cdot xH_2O$—is thermodynamically possible (Fig. 6.5). The compositions above the horizontal arrows denote the surface stoichiometries (in oxidized layers 5–8 nm thick) determined by X-ray photoelectron spectroscopy (XPS) after potentiostatic oxidation at potentials within the ranges indicated [112,113].

The flatband potential (E_{fb}) determined from the onset of p-type photocurrent on a UO_2 electrode (from plots of I_{Ph}^2 versus E) [114] is also shown in Fig. 6.8: the measured value (~ -1.3 V versus SCE) is in reasonable agreement with that predicted by Eq. (6.3). For an ideal wide-bandgap p-type semiconductor with a low

Table 6.1 Some Important Uranium Complexation and Solubility Equilibria[a,b]

Complexation Process	Equilibrium Constant
$U^{4+}_{(aq)} + Cl^-_{(aq)} \rightleftharpoons UCl^{3+}_{(aq)}$	$\log K^0 = 1.72$
$U^{4+}_{(aq)} + F^-_{(aq)} \rightleftharpoons UF^{3+}_{(aq)}$	$\log K^0 = 9.28$
$U^{4+}_{(aq)} + 2F^-_{(aq)} \rightleftharpoons UF^{2+}_{(aq)}$	$\log \beta^0_2 = 16.23$
$U^{4+}_{(aq)} + 4F^-_{(aq)} \rightleftharpoons UF_{4(s)}$	$-\log K_{s,0} = 33.5$
$U^{4+}_{(aq)} + SO_4^{2-}_{(aq)} \rightleftharpoons USO_4^{2+}_{(aq)}$	$\log K^0 = 6.58$
$U^{4+}_{(aq)} + 2SO_4^{2-}_{(aq)} \rightleftharpoons U(SO_4)_{2(aq)}$	$\log \beta^0_2 = 10.51$
$U^{4+}_{(aq)} + 4CO_3^{2-}_{(aq)} \rightleftharpoons U(CO_3)_4^{4-}_{(aq)}$	$\log \beta^0_4 = 35.1$
$U^{4+}_{(aq)} + 5CO_3^{2-}_{(aq)} \rightleftharpoons U(CO_3)_5^{6-}_{(aq)}$	$\log \beta^0_5 = 34.0$
$U^{4+}_{(aq)} + OH^-_{(aq)} \rightleftharpoons U(OH)^{3+}_{(aq)}$	$\log K^0 = 13.45$
$U^{4+}_{(aq)} + 4OH^-_{(aq)} \rightleftharpoons U(OH)_{4(aq)}$	$\log \beta^0_4 = 51.41$
$U^{4+}_{(aq)} + 5OH^-_{(aq)} \rightleftharpoons U(OH)_5^-_{(aq)}$	$\log \beta^0_5 = 53.40$
$UO_2^{2+}_{(aq)} Cl^-_{(aq)} \rightleftharpoons UO_2Cl^+_{(aq)}$	$\log K^0 = 0.2$
$UO_2^{2+}_{(aq)} F^-_{(aq)} \rightleftharpoons UO_2F^+_{(aq)}$	$\log K^0 = 5.1$
$UO_2^{2+}_{(aq)} 2F^-_{(aq)} \rightleftharpoons UO_2F_{2(aq)}$	$\log \beta^0_2 = 8.6$
$UO_2^{2+}_{(aq)} SO_4^{2-}_{(aq)} \rightleftharpoons UO_2SO_{4(aq)}$	$\log K^0 = 3.15$
$UO_2^{2+}_{(aq)} + 2SO_4^{2-}_{(aq)} \rightleftharpoons UO_2(SO_4)_2^{2-}_{(aq)}$	$\log \beta^0_2 = 4.14$
$UO_2^{2+}_{(aq)} + PO_4^{3-}_{(aq)} \rightleftharpoons UO_2PO_4^-_{(aq)}$	$\log K^0 = 13.23$
$UO_2^{2+}_{(aq)} + HPO_4^{2-}_{(aq)} \rightleftharpoons UO_2HPO_{4(aq)}$	$\log K^0 = 7.24$
$UO_2^{2+}_{(aq)} + HPO_4^{2-}_{(aq)} + 4H_2O_{(l)} \rightleftharpoons UO_2HPO_4 \cdot 4H_2O_{(s)}$	$-\log K_{s,0} = 11.9$
$3UO_2^{2+}_{(aq)} + 2HPO_4^{2-}_{(aq)} + 4H_2O_{(l)} \rightleftharpoons (UO_2)_3(PO_4)_2 \cdot 4H_2O_{(s)} + 2H^+_{(aq)}$	$-\log K_{s,0} = 24.7$
$UO_2^{2+}_{(aq)} + CO_3^{2-}_{(aq)} \rightleftharpoons UO_2CO_{3(aq)}$	$\log K^0 = 9.68$
$UO_2^{2+}_{(aq)} + 2CO_3^{2-}_{(aq)} \rightleftharpoons UO_2(CO_3)_2^{2-}_{(aq)}$	$\log \beta^0_2 = 16.94$
$UO_2^{2+}_{(aq)} + 3CO_3^{2-}_{(aq)} \rightleftharpoons UO_2(CO_3)_3^{4-}_{(aq)}$	$\log \beta^0_3 = 21.6$
$UO_2^{2+}_{(aq)} + CO_3^{2-}_{(aq)} \rightleftharpoons UO_2CO_{3(s)}$	$-\log K_{s,0} = 14.47$
$UO_2^{2+}_{(aq)} + OH^-_{(aq)} \rightleftharpoons UO_2OH^+_{(aq)}$	$\log K^0 = 8.80$
$UO_2^{2+}_{(aq)} + 2OH^-_{(aq)} \rightleftharpoons UO_2(OH)_{2(aq)}$	$\log \beta^0_2 = 17.68$
$UO_2^{2+}_{(aq)} + 3OH^-_{(aq)} \rightleftharpoons UO_2(OH)_3^-_{(aq)}$	$\log \beta^0_3 = 22.77$
$UO_2^{2+}_{(aq)} + 4OH^-_{(aq)} \rightleftharpoons UO_2(OH)_4^{2-}_{(aq)}$	$\log \beta^0_4 = 22.95$

[a] Equilibrium constants are from the NEA review [105] for zero-ionic-strength aqueous solution.
[b] There are no reliable data for the formation of soluble U^{4+} complexes with phosphate nor for β_2 and β_3 with OH^-.

density of mobile holes, the potential drop would be across a space-charge layer within the solid for $E < E_{fb}$, and only very small cathodic dark currents should be observed. The large current observed for the reduction of water suggests that a substantial portion of the potential is dropped across the Helmholtz layer and hence available to drive electrochemical reductions. An identical argument, based on the approach outlined by Myamlin and Pleskov [115], was used by Tench and Yeager to explain a similar behavior for lithium-doped p-type NiO [103]. A possible alternative interpretation is that electron transfer occurs via quantum-mechanical tunneling through the space-charge depletion layer [116–118]. Both mechanisms would be favored by the high charge-carrier density in (even slightly) hyperstoichiometric UO_2; however, the effects of the low electron mobility and of hole trapping are less well understood.

Oxidation of the UO_2 surface at peaks I and II (Fig. 6.8) is achieved at potentials below those at which bulk oxidation of the electrode becomes thermodynam-

ELECTROCHEMISTRY OF UO$_2$ NUCLEAR FUEL

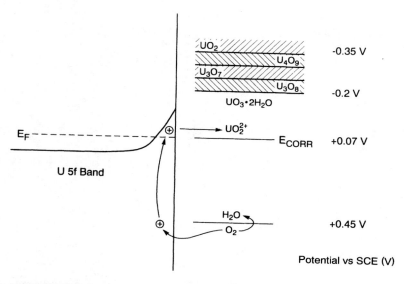

Figure 6.7. Simplified energy-level diagram (not to scale) for UO$_2$ to illustrate the expected behavior of UO$_2$ exposed to aerated solutions in neutral to alkaline solutions. The corrosion potential E_{CORR} is from Ref. 111.

ically possible. That surface oxidation is observed at such negative potential is not surprising, considering the affinity of clean UO$_2$ surfaces for oxygen under low-temperature, high-vacuum conditions [119–121]. Assuming that the oxidation process involved is U^{IV} to U^{VI}, then no more than a monolayer of charge is injected into the surface at peaks I and II, and potential scans reversed before the foot of peak III show these surface oxidation processes to be reversible with 100% charge recovery. The absence of irreversible oxidation of the surface in this potential region is demonstrated by photocurrent spectroscopy (Fig. 6.9), which indicates no loss of p-type photocurrent response after a scan to -0.4 V (versus SCE) and back. The fact that the photocurrent and the dark current were recorded under slightly different conditions is not significant, since neither response changes noticeably over the pH range from 7 to 10.5 or between sulfate and perchlorate solutions.

It is tempting to attribute peaks I and II to the redox transformations U$^{IV} \to$ U^{V} and U$^{V} \to$ U^{VI}, respectively, in the surface of the electrode, after a similar interpretation of surface states on (Li)NiO electrodes [103]. Although reproducibility is achieved from scan to scan on the same specimen, the definition of these peaks varies from specimen to specimen; commonly, only a plateau as opposed to two peaks can be resolved. It is more likely that they correspond to oxidation at surface sites of different energy, which might reflect differences in reactivity between grains and grain boundaries. Electrochemical reactivity should be concentrated at sites where electrical conductivity is highest, and high-frequency measurements on polycrystalline UO$_{2+x}$ (with $x = \sim 0.02$) have provided clear evidence of conductivity differences at a microscopic level. These differences were interpreted in terms of

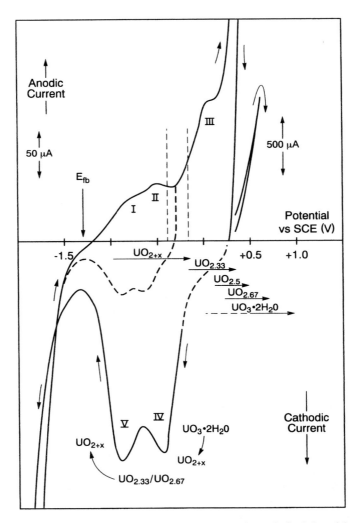

Figure 6.8. Voltammetric response of a stationary UO_2 electrode in 0.5 mol·L^{-1} Na$_2$SO$_4$ (pH = 10.5) recorded at a scan rate of 20 mV · s^{-1}; ——— scan reversed at +0.6 V; ——— scan reversed at −0.3 V [112]. The two vertical dashed lines mark the potential range over which oxidation of UO_2, through the sequence of phases $U_4O_9/U_3O_7/U_3O_8/UO_3 \cdot 2H_2O$, is thermodynamically possible. Composition ranges indicated by arrows represent the surface stoichiometries established by XPS [112,113]. E_{fb} is the flatband potential from photocurrent measurements [114].

distinct intragranular and grain-boundary contributions to macroscopic electron transport [86,87]. Both peaks I and II (Fig. 6.8) disappear for pH < 5 but increase in height and shift to more negative potentials as the pH increases. For pH < 5, hydrolysis of U^{VI} species becomes negligible [105], suggesting that the formation of these surface states may be stabilized by partial charge transfer to OH$^-$. A similar

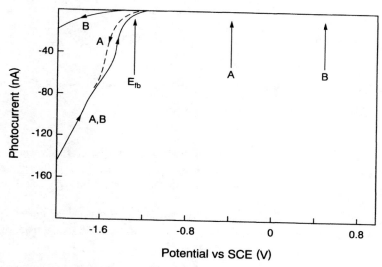

Figure 6.9. Photocurrents recorded on a UO_2 electrode (resistivity = 4 kΩ · cm) during potential scans to two different anodic limits (A, −0.4 V; B, +0.5 V) and back at 20 mV · s^{-1} in 0.2 mol · L^{-1} Na_2SO_4 solution (pH = 10). The electrode was illuminated with white light during the scan. E_{fb} indicates the flatband potential determined from plots of I_{Ph}^2 versus E [74,114].

adsorption/desorption of OH^- ions was observed on cobaltite (Co_3O_4) in alkaline solutions [122].

Polarization to more positive potentials causes irreversible oxidation of the electrode surface (Fig. 6.8), and XPS analysis indicates that this occurs in stages [112,113]. Peak III, which is often not as prominent as that shown in Fig. 6.8, can be attributed to incorporation of O^{2-} ions at interstitial sites in the fluorite lattice until a limiting stoichiometry of $UO_{2.33}$ (U_3O_7) is attained:

$$UO_2 + 0.33H_2O \rightarrow UO_{2.33} + 0.66H^+ + 0.66e \qquad (6.4)$$

For potentials at which dissolution is minor (≤0.1 V versus SCE), or at short times at more positive potentials, the logarithm of the current for film growth is a linear function of the logarithm of time, with a slope close to −1, indicating growth via a high-field ion-conduction process [123].

Because $UO_{2.33}$ is effectively the end of the fluorite-structure range, further oxidation of the surface involves dissolution as uranyl ions and/or recrystallization of the densely packed $UO_{2.33}$ to the more open orthorhombic structure associated with $U_3O_8(UO_{2.67})$. Analysis by XPS has also tentatively identified $U_2O_5(UO_{2.5})$ as a temporary surface composition in the conversion of $UO_{2.33}$ to $UO_{2.67}$ [112,113]. This is consistent with the fact that phases with a composition of $UO_{2.5}$ appear to have structures intermediate between $UO_{2.33}$ and the layered structures encountered in the higher oxides [3,46]. It has been proposed that the recrystallization involves an intermediate adsorbed uranyl species formed by further oxidation of $UO_{2.33}$ [11,112,124]:

$$UO_{2.33} + 0.33H^+ \rightarrow (UO_2^{2+})_{ads} + 0.33OH^- + 1.33e \qquad (6.5)$$

which may dissolve in solutions containing complexing anions:

$$(UO_2^{2+})_{ads} + nA^- \rightarrow UO_2A_n^{(n-2)-} \qquad (6.6)$$

or become incorporated into the recrystallized orthorhombic structure:

$$3UO_{2.33} + (UO_2^{2+})_{ads} + H_2O \rightarrow 4UO_{2.5} + 2H^+ \qquad (6.7)$$

$$2UO_{2.5} + (UO_2^{2+})_{ads} + H_2O \rightarrow 3UO_{2.67} + 2H^+ \qquad (6.8)$$

These recrystallizations are favored at high potentials ($\geq +0.3$ V versus SCE), where coverage by the adsorbed intermediate will be high. In the presence of complexing anions, reaction (6.6) predominates and reactions (6.7) and (6.8) are preempted [123].

For sufficiently positive potentials at stationary electrodes in noncomplexing solutions, local supersaturation leads to the formation of surface deposits of uranyl hydrates:

$$UO_2^{2+} + (z+1)H_2O \rightarrow UO_3 \cdot zH_2O + 2H^+ \qquad (6.9)$$

Schoepite ($UO_3 \cdot 2H_2O$) is the stable phase near room temperature [44,45]. These deposited layers are not formed on rotating disk electrodes; when present, they are reduced at peak IV on the cathodic scan, whereas the underlying $UO_{2.33}/UO_{2.67}$ layer is reduced at peak V (Fig. 6.8).

The reduced surface obtained by electrochemical reduction of the anodic film remains partially oxidized as indicated by XPS analyses and confirmed by photocurrent spectrocopy (Fig. 6.9) [114]. The number density of charge carriers N_h in the reformed surface [proportional to the degree of hyperstoichiometry via Eq. (6.1)] is sufficiently high that the formation of a space-charge layer for $E < E_{fb}$ is inhibited, and only a very weak photocurrent is observed.

In alkaline solutions (pH ≥ 12), the current due to uranyl-ion dissolution is inhibited by the formation of a surface layer, predominantly U^{VI} in composition (Fig. 6.10) [124]. This layer may be $UO_2(OH)_2$, as suggested earlier, or α-UO_3, which is close in structure to $U_3O_8(UO_{2.67})$ and could be formed if $UO_{2.67}$ is an intermediate phase in the overall oxidation process [112,113]. The surface is not totally passivated by this layer, suggesting that it can be regarded as a porous membrane through which ionic transport can maintain dissolution.

In acidic solutions [125], the extent of surface oxidation is much less marked, presumably because of the much higher solubility of U^{VI}, which renders thin films such as $UO_{2.33}/UO_{2.5}/UO_{2.67}$ very unstable [105]. Dissolution is still preceded by film formation (Fig. 6.11), but the characteristics of this film are very different from those of the films formed in neutral to alkaline solutions and show similarities to the hydrous gel-like layers formed on noble metals, such as iridium and platinum, in acidic solutions. Such layers on noble metals are thought to comprise multiple oxidation states (e.g., Ir^{III}/Ir^{IV}) in hydrolyzed aggregates, and electron transfer to reduce them involves proton transfer both into the film and between oxidized and reduced metal centers within the film [92,126,127]. The oxidation and reduction of

ELECTROCHEMISTRY OF UO$_2$ NUCLEAR FUEL

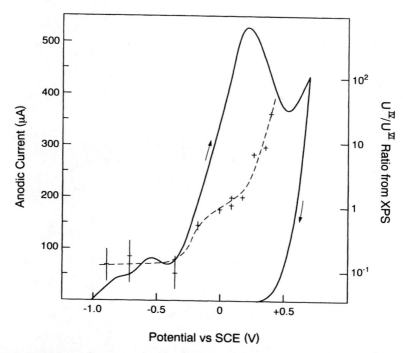

Figure 6.10. Voltammetric response recorded on a rotating UO$_2$-disk electrode (f = 16.7 Hz) in 0.5 mol · L^{-1} Na$_2$SO$_4$ (pH = 12.0) at a scan rate of 20 mV · s^{-1}. (Only anodic currents are shown.) The data points (connected by the dashed line) indicate the stoichiometries of the electrode surface after potentiostatic oxidation for 10 min at different potentials [124].

these layers is reversible on noble metals (Fig. 6.12), and the peak potentials show a pH dependence of ~88 mV/pH-unit. Their size is enhanced by potential cycling. For UO$_2$ the peaks appear close to reversible and may involve the formation of both UO$_2^+$ and UO$_2^{2+}$ species at peak I (Fig. 6.12). A pH-dependence of ~80 mV/pH-unit has been observed for peak II. The UO$_2$ peaks are not enhanced by repetitive cycling because simultaneous dissolution occurs (Fig. 6.11). Conversely, for iridium both peaks are distinctly separated from the dissolution region, which occurs at potentials above +1.5 V versus RHE (Fig. 6.12) [92,127].

6.8 Anodic Dissolution

Irrespective of the solution composition and pH, anodic dissolution of UO$_2$ occurs as U^{VI}, consistent with thermodynamic expectations (Fig. 6.6B) [128,129]. Steady-state dissolution currents for UO$_2$ in acidic solutions (pH ≤ 2) show two Tafel regions with slopes of 35–45 mV/decade at low anodic potentials and 140–160 mV/decade at higher potentials (Fig. 6.13). The curves show a cathodic shift of 30–

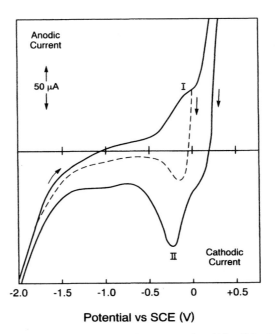

Figure 6.11. Voltammetric response recorded on a rotating UO$_2$-disk electrode (f = 16.7 Hz) in 0.5 mol · L^{-1} Na$_2$SO$_4$ (pH = 2.9) at a scan rate of 20 mV · s^{-1}; --- anodic limit 0 V; ——— anodic limit 0.3 V.

40 mV per increasing unit of pH. These data have been interpreted by Nicol and Needes [128,129] according to the following reaction scheme:

$$UO_2 + H_2O \underset{k_{-1}}{\overset{k_1}{\rightleftharpoons}} UO_2OH + H^+ + e \tag{6.10}$$

$$UO_2OH \overset{k_2}{\rightarrow} UO_3 + H^+ + e \tag{6.11}$$

$$UO_3 + 2H^+ \overset{k_3}{\rightarrow} UO_2^{2+} + H_2O \tag{6.12}$$

in which the chemical dissolution step, reaction (6.12), is very rapid in acidic solutions. At low pH and low anodic potentials, surface coverage by the U^V intermediate (UO$_2$OH) will be very small ($k_{-1} \gg k_1$) and the dissolution current given by

$$I_D = 2Fk_1k_2 \{k_{-1} [H^+]\}^{-1} \exp[(2 - \beta_2)EF(RT)^{-1}] \tag{6.13}$$

which predicts a Tafel slope of 40 mV/decade ($\beta_2 = 0.5$) and an inverse acid dependence with a cathodic shift of 40 mV per increasing unit of pH. At low pH but

ELECTROCHEMISTRY OF UO$_2$ NUCLEAR FUEL

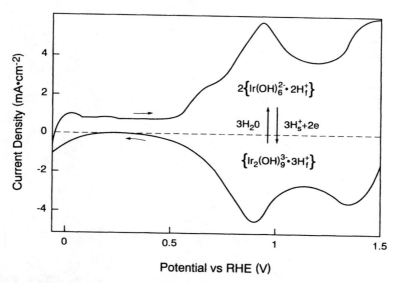

Figure 6.12. Voltammetric response of iridium in 1 mol · L^{-1} H$_2$SO$_4$ at 25°C at a scan rate of 3 V · s^{-1} from −0.05 V to +1.5 V versus RHE (in this solution the zero point of the RHE scale is at +0.018 V versus SHE or −0.223 V versus SCE). The profile shown is for the 120th potential scan (reproduced using data from Ref. 126). The nature of the oxidation/reduction reaction involved is indicated. H$_f^+$ and H$_s^+$ denote protons exchanged within the film (H$_f^+$) and between the film and bulk solution (H$_s^+$) during the redox reaction.

high anodic potentials, coverage by the U^V intermediate will be high ($k_{-1} \ll k_2$) and (for $\beta_1 = \beta_2 = \beta$)

$$I_D = 2Fk_1k_2(k_1 + k_2) \exp[(1 - \beta)EF(RT)^{-1}] \quad (6.14)$$

which predicts a Tafel slope of 120 mV/decade ($\beta = 0.5$) and no pH dependence (Fig. 6.13).

For higher pH values (3–9) the distinction between the two individual electron-transfer steps appears to be lost and a single Tafel slope of 80–90 mV/decade [128,129] or (at pH = 9.5) ~60 mV/decade [111] is obtained. Between pH 3 and 5 the current-potential curves exhibit an anodic shift of ~100 mV but are independent of pH near the neutral region (pH = 5–9). For higher pH values, a Tafel region becomes difficult to define and the dissolution currents become independent of potential at very anodic potentials (pH ≥ 12) [128,129], consistent with chemical control of the dissolution by reaction (6.12). This last possibility was discussed earlier when considering the nature of the surface films formed in alkaline solutions (see Fig. 6.10).

Although this simple reaction mechanism accounts for most of the available electrochemical data, it remains unproven in the absence of a more thorough analysis. Nonuniformity in the characteristics of the electrode material used adds further

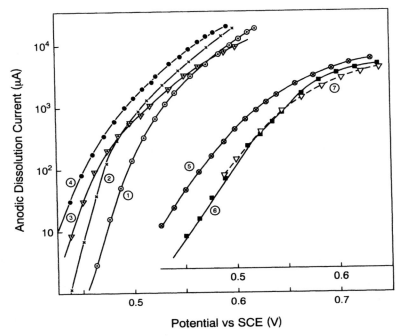

Figure 6.13. IR-compensated steady-state dissolution currents recorded on a $UO_{2.14}[UO_{2+x}(UO_{2.25})]$ rotating-disk electrode [resistivity (ρ) $\sim$40 $\Omega \cdot$ cm] in various solutions of different pH values. The electrode rotation rate was sufficient to eliminate solution transport effects.
(1) $- \odot -$ 1 mol $\cdot$ L^{-1} HClO$_4$;
(2) $- x -$ 1 mol $\cdot$ L^{-1} NaClO$_4$, pH = 1.0;
(3) $- \triangledown -$ 1 mol $\cdot$ L^{-1} NaClO$_4$, pH = 2.0;
(4) $- \bullet -$ 0.5 mol $\cdot$ L^{-1} Na$_2$SO$_4$, pH = 2.0;
(5) $- \otimes -$ 1 mol $\cdot$ L^{-1} NaClO$_4$, pH = 3.0;
(6) $- \blacksquare -$ 1 mol $\cdot$ L^{-1} NaClO$_4$, pH = 5.0;
(7) $- \triangledown -$ 1 mol $\cdot$ L^{-1} NaClO$_4$, pH = 9.0.
The curves marked (5), (6), and (7) have been shifted by 0.1 V to avoid confusion. (Replotted from data in Ref. 128.)

uncertainty. Nicol and Needes [128,129] recorded their data on an electrode with an overall composition of $UO_{2.14}$ but apparently composed of a mixture of UO_{2+x} and considerable amounts of U_4O_9 {designated $UO_{2.14}[UO_{2+x}(UO_{2.25})]$}. By contrast the electrode material used by Shoesmith et al. [111] was very close to stoichiometric UO_2.

Distinct effects on anodic dissolution of both anions and cations have been observed in the pH range 0–5 [128]—consistent with the prior formation of a film composed of hydrolyzed aggregates. The state of hydrolysis of such aggregates would be affected by the incorporation of anions, especially ones that complex U^V/U^{VI} species. Similarly, cation injection into these films would be expected to compete with proton injection, especially for small polarizable cations. Thus anodic dissolution currents increase with change in cation species in sulfate solutions in the following order:

$$[I_D]_{K^+} < [I_D]_{Na^+} < [I_D]_{Mg^{2+}} \quad (6.15)$$

over the pH range from 0 to 5, with $[I_D]_{Mg^{2+}}$ up to one order of magnitude greater than $[I_D]_{K^+}$.

Relative to perchlorate, sulfate enhances dissolution for pH $\geq$ 2 and inhibits it for pH < 2, an effect attributed to activation by SO_4^{2-} and inhibition by HSO_4^- [128]. This interpretation is incomplete, however, since it cannot explain either the change in Tafel slope (55 ± 5 mV/decade for sulfate compared with 40 ± 5 mV/decade in perchlorate) or the higher current densities at more anodic potentials, Fig. 6.13 [curves (3) and (4)]. A possible explanation for this higher Tafel slope at low positive potentials in sulfate is that the assumption of low coverage by the U^V intermediate used to obtain Eq. (6.13) no longer applies. Such a situation would be achieved if sulfate stabilized the adsorbed U^V intermediate ($0 < \theta < 1$) and accelerated the dissolution of adsorbed U^{VI} species. The reaction scheme proposed in perchlorate solutions could be rewritten for sulfate solutions as follows:

$$UO_2 + (HSO_4^-)_{ads} \rightleftharpoons (UO_2HSO_4)_{ads} + e \quad (6.16)$$

$$(UO_2HSO_4)_{ads} + OH^- \rightarrow (UO_2SO_4)_{ads} + e + H_2O \quad (6.17)$$

$$(UO_2SO_4)_{ads} \rightarrow UO_2SO_4 \quad (6.18)$$

$$UO_2SO_4 + SO_4^{2-} \rightleftharpoons UO_2(SO_4)_2^{2-} \quad (6.19)$$

A similar mechanism to that proposed for sulfate appears to apply for anodic dissolution in alkaline carbonate solutions (8 $\leq$ pH $\leq$ 11) [123,128,130,131]. In this pH range, dissolution occurs from an underlying film of $UO_{2.33}$ (~5 nm thick), whose presence has been confirmed by XPS [123]:

$$UO_{2.33} + 0.33H_2O + HCO_3^- \rightleftharpoons (UO_2HCO_3)_{ads} + 0.66OH^- + 0.33e \quad (6.20)$$

$$(UO_2HCO_3)_{ads} + OH^- \rightarrow (UO_2CO_3)_{ads} + H_2O + e \quad (6.21)$$

$$(UO_2CO_3)_{ads} + HCO_3^- \rightarrow UO_2(CO_3)_2^{2-} + H^+ \quad (6.22)$$

$$UO_2(CO_3)_2^{2-} + HCO_3^- \rightleftharpoons UO_2(CO_3)_3^{4-} + H^+ \quad (6.23)$$

Compared with curves recorded in perchlorate and sulfate in this pH region, current-potential curves in carbonate are cathodically shifted by up to 300 mV (depending on concentration), owing to the effect of complexation reactions (Table 6.1) on the equilibrium potential of the dissolution half-reaction [111,128,130].

Electrochemical studies in carbonate solution have also been performed on uranium-oxide electrodes with distinctly different characteristics, and although the general features of steady-state current-potential curves are the same, differences in slope and pH dependencies exist [128,130,131]. For specimens close to stoichiometric UO_2 [111,131], two distinct Tafel slopes are observed; by contrast, on

$UO_{2.14}[UO_{2+x}(UO_{2.25})]$ electrodes only one Tafel region was found [130]. For near stoichiometric UO_2 at low potentials, Tafel slopes between 62 and 75 mV/decade suggest that the conditions required to generate a response of the form of Eq. (6.13) (Tafel slope = 40 mV/decade) do not apply. Nicol and Needes claimed that a two-electron transfer process was rate-determining [130], but this seems highly unlikely. A more probable explanation is that the two one-electron transfers ($U^{IV} \rightarrow U^V$) [reaction (6.20)] and $U^V \rightarrow U^{VI}$ [reaction (6.32)] are kinetically similar at low potentials (i.e., $k_{IV/V} \sim k_{V/VI}$) and the approximations embodied in the derivation of Eq. (6.13) are not justified. At higher potentials on near stoichiometric UO_2, Tafel slopes between 95 mV/decade ($[CO_3]_T \geq 0.1$ mol · L^{-1}) and 155 mV/decade ($[CO_3]_T \leq 10^{-2}$ mol · L^{-1}) were obtained, along with a first-order dependence on carbonate concentration [131], consistent with control of the rate by reaction (6.20). That subsequent reactions, including electron transfer between carbonate-complexed U^V and U^{VI} species [reaction (6.21)], should be relatively rapid compared to the first electron transfer would not be surprising considering the strong carbonate complexes formed by the uranyl ion (see Table 6.1). The U^V ion, UO_2^+, is also stabilized by complexation with carbonate [106], although apparently only in the adsorbed state during anodic dissolution of UO_2. At a combination of high carbonate concentration and high anodic potential, the dissolution current becomes independent of potential and the rate is controlled by the chemical dissolution of UO_2CO_3 [reaction (6.22)], a species known to have limited solubility (see Table 6.1).

The a.c. impedance data displayed in Figs. 6.14A and 6.14B, which can be interpreted in terms of the equivalent circuit depicted at the top of Fig. 6.14B, provide further evidence that adsorbed carbonate species are involved in the anodic dissolution of UO_2. The high-frequency response (A in Fig. 6.14A) is independent of electrochemical and chemical parameters and is tentatively assigned to the parallel coupling of the electrode resistance (R_{UO_2}) and an associated geometric capacitance (C_g), although a contribution due to impedance within the capillary leading to the reference electrode may also be important. The response at intermediate frequencies (B) is due to the coupling of the charge-transfer resistance (R_{CT}) with a series combination of the double-layer capacitance (C_{dl}) and that due to the underlying film of $UO_{2.33}$ ($C_{UO_{2.33}}$). Variations in R_{CT}^{-1} with potential calculated from this

---

Figure 6.14. A.c. impedance spectra for the anodic dissolution of near stoichiometric UO_2 at room temperature in carbonate-containing solutions:
 (A) Effect of potential (vs. SCE) in 0.1 mol · L^{-1} $NaClO_4$ + 0.2 mol · L^{-1} Na_2CO_3 (pH = 9.5).
 (B) Effect of carbonate concentration in 0.1 mol · L^{-1} $NaClO_4$ (pH = 9.5) at an applied potential of 300 mV (versus SCE);
  – x – $[CO_3]_T = 5 \times 10^{-3}$ mol · L^{-1};
  – ■ – $[CO_3]_T = 3 \times 10^{-2}$ mol · L^{-1};
  – ○ – $[CO_3]_T = 10^{-1}$ mol · L^{-1};
  – ● – $[CO_3]_T = 2 \times 10^{-1}$ mol·L^{-1}.

The data points for a number of specific frequencies are noted. In (B) the high-frequency response (>10^2 Hz) has been omitted for clarity. An appropriate equivalent circuit is shown in (B). (See the text for the symbols in the circuit.)

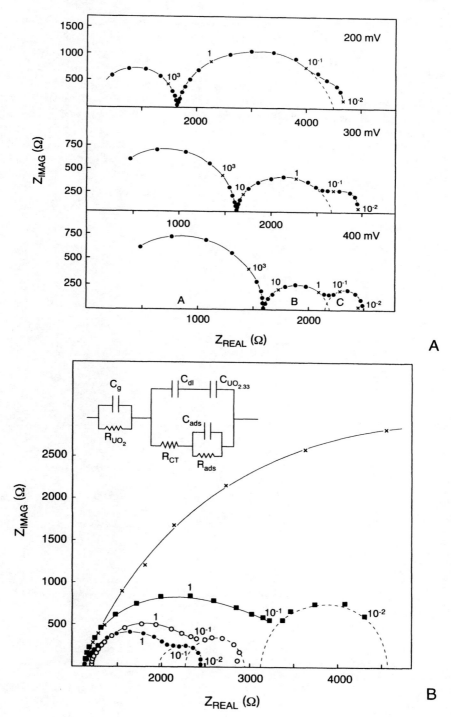

response are identical to those of the steady-state dissolution current. The low-frequency semicircle (C) is dependent on carbonate concentration and can be attributed to the relaxation of an adsorbed intermediate (probably UO_2HCO_3 with C_{ads} and R_{ads}).

In phosphate solutions a Tafel slope of 60 mV/decade is definitely observed (Fig. 6.15), and the steady-state current is limited by a chemical dissolution step to values two orders of magnitude less than those observed for carbonate [132]. This is not surprising, considering the insolubility of uranyl phosphate phases (see Table 6.1). The 60 mV/decade Tafel slope is probably a consequence of the relative stabilities of U^V and U^{VI} species complexed with phosphate and implies that U^V intermediates are not involved in the dissolution process. The chemical-dissolution step is first order in phosphate, suggesting the possible overall dissolution process [132]:

$$UO_2 + HPO_4^{2-} \rightarrow (UO_2HPO_4)_{ads} + 2e \qquad (6.24)$$

$$(UO_2HPO_4)_{ads} + HPO_4^{2-} \rightarrow UO_2(HPO_4)_2^{2-} \qquad (6.25)$$

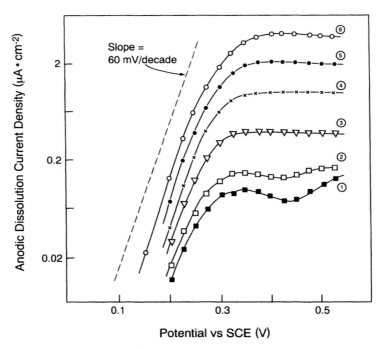

Figure 6.15. Steady-state dissolution currents recorded on a rotating UO_2-disk electrode (f = 16.7 Hz) in 0.1 mol · L^{-1} NaClO$_4$ (pH = 9.5) containing various concentrations of Na$_2$HPO$_4$: (1) ■ 10^{-2} mol · L^{-1}; (2) □ 2×10^{-2} mol · L^{-1}; (3) ▽ 5×10^{-2} mol · L^{-1}; (4) χ 10^{-1} mol · L^{-1}; (5) ● 2×10^{-1} mol · L^{-1}; and (6) ○ 5×10^{-1} mol · L^{-1}. (Data from Ref. 132.)

6.9 Redox Reactions on UO_2 Surfaces

The mechanism and kinetics of redox reactions on UO_2 surfaces have been studied with two principal aims in mind: (1) to define the most appropriate reagent and optimum conditions for the hydrothermal leaching of uranium from ores, and (2) to determine the dissolution mechanism of used fuel and predict its corrosion behavior under various proposed waste-disposal conditions.

The general properties of redox reactions at oxide surfaces have been reviewed in detail [92,133–135]. The kinetics are affected by the following parameters:

1. The density, mobility, and type of charge carriers in the oxide. Electrical conductivity varies from essentially metallic (e.g., RuO_2) through semiconductive (e.g., p-type NiO and n-type SnO_2) to quite insulating (e.g., SiO_2).
2. The potential distribution at the oxide–electrolyte interface. Which redox reactions can be sustained at a specific oxide electrode will depend on the conductivity type, and for semiconductive oxides, on the position of the oxide flatband potential relative to the equilibrium potential for that redox couple. Oxides that are n-type semiconductors will normally block current flow when polarized positive to their flatband potential, whereas p-type oxides should exhibit a similar effect when polarized negatively.
3. The presence of surface states. Many oxide electrodes can sustain redox transition on/in their surface that mediate electron transfer to redox reagents in solution. This is particularly true of transition-metal spinels and perovskites (mixed rare-earth/transition-metal oxides).

The unusually negative flatband potential for UO_2 (-1.3 V versus SCE; see Figs. 6.3, 6.4, and 6.9) allows the attainment of pseudometallic properties on the surface of the oxide over a wide potential range in aqueous solutions. However, the instability of the oxide with respect to anodic dissolution limits the achievable positive potentials, and the major changes in surface composition with potential (see Figs. 6.1 and 6.8) make the interpretation of mechanism and measurement of kinetics for redox reactions difficult. Differences in solid-state properties, such as conductivity, also exert a substantial influence on the kinetics of various redox reactions, and careful characterization of individual electrodes is essential.

Simple redox reactions involving only electron-transfer steps, especially those involving outer-sphere electron transfer in which interactions with the electrode surface are minimal, are capable of probing the electronic levels at the surface of the oxides. Redox couples such as $Fe^{III}(CN)_6^{3-}/Fe^{II}(CN)_6^{4-}$ and Fe^{III}/Fe^{II} are normally presumed to satisfy these requirements. The measured formal potentials (versus SCE) at a UO_2 electrode surface for these two couples are 0.276 V [$Fe^{III}(CN)_6^{3-}/Fe^{II}(CN)_6^{4-}$; pH = 1.5] and 0.443 V (Fe^{III}/Fe^{II}; pH = 1), making it easy to study the cathodic reaction on UO_2 but difficult to study the anodic reaction, which occurs simultaneously with anodic dissolution of the oxide (see Fig 6.8) [136]. Nicol and Needes separated these two anodic processes using a rotating gold-ring/UO_2-disk electrode [136]. For both redox couples the anodic and cathodic reactions were first order. For the Fe^{III}/Fe^{II} couple the cathodic Tafel slope was ~ 160 mV/decade compared with an anodic slope of ~95 mV/decade. No slopes were quoted for the $Fe^{III}(CN)_6^{3-}/Fe^{II}(CN)_6^{4-}$ couple, but the symmetry factors given (β = 0.5 to 0.7) suggest a lower cathodic than anodic slope. As expected, neither reaction depended significantly on the properties of the oxide {either $UO_{2.01}$ with resistivity = 2 k$\Omega \cdot$ cm or $UO_{2.14}[UO_{2+x}(UO_{2.25})]$ with resistivity = 40 $\Omega \cdot$ cm}. For both couples the rate constant k_0 (at $E = E^e$) was similar to that measured for the same reaction on carbon but ~10^2-fold lower than that on platinum. The rate constant for $Fe^{III}(CN)_6^{3-}/Fe^{II}(CN)_6^{4-}$ on UO_{2+x} varied with the nature of the cation, indicating electron transfer via a cation-bridged outer sphere intermediate with

$$(k_0)_{K^+} > (k_0)_{Na^+} > (k_0)_{Li^+} \tag{6.26}$$

as postulated for this reaction occurring homogeneously in solution [137]. In acidic sulfate solutions [136], the maximum value of k_0 for Fe^{III} reduction was obtained for $[Fe^{III}] = [SO_4^{2-}]$, suggesting that $FeSO_4^+$ is the favored electroactive species. In the absence of sulfate (i.e., in perchlorate solutions), the current for Fe^{III} reduction was independent of pH over the 0.5–2 range, implying that the electroactivity of the various hydrolyzed Fe^{III} species is approximately the same. The reduction current showed a peak around pH = 1.5 in sulfate solutions, which is attributed to the decrease in concentration of the species $FeSO_4^+$ at high H^+ activity, where the noncomplexing HSO_4^- predominates over SO_4^{2-}, and production of less electroactive hydrolyzed Fe^{III} species at higher pH values.

At sufficiently positive potentials in acidic solutions, redox reactions are blocked on UO_2 [136]. For reactions such as Ti^{III} oxidation, V^V reduction, and I_3^- reduction, this is attributed to the formation of the U^{VI} films, which also limit the rate of anodic dissolution of the electrode [UO_3 in reaction (6.11)]. Inhibition occurs gradually over the same potential range for which the anodic-dissolution curves become potential-independent (see Fig. 6.13). For the oxidation of $Fe^{II}(CN)_6^{4-}$, inhibition occurs more suddenly and has been attributed to precipitation of the insoluble salt $(UO_2)_2Fe(CN)_6$, although no analytical proof was offered [136].

The two redox reactions studied most extensively on UO_2 electrodes are the reduction of oxygen and the reduction/oxidation of H_2O_2, because they are important not only for the hydrothermal extraction of ores but also for the oxidative dissolution of natural uraninite and the corrosion of used nuclear fuel.

Equilibrium potentials for a number of redox equilibria involving oxygen species are listed in Table 6.2 [104,138,139]. Although both O_2 and H_2O_2 have quite positive standard electrode potentials for their complete reduction to OH^- or H_2O [reactions (6) and (7) in Table 6.2], the first electron-transfer step in each case [reactions (1) and (3) in Table 6.2] is much less favorable. However, two factors increase the facility of reactions (1) and (3) in Table 6.2:

1. The extreme reactivity of OH and O_2^- means that only low concentrations of these species can be achieved, which causes E^e to lie above E^o for this first step [reactions (1) and (3)].
2. Strong adsorption of the reaction intermediates on the electrode surfaces.

Table 6.2 Redox Equilibria for Oxygen Reactions[a]

Reduction Reaction[b]	E^0_{SHE} (V)
1. $O_{2(g)} + e \rightleftharpoons O^-_{2(aq)}$	-0.330
2. $O^-_{2(aq)} + H_2O_{(l)} + e \rightleftharpoons HO^-_{2(aq)} + OH^-_{(aq)}$	0.201
3. $HO^-_{2(aq)} + H_2O_{(l)} + e \rightleftharpoons HO_{(aq)} + 2OH^-_{(aq)}$	-0.166
4. $OH_{(aq)} + e \rightleftharpoons OH^-_{(aq)}$	1.900
5. $O_{2(g)} + H_2O_{(l)} + 2e \rightleftharpoons HO^-_{2(aq)} + OH^-_{(aq)}$	-0.065
6. $HO^-_{2(aq)} + H_2O_{(l)} + 2e \rightleftharpoons 3OH^-_{(aq)}$	0.867
7. $O_{2(g)} + 2H_2O_{(l)} + 4e \rightleftharpoons 4OH^-_{(aq)}$	0.401

[a] Standard reduction potentials calculated from thermodynamic data [104,138,139].
[b] Written in terms of the species that predominate for unit OH^- activity, i.e., O^-_2 and HO^-_2, based on the acid dissociation constants for hydrogen peroxide and the perhydroxyl radical:

$$H_2O_2 \rightleftharpoons H^+ + HO^-_2 \quad pK = 11.7$$
$$HO_2 \rightleftharpoons H^+ + O^-_2 \quad pK = 4.8$$

On oxide electrodes this is accomplished by the coupling of reduction to changes in the valence state of the oxide cations [140]. Consequently, we would expect the kinetics for O_2 reduction on UO_2 to change significantly with the solid-state properties of the oxide and the degree of oxidation of its surface (see Fig. 6.8). The latter is illustrated by the current-voltage plots for O_2 reduction recorded on a slightly hyperstoichiometric UO_2 electrode before and after surface oxidation in air-saturated, carbonate-free solutions for a period of four days [93,114] (Fig. 6.16A). Figure 6.16B shows a voltammogram recorded on the same electrode. The inflection points (E_1, E_2, E_3) in the current-potential curves correlate well with qualitative changes in composition of the electrode surface. For the prereduced electrode, deviations from the Tafel region are apparent for $E \lesssim -0.7$ V versus SCE. At potentials more negative than shown in Fig. 6.16A, the reduction current becomes independent of potential and may even decrease [93]. Unfortunately, the IR compensation required at these potentials and the need to correct for the current due to water reduction make it difficult to be certain about the validity of such a decrease. Since the flatband potential for this oxide is more than 0.5 V negative to the potential at which deviations from Tafel behavior start (Fig. 6.16A), it is extremely unlikely that the development of a space-charge layer in the oxide can account for this potential-independent current. The correlation between the initial stages of surface oxidation and the cathodic end of the Tafel region (Figs. 6.16A and 6.16B) suggests instead that changes in surface composition are the cause of the curvature. Because the state of oxidation of the surface is changing throughout the Tafel region, the Tafel slope of ~ 190 mV/decade may result partially from a potential-dependent change in the number of available reduction sites on the electrode surface [93].

The O_2-reduction current is increased by up to an order of magnitude when

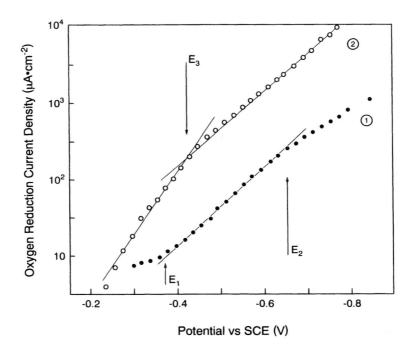

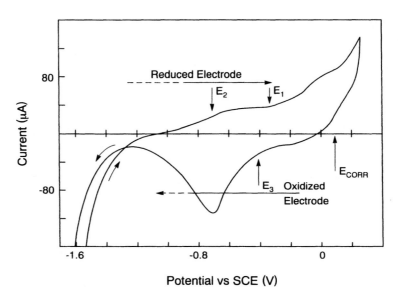

natural corrosion in aerated solution to a steady-state open-circuit potential of ~+0.1 V versus SCE is allowed (see Fig. 6.16A). Under these conditions, the surface will be covered by a layer with a composition of $UO_{2.33}$ (~5–8 nm thick) with possibly a thin film of U^{VI} species at the electrode–electrolyte interface [111]. Oxidation in this manner obviously increases the number density of sites available for O_2 reduction. Before reduction of the oxidized surface layer commences at E_3 in Fig. 6.16, a Tafel slope of 120 mV/decade is achieved, demonstrating that, with the first electron-transfer step rate-controlling, the reduction process is not limited by the availability of reactive sites.

The theory developed by Presnov and Trunov [141] for O_2 reduction on transition-metal oxides with p-type semiconductivity is readily adaptable to explain this behavior on UO_{2+x}. Electron-transfer reactions are assumed to occur exclusively at active sites formed by adjacent cations in different valence states. The cation at the surface must be in a higher oxidation state than its nearest neighbor below the surface, thought most likely to be U^{VI} and U^V, respectively, although U^V/U^{IV} and U^{VI}/U^{IV} combinations would also be suitable. In the lower valence states the uranium cation possesses electrons in f-orbitals capable of interacting with sp^2 orbitals on molecular oxygen. Thus these active sites function as electron acceptors relative to the oxide bulk and electron donors relative to the adsorbed molecular oxygen.

A reaction mechanism consistent with experimental observations is the following [93]:

$$\{U^{VI}\} + e \rightarrow \{U^V\} \quad (6.27)$$

$$\{U^V\} + O_2 \rightleftharpoons \{U^V\}O_2 \quad (6.28)$$

$$\{U^V\}O_2 \rightarrow \{U^{VI}\}O_2^- \quad (6.29)$$

where {U} represents a donor–acceptor site in the electrode surface and $\{U\}O_2$ an oxygen species adsorbed onto such a site. Weak adsorption of O_2 at donor–acceptor sites would yield a low coverage of the adsorbed species, $\{U^V\}O_2$, and a first-order dependence on O_2 concentration with reaction (6.29) rate-determining. Protonation during or prior to the rate-determining step can be discounted, since only a very

Figure 6.16. (A) Transport- and IR-compensated current-potential plots for O_2 reduction on UO_2 ($\rho \approx 4$ k$\Omega \cdot$ cm) in aerated 0.1 mol $\cdot$ L^{-1} NaClO$_4$ (pH = 9.5). E_1, E_2, and E_3 indicate potentials at which the curves either deviate from linear behavior (E_1, E_2) or change slope (E_3). The electrode was polished to a 600-grit (18 μm) finish and electrochemically reduced in argon-purged solution before the experiment. (1) Data recorded from the most negative to most positive potential; (2) data recorded from the most positive to most negative potential after natural corrosion in aerated 0.1 mol $\cdot$ L^{-1} NaClO$_4$ (pH = 9.5) until a steady-state corrosion potential was achieved (~4 days) [93,114].

(B) Voltammetric response (IR-compensated) recorded on the same electrode in argon-purged 0.1 mol $\cdot$ L^{-1} NaClO$_4$ (pH = 9.5) at 10 mV $\cdot$ s^{-1}. The horizontal arrows indicate the potential ranges over which O_2-reduction currents were measured (see Fig. 6.16A). E_{CORR} indicates the corrosion potential achieved prior to recording curve 2 in A [114].

weak pH dependence is observed over the range $8.5 \leq \text{pH} \leq 10.5$. Further reduction of the adsorbed superoxide ion (O_2^-) then proceeds via peroxide, with protonation and electron transfer occurring either sequentially or simultaneously,

$$\{U^{VI}\}O_2^- + e \rightarrow \{U^V\}O_2^- \tag{6.30}$$

$$\{U^V\}O_2^- + H_2O \rightarrow \{U^{VI}\}O_2H^- + OH^- \tag{6.31}$$

No peroxide is detected at the gold ring of a rotating ring-disk electrode, unless the electrode surface is either highly polished (1 μm) or the solution contains large quantities of carbonate ion (>0.1 mol · L^{-1}). In the former case, a small fraction of the disk current yields peroxide for low–intermediate overpotentials, whereas in the latter case, large amounts of peroxide are observed for all potentials. The effect of polishing suggests that the relative rates of peroxide reduction and desorption are subtly affected by the state of the surface. The presence of carbonate also causes a decrease in the total O_2-reduction current by approximately one order of magnitude, which can be attributed to blocking of donor–acceptor sites by adsorbed carbonate [93]. Although incompletely understood, these observations indicate that O_2 reduction proceeds, at least partly, through an adsorbed peroxide species, and a further sequence of electron- and proton-transfer reactions involving the donor–acceptor sites would complete the four-electron reduction process to hydroxide ion [93].

The effect of the solid-state properties of UO_2 on the kinetics of O_2 reduction are illustrated by the results in Fig. 6.17 [93,114], which shows current-potential curves recorded on two different specimens of UO_2 and on a UO_2 specimen doped with a number of elements in an attempt to replicate the chemical impact of fuel irradiation during reactor operation (i.e., SIMFUEL) [142]. The resistivity (ρ) of the doped UO_2 (B in Fig. 6.17) is four times lower than that of the typical undoped UO_2 (A in Fig. 6.17), because of the substitution of trivalent ions (e.g., Y^{III}, La^{III}, and Nd^{III}) for U^{IV} in the uraninite lattice, which creates mobile holes in the U 5f band even in the absence of interstitial O^{2-} ions. Charge carrier trapping at substitution sites may limit the contribution of the dopants to the bulk electrical conductivity at ambient temperatures. Nonetheless, this doped UO_2 possesses a high density of surface donor–acceptor sites, leading to a superior reactivity for the cathodic reduction of O_2. The Tafel slope for the doped UO_2 is close to that for undoped UO_2, and the current is first order in O_2 concentration, indicating that O_2 reduction proceeds by the same mechanism on both specimens, reactions (6.27) to (6.31).

By contrast, a larger Tafel slope is obtained on the second undoped specimen of UO_2 (C in Fig. 6.17), and the reaction order with respect to O_2 concentration is 0.5. The resistivity of this material is an order of magnitude lower than that of standard undoped UO_2 (A in Fig. 6.17). Furthermore, it exhibits a positive photocurrent response (compared with the negative response in Fig. 6.9) under illumination by white light despite a dark-current voltammogram almost identical to that of standard p-type UO_2 [114]. This positive photocurrent response can be understood in terms of the energy-level diagram of Fig. 6.3, provided there are sufficient vacancies in the occupied U 5f band. For a high degree of hyperstoichiometry, an indirect charge transfer appears to occur in which excitation of a U 5f electron into the conduction

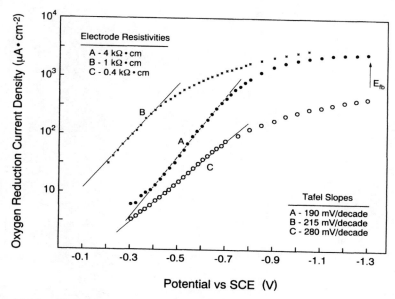

Figure 6.17. Transport- and IR-compensated plots for O_2 reduction on two different specimens of UO_2 (A and C) and a UO_2 specimen (B) doped with a number of elements (SIMFUEL [142]) recorded in aerated 0.1 mol · L^{-1} $NaClO_4$ solution (pH = 9.5). Each electrode was polished to a 600-grit finish and electrochemically reduced in argon-purged solution before the experiment [93,114]. E_{fb} is the flatband potential for an electrode with properties similar to electrode A (see Fig. 6.9).

band leads to the capture of an O 2p electron by the U^V or U^{VI} state. This would leave a highly reactive hole in the valence band capable of extracting electrons from species in solution over the whole potential range (see Fig. 6.18). A detailed understanding of the variation in photoresponse with potential remains to be achieved.

Significant hyperstoichiometry would be expected to yield a large number of donor-acceptor sites in the surface of the electrode and hence a similar enhancement of the O_2-reduction current to that observed for rare-earth-doped UO_2 (B in Fig. 6.17). However, the hyperstoichiometry appears to be nonuniformly distributed, possibly crowded into highly conductive grain boundaries. Strong interactions between such a network of closely spaced donor-acceptor sites could lead to O_2 adsorption under Temkin conditions and a strong coupling between the adsorption step [reaction (6.28)] and the first electron transfer [reaction (6.29)]. Following arguments made by Calvo and Schiffrin for O_2 reduction on iron oxide [143], this can be shown to yield a near square-root dependence on O_2 concentration and a doubling of the expected Tafel slope to 240 mV/decade.

Needes and Nicol observed quite different behavior for O_2 reduction on $UO_{2.14}$ [$UO_{2+x}(UO_{2.25})$] [136]. Current-potential curves exhibited two distinct stages of reduction compared with the single stage observed on a single phase UO_{2+x} elec-

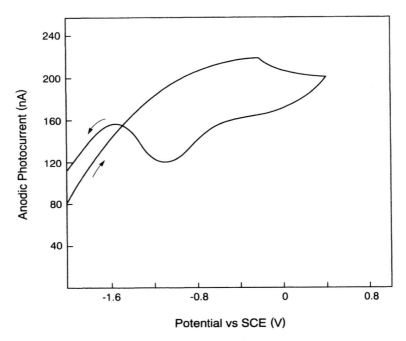

Figure 6.18. Photocurrent recorded on electrode C (Fig. 6.17) during a potential scan from −2.0 V to +0.4 V and back, at 20 mV · s^{-1} in 0.2 mol · L^{-1} Na$_2$SO$_4$ solution (pH = 10.0). The electrode was illuminated with white light during the experiment [114].

trode under the same conditions (Fig. 6.19). The second stage at more negative potentials appears similar in mechanism to that described earlier, since the Tafel slope, reaction order with respect to O$_2$, and lack of pH dependence are identical. To explain the stage at less negative potentials, they found it necessary to invoke a step involving protonation of surface sites to account for the apparent pH dependence of the overall reaction. It is tempting to view the two stages, however, as O$_2$ reduction occurring on the two distinct phases: on the more oxidized UO$_{2.25}$ phase at less negative potentials and on the UO$_{2+x}$ phase at higher potentials. Such an interpretation would be consistent with the evidence in Fig. 6.16A showing that O$_2$ reduction was enhanced on more oxidized surfaces.

A definitive study of the oxidation and reduction of H$_2$O$_2$ on well-characterized UO$_2$ has not yet been performed. On noble-metal electrodes, the oxidation of H$_2$O$_2$ appears to be kinetically more facile than its reduction (144–148). For H$_2$O$_2$ oxidation on gold in alkaline solutions Zurilla et al. [148] observed a Tafel slope of 120 mV/decade and reaction orders of 0.5 with respect to peroxide and hydroxide ion, and interpreted these observations by a reaction sequence involving a dismutation equilibrium prior to the rate-determining one-electron oxidation of an adsorbed superoxide ion:

ELECTROCHEMISTRY OF UO_2 NUCLEAR FUEL

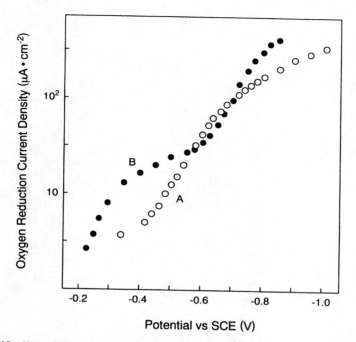

Figure 6.19. IR-compensated plots for O_2 reduction on different specimens of UO_2:
(A) Data (○) recorded on an electrode similar to electrode A (Fig. 6.17) rotating at 41.67 Hz in an aerated solution containing 0.5 mol · L^{-1} Na_2CO_3 (pH = 9.5) from Ref. 93;
(B) Data (●) recorded on a $UO_{2.14}[UO_{2+x}(UO_{2.25})]$ (ρ = 40 Ω · cm) electrode rotating at an unspecified rate in oxygenated 0.5 mol · L^{-1} Na_2CO_3 solution (pH = 9.5) from Ref. 136.

$$O_2 + HO_2^- + OH^- \rightleftharpoons 2(O_2^-)_{ads} + H_2O \quad (6.32)$$

$$(O_2^-)_{ads} \rightarrow O_2 + e \quad (6.33)$$

Similar Tafel slopes and reaction orders were observed by Nicol and Needes on $UO_{2.14}$ $[UO_{2+x}(UO_{2.25})]$, the oxidation of H_2O_2 having been separated from the anodic dissolution of UO_2 using a rotating gold-ring/UO_2-disk electrode [136]. Based on suggestions by Bockris and others to explain similar behavior on platinum and gold [144–146], a mechanism was invoked involving the dissociation of peroxide into adsorbed radical species before electron transfer; however, the sequence of reactions (6.32) and (6.33) appears at least as plausible an explanation. The potentials required to inhibit peroxide oxidation are not as positive as for other reactions (Fig. 6.20), which implies that H_2O_2 oxidation is more sensitive to surface composition than, say, Ti^{III} oxidation. Because both oxidation and reduction sites are required for the dismutation reaction (6.32), it is possible that the simple elimination of catalytic sites involving U^{IV} as well as U^V/U^{VI} is sufficient to block peroxide

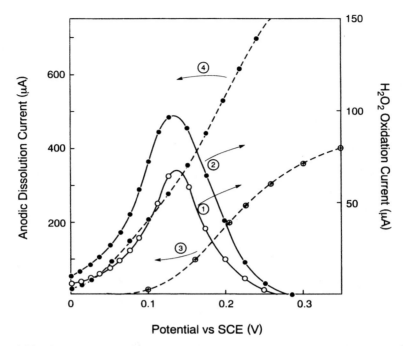

Figure 6.20. Currents for the anodic oxidation of H_2O_2 [curves (1) and (2)] and the anodic dissolution of UO_2 [curves (3) and (4)] recorded on a rotating $UO_{2.14}[UO_{2+x}(UO_{2.25})]$-disk/gold-ring electrode in 0.5 mol · L^{-1} Na_2CO_3 (pH = 11.3):
(1) ○ $[H_2O_2] = 4.76 \times 10^{-3}$ mol · L^{-1};
(2) ● $[H_2O_2] = 1.67 \times 10^{-2}$ mol · L^{-1};
(3) ⊕ $[H_2O_2] = 0.0$ mol · L^{-1};
(4) ● $[H_2O_2] = 1.67 \times 10^{-2}$ mol · L^{-1}
(Data from Ref. 136.)

oxidation. The blocking of other oxidations not involving specific site interactions appears to require a thicker, more insulating layer. Such an explanation must be considered tentative, since the blocking of peroxide oxidation is accompanied by an enhancement in the current for the anodic dissolution of UO_2, because of a cathodic shift in the current-potential curve (see Fig. 6.20) [136]. Nicol and Needes attributed this shift to the effect of the complexation equilibrium;

$$UO_2(CO_3)_3^{4-} + H_2O_2 \rightleftharpoons UO_2(CO_3)_2O_2H^{3-} + HCO_3^- \qquad (6.34)$$

on the equilibrium potential for the anodic dissolution half-reaction.

On both platinum and gold, the current for peroxide reduction in alkaline solution varies little with potential, suggesting rate control by a chemical step prior to the first electron transfer. Zurilla et al. [148] tentatively proposed a rate-determining dissociation step;

$$HO_2^- + H_2O \rightarrow 2(OH^·)_{ads} + OH^- \qquad (6.35)$$

equivalent to the step;

$$H_2O_2 \rightarrow 2(OH^\cdot)_{ads} \quad (6.36)$$

in less alkaline solutions ($pK_a = 11.7$ for $H_2O_2 \rightleftharpoons HO_2^- + H^+$). As an alternative, Genshaw et al. [147] and others [144–146] proposed rate control by chemical-dissociation steps to yield surface O species. Nicol and Needes [136] suggested that a similar sequence of events could explain their large Tafel slopes (~520 mV/decade) and first-order dependence on peroxide concentration for $UO_{2.14}[UO_{2+x}$ ($UO_{2.25}$)] electrodes. Since their Tafel slope was maintained at large negative potentials without any evidence for increasing chemical control, however, they preferred a mechanism that allowed rate control by a first electron-transfer process such as:

$$H_2O_2 + e \rightarrow OH^\cdot + OH^- \quad (6.37)$$

Neither possibility appears to be a totally satisfactory explanation, and a more complete study involving UO_{2+x} electrodes with different properties will be required before a clear mechanistic understanding can be claimed.

6.10 Electrochemical Reactivity

Various solid-state parameters (e.g., conductivity, distribution of grain sizes, inhomogeneities in composition) have a major impact on the oxidation and dissolution of UO_2. Unfortunately, many of these effects are not understood, despite their critical importance in determining the rate of release of radionuclides from used nuclear fuel [10]. Of particular importance is the relative reactivity of grains and grain boundaries, since the latter provide preferred sites for many radionuclides in used fuel. The effects of grain boundaries on the kinetics of anodic dissolution were illustrated by a comparison of the behavior of a single crystal ($UO_{2.03}$) and a sintered pellet ($UO_{2.01}$) with a mean grain size of 15 μm (3–30 μm), but also containing significant quantities of small (<1 μm) grains of unrecrystallized UO_2 [128]. The form of Tafel plots was the same, suggesting a similar dissolution mechanism; however, for a constant anodic potential of 0.4 V (versus SCE), the dissolution current for the sintered disk was $>10^3$ times that of the single crystal. Other specimens with various resistivities, compositions, and grain sizes gave intermediate results and no clear correlations were apparent. Nonetheless, the tendency of grain boundaries to dissolve preferentially, leading to pellet erosion, was demonstrated by extensive anodic-dissolution experiments in carbonate solutions [149].

The claim that the simple density of grain boundaries will be the major factor determining reactivity and oxidative-dissolution rates [128] is likely to prove incomplete. The corrosion rates of nuclear fuel [14,111,150] and naturally occurring uraninites (UO_{2+x}) [151] appear to be controlled by the kinetics of the cathodic reaction(s), O_2 and H_2O_2 reduction, which are subtly dependent on the properties of the UO_2 surface as demonstrated in this review. Detailed electrochemical studies of such reactions on UO_{2+x} surfaces should eventually yield a more complete understanding of the reactivity of UO_2, especially under geological and nuclear-waste-disposal conditions.

Acknowledgments

R. J. Lemire is thanked for calculating the data plotted in Figs. 6.5 and 6.6. We would also like to thank L. M. Peter (University of Bath, U.K.) in whose laboratory the a.c. impedance and photoelectrochemical data were recorded. SIMFUEL pellets were provided by P. G. Lucuta and R. A. Verrall (Chalk River Laboratories). The writing of this review was sponsored by the Canadian Nuclear Fuel Waste Management Program, which is jointly funded by Atomic Energy of Canada Limited and Ontario Hydro under the auspices of the CANDU Owners Group (COG). Constructive criticism from D. Mancey, C. Frost, and one reviewer is acknowledged.

References

1. Collier, J. G.; Hewitt, G. F. "Introduction to Nuclear Power." Springer-Verlag, Berlin, 1987.
2. Matzke, Hj. "Science of Advanced LMFBR Fuels." Elsevier, Amsterdam, 1986.
3. Smith, D. K.; Scheetz, B. E.; Anderson, C. A. F.; Smith, K. L. Uranium **1982**, *1*, 79.
4. Nicol, M. J.; Needes, C. R. S.; Finkelstein, N. P. In "Leaching and Reduction in Hydrometallurgy," Burkin, A. R., Ed. Institute of Mining and Metallurgy, London, 1975, p. 1.
5. Needes, C. R. S.; Nicol, M. J.; Finkelstein, N. P. In "Leaching and Reduction in Hydrometallurgy," Burkin, A. R., Ed. Inst. Min. Metall., London, 1975, p. 12.
6. Hiskey, J. B. Inst. *Min. Metall. Trans.* **1979**, *88*, C145.
7. Hiskey, J. B. *Inst. Min. Metall. Trans.* **1980**, *89*, C145.
8. Taylor, R. F.; Sharratt, E. W.; de Chazal L. E. M.; Logsdail, D. H. *J. Appl. Chem.* **1963**, *13*, 32.
9. Habashi, F.; Thurston, G. *Energia Nucleare* **1967**, *14*, 238.
10. Johnson, L. H.; Shoesmith, D. W. In "Radioactive Waste Forms for the Future," Lutze, W.; Ewing, R. C., ed., Elsevier, Amsterdam, 1988, p. 635.
11. Sunder, S.; Shoesmith, D. W. Atomic Energy of Canada Limited Report, AECL-10395, 1991.
12. Garisto, F.; Garisto, N. C. *Nucl. Sci. Eng.* **1985**, *90*, 103.
13. Lemire, R. J.; Garisto, F. Atomic Energy of Canada Limited Report, AECL-10009, 1989.
14. Shoesmith, D. W.; Sunder, S. *J. Nucl. Mater.* **1992**, *190*, 20.
15. Shoesmith, D. W.; Sunder, S. Atomic Energy of Canada Limited Report, AECL-10488, 1991.
16. Cramer, J. J. In "Proceedings of the International Canadian Nuclear Society Conference on Radioactive Waste Management," Canadian Nuclear Society, Toronto, 1986, p. 697.
17. Cramer, J. J.; Smellie, J. A. T. In "Proceedings of the Fifth Natural Analogue Working Group," CEC Publication EUR 15176 (in press).
18. Belle, J. Ed., "Uranium Dioxide: Properties and Nuclear Applications," AEC-USA, Washington, DC, 1980.
19. Matzke, Hj. *Adv. Ceram.* **1986**, *17*, 1.

20. Matzke, Hj. In "Plutonium and Other Actinides," Blank, H.; Lindner, R., eds. North-Holland, Amsterdam, 1976, p. 801.

21. Kleykamp, H. *J. Nucl. Mater.* **1985**, *131*, 221.

22. Matzke, Hj. *Radiat. Effects* **1982**, *64*, 3.

23. Matzke, Hj. *J. Chem. Soc. Faraday Trans. II*, **1987**, *83*, 1121.

24. Matzke, Hj. Nucl. Instrum. Methods Phys. Res. **1988**, *B32*, 455.

25. Matzke, Hj. *Radiat. Effects* **1980**, *53*, 219.

26. Matzke, Hj. *J. Chem. Soc. Faraday Trans.* **1990**, *86*, 1243.

27. Ball, R. G. J.; Grimes, R. W. *J. Chem. Soc. Faraday Trans.* **1990**, *86*, 1257.

28. Grimes, R. W.; Catlow, C. R. A. *Phil. Trans. R. Soc. London* **1991**, 335A, 609.

29. Thomas, L. E.; Charlot, L. A. *Ceram. Trans.* **1990**, *9*, 397.

30. Thomas, L. E. In "Fundamental Aspects of Inert Gases in Solids," Donnelly, S. E.; Evans, J. H. Ed. Plenum, New York, 1991, p. 431.

31. Une, K.; Tominaga, Y.; Kashibe, S. *J. Nucl. Sci. Technol.* **1991**, *28*, 409.

32. Matzke, Hj.; Ottaviani, J.; Pellottiero, D.; Rouault, J. *J. Nucl. Mater.* **1988**, *160*, 142.

33. Kleykamp, H. *J. Nucl. Mater.* **1979**, *84*, 109.

34. Kleykamp, H. *J. Nucl. Mater.* **1990**, *171*, 181.

35. Vollath, D. In "Uranium-Uranium Dioxide, UO_2, Preparation and Crystallographic Properties," Gmelin Handbook of Inorganic Chemistry, Vol. C4, Springer-Verlag, Heidelberg, 1984, p. 97.

36. Willis, B. T. M. *J. Chem. Soc. Faraday Trans.* **1987**, II, *83*, 1073.

37. Bevan, D. J. M.; Grey, I. E.; Willis, B. T. M. *J. Solid State Chem.* **1986**, *61*, 1.

38. Aronson, S.; Rulli, J. E.; Schaner, B. E. *J. Chem. Phys.* **1961**, *35*, 1382.

39. Ishii, T.; Naito, K.; Oshima, K. *J. Nucl. Mater.* **1970**, *36*, 288.

40. Naito, K.; Tsuji, T.; Matsui, T. *J. Nucl. Mater.* **1973**, *48*, 58.

41. Hoekstra, H. R.; Santoro, A.; Siegel, S. *J. Inorg. Nucl. Chem.* **1961**, *18*, 166.

42. Allen, G. C.; Tempest, P. A. *Proc. R. Soc. London* **1986**, A406, 325.

43. Hoekstra, H. R.; Siegel, S.; Gallagher, F. X. *J. Inorg. Nucl. Chem.* **1970**, *32*, 3237.

44. Taylor, P.; Lemire, R. J.; Wood, D. D. In "High Level Radioactive Waste Management," Vol. 2, Amer. Nucl. Soc., La Grange Park, Illinois, (1992) 1442.

45. O'Hare, P. A. G.; Lewis, B. M.; Nguyen, S. N. *J. Chem. Thermodyn.* **1988**, *20*, 1287.

46. Hoekstra, H. R.; Siegel, S. *J. Inorg. Nucl. Chem.* **1973**, *35*, 761.

47. Moore, J. P.; Fulkerson, W.; McElroy, D. L. *J. Amer. Ceram. Soc.* **1970**, *53*, 76.

48. Matsui, H.; Tamaki, M.; Nasu, S.; Kurasawa, T. *J. Phys. Chem. Solids* **1980**, *41*, 351.

49. Matsui, H.; Horiki, M.; Ohya, N.; Kato, T.; Osada, M. *J. Nucl. Mater.* **1983**, *115*, 128.

50. Keller, J.; Erbudak, M. *J. Physique Coloq.* **1979**, *C4*, 22.

51. Weinberger, P.; Podloucky, R.; Mallett, C. P.; Neckel, A. *J. Phys. C.* **1979**, *12*, 801.

52. Killeen, J. C. *J. Nucl. Mater.* **1980**, *88*, 185.

53. Dudney, N. J.; Coble, R. L.; Tuller, H. L. *J. Amer. Ceram. Soc.* **1981**, *64*, 627.
54. Hyland, G. J.; Ralph, J. *High Temp.–High Press.* **1983**, *15*, 179.
55. Winter, P. W. *J. Nucl. Mater.* **1989**, *161*, 38.
56. Fournier, J. M.; Manes, L. In "Structure and Bonding 59/60, Actinides—Chemistry and Physical Properties," Manes, L., Ed., Chapter A, Springer-Verlag, Berlin, 1985, p. 1.
57. Honig, J. M. *J. Chem. Educ.* **1966** *43*, 76.
58. Bosman, A. J.; van Daal, H. J. *Adv. Phys.* **1970**, *19*, 1.
59. Tuller, H. L. In "Nonstoichiometric Oxides," Sorensen, O. T., Ed., Academic Press, New York, 1981, Chapter 6, p. 271.
60. Kelly, P. J., Brooks, M. S. S. *J. Chem. Soc. Faraday Trans.* II, **1987**, *83*, 1189.
61. Baer, Y. In "Handbook on the Physics and Chemistry of the Actinides," Freeman, A. J.; Lander, G. H. eds. Elsevier, Amsterdam, 1984, Chapter 4, p. 271.
62. Naegele, J. R.; Ghijsen, L.; Manes, L. In "Structure and Bonding 59/60, Actinides—Chemistry and Physical Properties," Manes, L., ed. Springer-Verlag, Berlin, 1985, Chapter E, 197.
63. Baer, Y.; Schoenes, J. *Solid State Commun.* **1980**, *33*, 885.
64. Schoenes, J. *Phys. Rep.* **1980**, *63*, 301.
65. Schoenes, J. *J. Chem. Soc. Faraday Trans. II* **1987**, *83*, 1205.
66. Evans, S. *J. Chem. Soc. Faraday Trans. II*, **1977**, *73*, 1341.
67. Veal, B. W. *J. Phys. Colloq.* **1979**, *C4*, 163.
68. Naegele, J. *J. Phys. Colloq.* **1979**, *C4*, 169.
69. Chauvet, G.; Baptist, R. *Solid State Commun.* **1982**, *43*, 793.
70. Cox, L. E.; Ellis, W. P.; Cowan, R. D.; Allen, J. W.; Oh, S. J.; Lindau, I.; Pate, B. B.; Arko, A. J. *Phys. Rev.* **1987**, *B35*, 5761.
71. Ellis, D. E.; Goodman, G. L. *Int. J. Quant. Chem.* **1984**, *25*, 185.
72. Hiernaut, J. P.; Magill, J.; Ohse, R. W.; Tetenbaum, M. *High Temp.–High Press.* **1985**, *17*, 633.
73. McLean, W.; Chen, H. L. *J. Appl. Phys.* **1985**, *58*, 4679.
74. Memming, R. In "Comprehensive Treatise of Electrochemistry," Conway, B. E.; Bockris, J. O'M.; Yeager, E.; Khan, S. V. M.; White, R. E., Eds. Plenum, New York, 1983, Vol. 7, p. 529.
75. Gubanov, V. A.; Rosen, A.; Ellis, D. E. *Solid State Commun.* **1977**, *22*, 219.
76. Cox, L. E. *J. Electron Spectrosc. Relat. Phenom.* **1982**, *26*, 167.
77. Cox, L. E.; Ellis, W. P. *Solid State Commun.* **1991**, *78*, 1033.
78. Bates, J. L.; Hinman, C. A.; Kawada, T. *J. Amer. Ceram. Soc.* **1967**, *50*, 652.
79. Nagels, P.; Devreese, J.; Denayer, M. *J. Appl. Phys.* **1964**, *35*, 1175.
80. DeConinck, R.; Devreese, J. *Physica Status Solidi* **1969**, *32*, 823.
81. Lee, H. M. *J. Nucl. Mater.* **1975**, *56*, 81.
82. Tateno, J. *J. Chem. Phys.* **1984**, *81*, 6130.
83. Hampton, R. N.; Saunders, G. A.; Stoneham, A. M. *J. Nucl. Mater.* **1986**, *139*, 185.

ELECTROCHEMISTRY OF UO$_2$ NUCLEAR FUEL 335

84. Hampton, R. N.; Saunders, G. A.; Harding, J. H.; Stoneham, A. M. *J. Nucl. Mater.* **1987**, *150*, 17.

85. Hampton, R. N.; Saunders, G. A.; Stoneham, A. M.; Harding, J. H. *J. Nucl. Mater.* **1988**, *154*, 245.

86. Collier, I. T.; Hampton, R. N.; Saunders, G. A.; Stoneham, A. M. *J. Nucl. Mater.* **1989**, *168*, 268.

87. Collier, I. T.; Hampton, R. N.; Saunders, G. A.; Stoneham, A. M. *High Temp.–High Press.* **1989**, *21*, 199.

88. Finklea, H. O. In "Semiconductor Electrodes," Finklea, H. O. Ed. Elsevier, Amsterdam, 1988, p. 1.

89. Butler, M. A. *J. Appl. Phys.* **1977**, *48*, 1914.

90. Butler, M. A.; Ginley, D. S. *J. Electrochem. Soc.* **1978**, *125*, 228.

91. Freese, Jr., K. W. *J. Vac. Sci. Technol.* **1979**, *16*, 1042.

92. O'Sullivan, E. J. M.; Calvo, E. J. In "Comprehensive Chemical Kinetics," Vol. 27, "Electrode Kinetics," Compton, R. G. Ed. Elsevier, Amsterdam, 1987, p. 247.

93. Hocking, W. H.; Betteridge, J. S.; Shoesmith, D. W. Atomic Energy of Canada Limited Report, AECL-10402, 1991.

94. Li, J.; Peat, R.; Peter, L. M. *J. Electroanal. Chem.* **1984**, *165*, 41.

95. Gleria, M.; Memming, R. *J. Electroanal. Chem.* **1975**, *65*, 163.

96. Mollers, F.; Memming, R. *Ber. Bunsenges. Phys. Chem.* **1972**, 76, 469.

97. Finklea, H. O. In "Semiconductor Electrodes," Finklea, H. O. ed. Elsevier, Amsterdam, 1988, p. 43.

98. Anderman, M.; Kennedy, J. H. In "Semiconductor Electrodes," Finklea, H. O., ed. Elsevier, Amsterdam, 1988, p. 147.

99. Koffybery, F. P.; Benko, F. A. *J. Electrochem. Soc.* **1981**, *128*, 2476; **1982**, *129*, 2880.

100. Dare-Edwards, M. P.; Goodenough, J. B.; Hamnett, A.; Nicholson, N. D. *J. Chem. Soc. Faraday Trans.* II, **1981**, *77*, 643.

101. Sawatzky, G. A.; Allen, J. W. *Phys. Rev.* **1984**, *53*, 2339.

102. Duffy, J. A. "Bonding, Energy Levels and Bands in Inorganic Solids." John Wiley, New York, 1990.

103. Tench, D. M.; Yeager, E. *J. Electrochem. Soc.* **1973**, *120*, 164.

104. Lide, D. R., ed. "CRC Handbook of Chemistry and Physics," 71st ed. CRC Press, Boca Raton, Florida, 1990.

105. Grenthe, I.; Fuger, J.; Konings, R. J. M.; Lemire, R. J.; Muller, A. B.; Nguyen-Trung, C.; Wanner, H. "Chemical Thermodynamics of Uranium," Vol. 1, "Chemical Thermodynamics," Wanner, H.; Forest, I. Eds. North Holland, Amsterdam, 1992.

106. Mueller, T. R.; Petek, M. In "Encyclopedia of Electrochemistry of the Elements," Bard, A. J., Ed. Dekker, New York, 1986, Vol. IX, Part B.

107. Paquette, J.; Lemire, R. J. *Nucl. Sci. Eng.* **1981**, *79*, 26.

108. Lemire, R. J.; Tremaine, P. R. *J. Chem. Eng. Data* **1980**, *25*, 361.

109. Lemire, R. J.; Garisto, F. Atomic Energy of Canada Limited Report, AECL-10009, 1989.

110. Lemire, R. J. Atomic Energy of Canada Limited Report, AECL-9549, 1988.
111. Shoesmith, D. W.; Sunder, S.; Bailey, M. G.; Wallace, G. J. *Corros. Sci.* **1989**, *29*, 1115.
112. Sunder, S.; Shoesmith, D. W.; Bailey, M. G.; Stanchell, F. W.; McIntyre, N. S. *J. Electroanal. Chem.* **1981**, *130*, 163.
113. McIntyre, N. S.; Sunder, S.; Shoesmith, D. W.; Stanchell, F. W. *J. Vac. Sci. Technol.* **1981**, *18*, 714.
114. Hocking, W. H.; Shoesmith, D. W.; Betteridge, J. S. *J. Nucl. Mater.* **1992**, *190*, 36.
115. Myamlin, V. A.; Pleskov, Y. V. "Electrochemistry of Semiconductors." Plenum, New York, 1967.
116. Morrison, S. R. "The Chemical Physics of Surfaces." Plenum, New York, 1990.
117. Memming, R.; Möllers, F. *Ber. Bunsenges. Phys. Chem.* **1974**, *78*, 450.
118. Pettinger, B.; Schöppel, H. R.; Gerischer, H. *Ber. Bunsenges. Phys. Chem.* **1974**, *78*, 450.
119. Roberts, L. E. J. *J. Chem. Soc.* **1954**, 3332.
120. Ferguson, I. F.; McConnell, J. D. M. *Proc. R. Soc.* (London) A **1957**, *241*, 67.
121. McConnell, J. D. M. *J. Chem. Soc.* **1958**, 947.
122. Efremov, B. N.; Zakharkin, G. I.; Zhukov, S. R.; Tarasevich, M. R. *Electrokhimiya* **1978**, *14*, 937.
123. Shoesmith, D. W.; Sunder, S.; Bailey, M. G.; Wallace, G. J.; Stanchell, F. W. *Applic. Surf. Sci.* **1984**, *20*, 39.
124. Sunder, S.; Shoesmith, D. W.; Bailey, M. G.; Wallace, G. J. *J. Electroanal. Chem.* **1983**, *150*, 217.
125. Shoesmith, D. W.; Sunder, S.; Bailey, M. G. Unpublished results.
126. Burke, L. D.; Scannell, R. A. *Platinum Met. Rev.* **1984**, *28*, 56.
127. Burke, L. D.; Whelan, D. P. *J. Electroanal. Chem.* **1984**, *162*, 121.
128. Nicol, M. J.; Needes, C. R. S. Nat. Inst. Met. Repub. S. Africa, Report No. 7079, 1973.
129. Nicol, M. J.; Needes, C. R. S. *Electrochim. Acta* **1975**, *20*, 585.
130. Nicol, M. J.; Needes, C. R. S. *Electrochim. Acta* **1977**, *22*, 1381.
131. Shoesmith, D. W.; Sunder, S.; Bailey, M. G.; Owen, D. G. In "Passivity of Metals and Semiconductors," Froment, M., ed. Elsevier, Amsterdam, 1983, p. 125.
132. Shoesmith, D. W.; Sunder, S.; Bailey, M. G.; Wallace, G. J. *Can. J. Chem.* **1988** *66*, 259.
133. Tarasevich, M. R.; Efremov, B. In "Electrodes of Conductive Metallic Oxides, Part A," Trasatti, S., Ed. Elsevier, Amsterdam, 1980, p. 221.
134. Tamura, H.; Yoneyama, H.; Matsumoto, Y. In "Electrodes of Conductive Metallic Oxides, Part, A," Trasatti, S. Ed. Elsevier, Amsterdam, 1980, p. 261.
135. Trasatti, S.; Lodi, G. In "Electrodes of Conductive Metallic Oxides, Part B," Trasatti, S., ed. Elsevier, Amsterdam, 1980.
136. Needes, C. R. S.; Nicol, M. J. Nat. Inst. Met. Repub. S. Africa, Report No. 7073, 1973.
137. Basolo, F.; Pearson, R. G. "Mechanisms of Inorganic Reactions." John Wiley, New York, 1967, p. 488.
138. Wardman, P. *J. Phys. Chem. Ref. Data* **1989**, *18*, 1637.

139. Hoare, J. P. In "Standard Potentials in Aqueous Solutions," Bard, A. J.; Parsons, R.; Jordan, J., eds. Marcel Dekker, New York, 1985, 49.
140. Bronoel, G.; Grenier, J. C.; Reby, J. *Electrochim. Acta* **1980**, *25*, 1015.
141. Presnov, V. A.; Trunov, A. M. *Elektrokhimiya* **1975**, *11*, 71, 77, 290.
142. Lucuta, P. G.; Verrall, R. A.; Matzke, Hj.; Palmer, B. J. *J. Nucl. Mater.* **1991**, *178*, 48.
143. Calvo, E. J.; Schiffrin, D. J. *J. Electroanal. Chem.* **1988**, *243*, 171.
144. Evans, D. H.; Lingane, J. J. *J. Electroanal. Chem.* **1963**, *6*, 283.
145. Bockris, J. O'M.; Oldfield, L. F. *Trans. Faraday Soc.* **1955**, *51*, 249.
146. Bianchi, G.; Mazza, F.; Mussini, J. *Electrochim. Acta* **1962**, *7*, 457.
147. Genshaw, M. A.; Damjanovic, A.; Bockris, J. O'M. *J. Electroanal. Chem.* **1967**, *15*, 163, 173.
148. Zurilla, R. W.; Sen, R. K.; Yeager, E. *J. Electrochem. Soc.* **1978**, *125*, 1103.
149. Johnson, L. H.; Shoesmith, D. W.; Lunansky, G. E.; Bailey, M. G.; Tremaine, P. R. *Nucl. Technol.* **1982**, *56*, 238.
150. Sunder, S.; Shoesmith, D. W.; Lemire, R. J.; Bailey, M. G.; Wallace, G. J. *Corros. Sci.* **1991**, *32*, 373.
151. Grandstaff, D. E. *Econ. Geol. Bull. Soc. Econ. Geol.* **1976**, *71*, 1493.

CHAPTER
7

Electrochemistry with Clays and Zeolites

M. D. Baker and Chandana Senaratne

7.1 Introduction

The intent of this chapter is to highlight recent developments in the field of zeolite and clay-modified electrodes. A brief introduction to the structures of zeolites and clays is included for the benefit of the nonspecialist. Although comprehensive coverage is, of course, impossible, the reader is directed to several key articles and books where appropriate. The background material concerning zeolites and clays is somewhat truncated because of the space requirements of this review article. The limited amount of material covered, particularly in the applications of zeolites and clays outside of the electrochemical theater, reflects to a certain extent the interests of the authors.

Several excellent reviews concerning zeolite-modified electrodes [1,2] and clay-modified electrodes [3] have appeared recently, attesting to the continued interest in this field. Applications envisaged include uses in electrocatalysis, molecular recognition—including size and chiral selectivity, amperometric determinations (electroanalysis), uses in photocatalytic processes plus electrochemical synthesis and assembly of intrazeolite conductive polymers ("molecular-wires"). Some of these areas of interest lie far from the "stereotypical view of the zeolite" [4] and clays, which together are sometimes "typecast" [4] into a limited number of chemical roles. Modification of electrode surfaces with clay and zeolite coatings, although not in the mainstream of their uses (i.e., catalysis and separation), provides an interesting new line of inquiry. A maturing of this field will occur only when fundamental issues concerning the electrochemistry of these modified electrodes as well as the role of intrazeolitic and intraclay processes (i.e. ion exchange and/or diffusion) have been addressed. In this review article the recent progress in the study of clay- and zeolite-modified electrodes will be highlighted.

7.2 Zeolites and Clays: Structure and Properties Pertaining to Electrode Modification

7.2.1 Zeolites: Definitions and Types

Zeolites are restricted to crystalline aluminosilicate materials. Since 1980, a number of new "zeolites" containing framework atoms other than silicon and aluminum have been synthesized [5]. Dyer [6] has suggested the term *zeotypes* for this class of materials. Within the umbrella of electrochemistry several groups have also reported "zeolitic films" that have been grown on electrode surfaces [7,8]. Although this review will cover only aluminosilicate zeolites, the structures of two recently synthesized zeotypes with extra large pores will be briefly described in Section 7.2.2. To our knowledge, there have been no electrochemical studies using zeotypes.

7.2.2 Structure of Zeolites

The primary building blocks of zeolites are SiO_4 and AlO_4 tetrahedra. These can link in several ways, resulting in an array of structurally distinct three-dimensional networks. Three different depictions of zeolite structures are shown in Fig. 7.1a using the cubooctahedron (or sodalite cage) as an example. The tetrahedral atoms or T-atoms (Si or Al in zeolites) are denoted by filled circles, and the open circles represent oxygen atoms. Thus in the commonly used representation of zeolite structures (Fig. 7.1a(iii)) the T-atoms are at the vertices and the lines connecting them represent T—O—T bonds. Zeolites are formed by linking these tetrahedra to produce three-dimensional anionic networks, as depicted schematically in Fig. 7.1b for the chabazite structure and in Fig. 7.2a and b for zeolites A and Y, respectively. There are no unshared oxygens in zeolite networks (i.e., they are tectoaluminosilicates) and therefore (Al + Si):O = 1:2. Note also that Al—O—Al linkages are absent. The presence of aluminum in the framework introduces a negative charge that is balanced by an extraframework cation. These are readily exchangeable with other ions, offering a convenient method for implanting a variety of electroactive cations into zeolite micropores. The framework silicon-to-aluminum ratio can be controlled in zeolite syntheses [9,10], although for some structure types it is not infinitely variable. The framework aluminum content affects such properties as unit cell dimensions, hydrophilicity/hydrophobicity, acid and thermal stability, and the cation-exchange capacity. The key feature of zeolites is their microporosity. The pores, which can be one-, two-, or three-dimensional in these materials (*vide infra*), span a range of sizes and shapes, and in conventional zeolites the pore diameters range from 3 to 7.4 Å.

Figure 7.1b shows the structure of chabazite, which can be envisioned as formed by the interconnection of double six-rings (D6R). Some of the other secondary

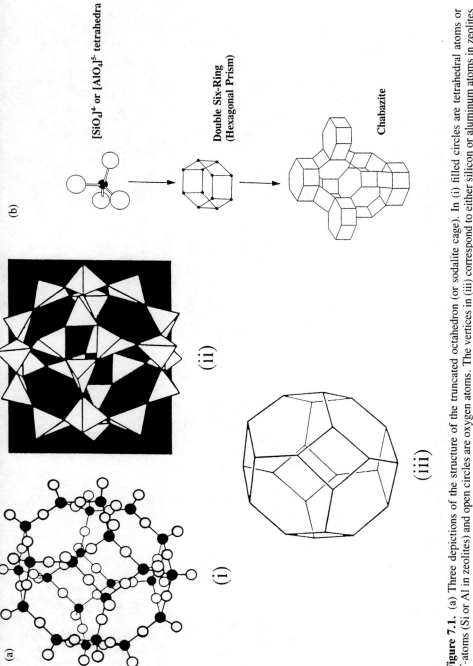

Figure 7.1. (a) Three depictions of the structure of the truncated octahedron (or sodalite cage). In (i) filled circles are tetrahedral atoms or T-atoms (Si or Al in zeolites) and open circles are oxygen atoms. The vertices in (iii) correspond to either silicon or aluminum atoms in zeolites and lines represent T—O—T bonds (the shared oxygen atoms of the tetrahedra are omitted for clarity). (Reproduced from Rabo, J. A., Ed. "Zeolite Chemistry and Catalysis." ACS Monograph 171. ACS: Washington, D.C., © 1976, American Chemical Society.) (b) A schematic representation of the formation of chabazite from primary (SiO_4 and AlO_4 tetrahedra) and secondary building units (SBUs). (Reprinted by

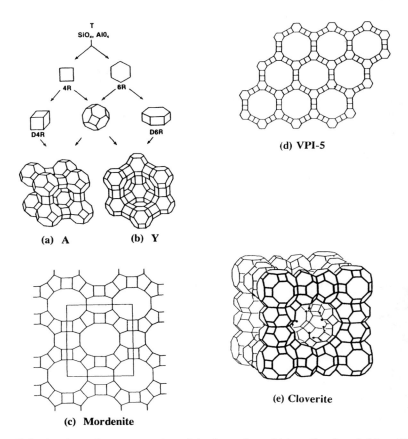

Figure 7.2. A schematic representation of the formation of (a) zeolite A and (b) zeolite Y from primary building units and the following SBUs; 4R—single four-rings, 6R—single six-rings, D4R—double four-rings, D6R—double six-rings. Digits refer to the number of T-atoms in the rings. Sodalite cage also shown. (Reproduced from Stucky, G. D.; Dwyer, F. G. "Intrazeolite Chemistry." ACS Symposium Series 218. ACS: Washington, D.C., © 1983, American Chemical Society.) The structures of (c) mordenite, showing the unit cell. (Reprinted with permission from Ref. 15.) (d) VPI-5. (Reprinted with permission from Ref. 13.) (e) Cloverite. (Reprinted with permission from Ref. 14, © 1991, Macmillan Magazines Ltd.)

building units (SBUs) encountered in zeolite structures together with the sodalite cage are shown in Fig. 7.2a and b (4R—single four-ring; 6R—single six-ring; D4R—double four-ring; D6R—double six-ring; digits refer to the number of T-atoms in the rings). The structures of zeolite A and Y can be considered as formed by linking sodalite cages through double four-rings and double six-rings, respectively. Zeolite A (Fig. 7.2a) is classified as a small-pore zeolite in which the Si/Al ratio equals 1.00. Depending on the nature of the extraframework cation, the pore diameter of the 8-ring opening into the large cavity (or supercage) varies from about 3 to 5 Å [11]. In certain circumstances the dimensions of this pore can be altered with a resolution of 0.1 Å [12]. Supercages in zeolite A are linked octahedrally to one another resulting in three-dimensional interconnected linear pores. Extraframework cations are located in both the supercages and sodalite cavities.

Zeolite Y (Fig. 7.2b) is a large-pore zeolite (pore diameter 7.4 Å) and has been synthesized with various framework Si/Al ratios [9]. The most commonly encountered type holds 56 extraframework cations per unit cell and thus possesses a Si/Al ratio of 2.4 [11]. In zeolite Y the supercages are linked tetrahedrally and thus the three-dimensional pores are not linear. In addition to the large-channel network formed by the interconnection of supercages (not shown in the figure), there is a small-channel network formed by the tetrahedrally interconnected sodalite cages. The structure of mordenite, which is also a large-pore zeolite with a Si/Al ratio in excess of 5, is shown in Fig. 7.2c. Although the pore system is classified as two-dimensional, molecules can freely diffuse only in one direction. The eight-membered ring channels, which are accessed from the main 12-membered ring elliptical channels (6.7 × 7.0 Å,) are blocked, making them more like pockets on the wall of the main channels.

Among the recent significant developments in molecular sieve science are the synthesis of extralarge-pore "zeotypes" [13,14]. Davis and co-workers were the first to break the longstanding 12-membered-ring-pore barrier when they synthesized the aluminophosphate VPI-5 in 1988 [13]. Molecular sieve VPI-5 consisting of 18-membered (18 T-atoms) rings (Fig. 7.2d) has a one-dimensional pore structure (free diameter 12–13 Å) without any ion-exchange capacity (i.e., Al/P = 1.0). The main SBUs in VPI-5 are single six-rings. Not long after the discovery of VPI-5 the synthesis of a gallophosphate with 20-membered rings, cloverite (Fig. 7.2e), was reported[14]. The 20-membered rings have an unusual clover-leaf shape with dimensions of 13.5 × 4.0 Å. This molecular sieve has three-dimensional pores and can be considered to be formed from double four-rings.

It is impossible within the framework of this review to cover the synthesis, structure, and characterization of zeolites in any depth. The novice is directed to the books by Barrer [15], Breck [11], and Dyer [6]. The latter in particular provides an extremely readable introduction to the world of zeolite molecular sieves.

7.2.3 Structure of Clays

Clays have long been recognized for their ability to intercalate guests between their sheetlike structures. Unlike zeolites, intercalation can lead to large changes in the interlayer spacings since the clay layers (*vide infra*) are held together by weak van der Waals and electrostatic forces. Electrostatic bonding occurs when charges arising on the sheets are countered by interlayer ions. The smectite clays described later in this review possess a single aluminum octahedral layer sandwiched between two silicon tetrahedral layers and, therefore, are classified as 2:1 phyllosilicates (materials in which all but one oxygen per tetrahedral or octahedral unit are shared). In the electroneutral mineral pyrophyllite two of the three octahedral sites are occupied by Al, whereas in talc all three are occupied by Mg. Partial substitution of Al(III) in pyrophyllite by Mg(II) and Mg(II) in talc by Li(I) results in montmorillonite and hectorite, respectively [16] (see Table 7.1). Since there is a charge deficiency in the octahedral sheets, hydrated cations are intercalated between the silicate layers. As with the zeolites these cations are exchangeable. The structure of montmorillonite is shown in Fig. 7.3 as an example. The idealized formulae of some important smectite clays are given in Table 7.1 [17].

Recently, several methods have been developed for *pillaring* clays. All of these essentially involve inserting polymeric cations between the clay sheets to prop open the layers. The interlayer distance is then determined by the dimensions of the pillar. For example, the polymeric aluminum-containing cation $[Al_{13}O_4(OH)_{24+x}(H_2O)_{(12-x)}]^{(7-x)+}$ can be intercalated by ion exchange followed by calcination. In addition to the aluminum polymer, polymeric cations such as $[Zr_4(OH)_{16-n}(H_2O)_{n+8}]^{n+}$, $[Fe_3(OCOC_3H_7)_7OH]^+$, $[Nb_6Cl_{12}]^{2+,3+}$, and $[Ta_6Cl_{12}]^{2+}$ have also been used as pillars [18–21].

Although the "cationic" clays described earlier predominate in nature, anionic clays can also be readily synthesized [22]. These are also layered materials except that the interlayer species are anionic. For example, in the structure of brucite (magnesium hydroxide; see Fig. 7.4), magnesium is octahedrally coordinated to six oxygens. The octahedra form infinite sheets via edge sharing of these oxygens. These sheets stack on top of each other via hydrogen bonds. In the anionic clay hydrotalcite Al(III) cations replace some of the Mg(II) cations. Anions that are located between the metal hydroxide sheets are now required for electrical neutrality (see Fig. 7.4b). There are several examples of the use of anionic clays as electrode modifiers, and these will be described later.

7.3 Fabrication of Electrodes

In this section we will briefly outline some of the methods used in the construction of zeolite- and clay-modified electrodes. Although the fields are closely related, we have chosen to deal with the materials separately for convenience.

7.3.1 Zeolite-Modified Electrodes

A recent review by Rolison [1] lists most of the methods used for the fabrication of zeolite-modified electrodes, and the reader is therefore directed primarily to this

Table 7.1 Idealized Formulae for Some Important Smectite Clays

Clay Type	Idealized Formula
Montmorillonite	$(Al_{2-y}Mg_y)Si_4O_{10}(OH)_2 \cdot nH_2O$
Beidellite	$Al_2(Si_{4-x}Al_x)O_{10}(OH)_2 \cdot nH_2O$
Nontronite	$Fe(III)_2(Si_{4-x}Al_x)O_{10}(OH)_2 \cdot nH_2O$
Saponite	$Mg_3(Si_{4-x}Al_x)O_{10}(OH)_2 \cdot nH_2O$
Hectorite	$(Mg_{3-y}Li_y)Si_4O_{10}(OH)_2 \cdot nH_2O$
Sauconite	$Zn_3(Si_{4-x}Al_x)O_{10}(OH)_2 \cdot nH_2O$

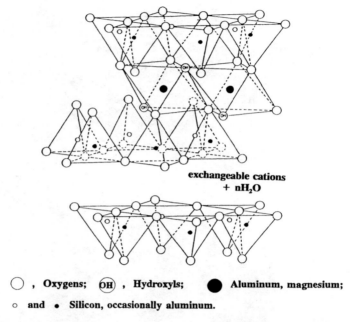

◯ , Oxygens; ⓞⒽ , Hydroxyls; ● Aluminum, magnesium;
o and • Silicon, occasionally aluminum.

Figure 7.3. The structure of montmorillonite (see text). (Reprinted with permission from Ref. 17.)

source. In this section we will describe each approach in general terms with representative examples for each method.

In general, zeolite-modified electrodes are produced by forming a film containing the modifier on the electrode surface [23–27]. Variations can consist of, for example, the inclusion of the zeolite in a conductive composite [27–35], tethering of the zeolite to the surface via a silane-functionalized electrode [36] and the use of conductive polymers as binders [2]. For example, zeolite-modified electrodes have been fabricated using the reductive coelectrodeposition of zeolite particles with an organic oxidant on a rotating disk electrode [37]. Subsequently this procedure was extended to allow the formation of uniform particle-polymer coatings on a disk electrode by electropolymerizing pyrrole with zeolite A [2]. The film formed by electrodeposition was stabilized by ion association between the resultant organic anions and cations present both in solution and on the zeolite, and therefore was suitable only for nonaqueous electrochemistry. Perhaps the simplest method of electrode fabrication is to form a zeolite–polymer film on an electrode surface by evaporating a suspension containing zeolite and the dissolved polymer [2,23–25]. In the method we use at Guelph a suspension of 100 mg of the zeolite in 2 mL of tetrahydrofuran (THF) containing 10 mg of polystyrene is used. Aliquots of 20 μL are then placed onto indium tin oxide (ITO) blanks and the electrode is allowed to dry in air [26]. If an equal amount of graphite powder is used in the mixture, enhanced faradaic currents are observed, presumably due to an increase in the effective surface area of the modified electrode. The main disadvantage of this

Figure 7.4. The relationship between brucite (Mg(OH)$_2$) and hydrotalcite (Mg$_6$Al$_2$(OH)$_{16}$ (CO$_3^{2-}$)·4H$_2$O). In hydrotalcite the isomorphous replacement of Mg^{2+} with Al^{3+} results in a cationic sheet. Intercalation of anions occurs to maintain electroneutrality. (Adapted with permission from Ref. 22, © 1986, American Chemical Society.)

approach is the lack of mechanical strength of the films, especially in stirred solutions. Note that superior films are formed on ITO compared to Pt due to poor wetting of ITO by the polymer [25,27]. These electrodes are compatible with common nonaqueous solvents such as acetonitrile (CH$_3$CN), N, N'-dimethylformamide (DMF), and dimethylsulfoxide (DMSO).

Zeolite composite electrodes with or without a conducting component have been employed in studies of a fundamental nature as well as in more applied systems [27–35]. In the majority of cases the conducting component of choice has been graphite, whereas the binders used include epoxy [28,29], petroleum jelly [30], mineral oil [32], paraffin oil [34], and a copolymer containing styrene and a cross-linking agent formed in situ [34]. Electrodes have also been fabricated by pressing a mixture of the zeolite and graphite on a metal grid [35].

Covalent tethering of a zeolite to a tin oxide electrode via a silane layer has been successfully demonstrated by Mallouk and co-workers [36]. The surface of the electrode, following reaction with a cationic silane (see Section 7.7), possesses free silanol groups that can bind the zeolite via surface hydroxyl groups. This is quite a general scheme and has also been employed effectively by this research group for the preparation of clay-modified electrodes (*vide infra*). Another useful technique for the fabrication of zeolite-modified electrodes, which employs photochemical polymerization, was recently reported [38]. Here the zeolite was exposed to triethoxyvinyl silane, which was then photochemically polymerized to form an electrode with good mechanical and electrochemical properties.

Suib and co-workers [39] have reported the growth of crystalline thin films of zeolite Y on Cu foil. The films were laterally uniform and mechanically stable and could therefore be useful in electrochemical measurements. Since the method described by Suib and co-workers can be extended to other zeolite types, coupled with reports of the growth of zeolites on other metallic substrates (e.g., titanium, aluminum, stainless steel) [40], this avenue of electrode modification holds much future promise, and applications in electrochemistry, sensors, and so on, could ultimately follow [41].

7.3.2 Clay-Modified Electrodes

The direct synthesis of clays on conductive substrates has not been reported in the open literature and thus the modification procedure relies on methods for tethering clay particles to the conductive substrate. The films first used by Ghosh and Bard (see Section 7.4) were formed by dropping a suspension of the clay onto an electrode and allowing the carrier solvent to evaporate. The addition of colloidal Pt to the mixture produced a more uniform film with fewer cracks. As with zeolites, electroactive species can be incorporated into clay coatings before or after the formation of the modified electrode. The technique used by Bard and co-workers is still popular and has been used in several laboratories [3,42–45].

Several other strategies have been adopted. Films produced in the presence of conductive polymers have been reported [46,47] in addition to surfactant-incorporated clay films [48] (see also Section 7.7). Recently, Villemure and Bard employed spin-coating techniques [49] to fabricate electrodes. Carbon paste clay-modified electrodes have also been studied [50,51]. As mentioned in the zeolite section, the covalent tethering of clay particles to the electrode surface via a silane has been described by Rong et al. [52].

7.4 Historical Perspective

The first reports concerning clay and zeolite layers as electrode modifiers were published in the early 1980s [35,37,53]. Ghosh and Bard [53] successfully demonstrated that several complex ions incorporated into sodium montmorillonite were electrochemically active. For example, Ru(bpy)$_3^{2+}$ showed diffusion-controlled electroactivity with an apparent diffusion coefficient of $\sim 10^{-11}$ cm^2 s^{-1}. Rolison and co-workers [37] demonstrated that molecular sieving properties of zeolite A were retained when immobilized at the electrode surface. Specifically, the cyclic voltammetry of 3A and 4A molecular sieves showed that oxygen reduction occurred even in an oxygen-free electrolyte in the case of the smaller-pore zeolite. This was ascribed to the fact that nitrogen (molecular diameter 3.15 Å), used as the purging gas, could displace oxygen from the 4A but not the 3A coating, where it remained trapped.

Pereira-Ramos et al. [35] investigated the electrochemical behavior of a silver ion–exchanged mordenite zeolite. They were able, through a combination of electrochemistry and scanning electron microscopy, to identify two kinds of silver deposits following electrochemical reduction of silver ions. These were ascribed to silver dendrites formed on the conductive part of the electrode that could be reoxidized and to silver particles that apparently migrated into the zeolite and could not be further oxidized.

7.5 Mechanism of Electrochemistry Occurring at Clay- and Zeolite-Modified Electrodes

The electrochemistry occurring at clay- and zeolite-modified electrodes showed much early promise. The retention of the structural integrity of the modifier and the observation of electroactive species incorporated into the modified electrode paved the way for further study. In this section we first examine the electron transfer events occurring at the modified electrode. The work of several groups that have played an important role in elucidating the manner in which electrical communication can be made with intrazeolite or intraclay moieties will be described. Several studies on clay-modified electrodes have shown that only a small fraction of an electroactive moiety intercalated into a clay layer is electroactive [42,43,49,53,54–57]. Similarly, it has been shown that intrazeolite cations and neutrals must in general diffuse to the conductive part of the electrode in order to participate in electron transfer [25,27,58,59]. We now review the electroactivity of intraclay and intrazeolite moieties.

King et al. [43] studied the electrochemistry of montmorillonite clays deposited on pyrolytic graphite. The clays were pre-exchanged with ML$_3^{2+}$ (M = Fe, Os, Ru, and L = bpy; M = Fe and L = phen) and methyl viologen and showed no electroactivity. The authors used this as evidence for rigorous inactivity of cations either intercalated in the clay galleries or bonded to the external surfaces of the clays. An example is shown in Fig. 7.5. This shows the cyclic voltammetry of a sodium montmorillonite film (A) in a solution of Os(bpy)$_3^{2+}$ and (B) pre-exchanged with Os(bpy)$_3^{2+}$. In (B) the complex is electrochemically silent. However, edge-adsorbed electroactive ion pairs present in excess of the ion-exchange capacity were observed. Kaviratna and Pinnavaia [60] have also demonstrated that mobile com-

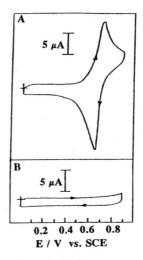

Figure 7.5. Cyclic voltammograms obtained for (A) a sodium montmorillonite film in 0.2 mM $Os(bpy)_3^{2+}$ solution; (B) a pre-exchanged $Os(bpy)_3^{2+}$–montmorillonite film (80% exchanged in pure electrolyte). Scan rate, 50 mV s^{-1} in 0.1 M Na_2SO_4. Potentials are with respect to SCE. (Adapted with permission from Ref. 43.)

plex ions in clay galleries are electrochemically addressable. Electrochemical silence of trisdiimine complexes intercalated in clay galleries was attributed to their inability to ion-exchange with electrolyte cations.

Villemure and Bard [49] also studied natural and reduced-charge samples of montmorillonite exchanged with $M(bpy)_3^{2+}$ (M = Ru, Os, and Fe). For the reduced-charge clay, both the total concentration of complex and the fraction that was electroactive was a factor of 10 smaller than in the normal sample, even though the cations were restricted to the external clay surfaces in the reduced-charge clay sample. The conclusion was therefore that a fraction of the cations on the external clay surfaces of the normal sample was also inactive. Results obtained as a function of the film preparation also pointed to the importance of defect sites in the clays. It was therefore proposed that cations could be electroactive if a defect in their vicinity provided access to the electrode surface.

Fitch et al. [42] studied the uptake of solution-phase $Cr(bpy)_3^{3+}$ into a montmorillonite clay-modified electrode. The results were interpreted in terms of the complex docking with the clay in a two-step process. Note that pre-exchanged $Cr(bpy)_3^{3+}$ had little influence on the electroactivity of the film. Cyclic voltammograms shown in Fig. 7.6 demonstrate how the uptake into the film was followed. The isopotential points in these data indicate the presence of two interchanging electroactive species. These were ascribed to edge (II) and face-adsorbed $Cr(bpy)_3^{3+}$, as shown in Fig. 7.6b. This interpretation invokes the reduction of $Cr(bpy)_3^{3+}$ adsorbed in the gallery sites. There is some evidence to indicate that the gallery cations are electrochemically silent [43,60]. Thus there is disagreement concerning the nature of the

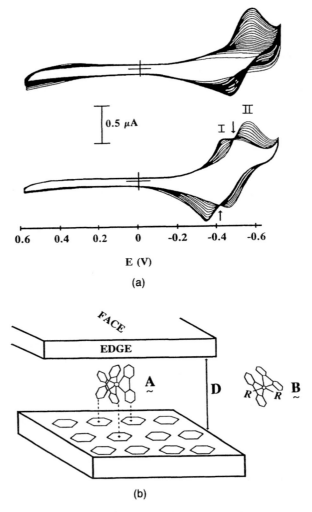

Figure 7.6. (a) Cyclic voltammograms of 0.15 mM $Cr(bpy)_3^{3+}$ in 0.1 M Na_2SO_4 at clay-modified electrodes. Potentials are with respect to Ag/AgCl. (b) Model used to simulate docking of trisdiimine complex with a clay-modified electrode (see text). (Adapted with permission from Ref. 42, © 1988, American Chemical Society.)

electroactive moieties as well as the electron transfer processes that occur [61]. A recent study by Petridis et al. [62], concerning clay-modified electrodes containing self-assembled $Fe(bpy)_3^{3+}/Fe(CN)_6^{4-}$ indicated that the electroactivity of the anionic $Fe(CN)_6^{3-/4-}$ couple was due to ion pairs of the iron complexes. They also demonstrated that $Fe(CN)_6^{3-}$ can act as a charge mediator shuttling electrons between the electrode and the molecules in the clay film. This paper again points to the rigorous electroinactivity of adsorbed cations.

Another interesting aspect of the nature and location of electron transfer processes occurring in clay-modified electrodes is the participation of framework iron, which can be present in the clay as Fe(II) or Fe(III). Xiang and Villemure [63] have recently demonstrated the effects of structural Fe(II) on electron transfer to Fe(bpy)$_3^{2+}$ incorporated into smectite clays. When the Fe(II) concentration in the clay was increased by deliberate chemical reduction of framework Fe(III), there was some indication that the Fe(II) produced was acting as an electron well. However, the results did not follow a logical trend with the known amount of Fe(II) in the clay.

The possible mechanisms for the electrochemical processes occurring at a zeolite-modified electrode, where the reduction of an electroactive ion (E^{m+}) that has been ion-exchanged into the zeolite occurs, were first discussed by Shaw et al. [27]. These are reproduced as follows:

1. $E^{m+}(z) + nC^+(s) + ne^-(s) \rightleftharpoons E^{m-n}(z) + nC^+(z)$
2. $E^{m+}(z) + mC^+(s) \rightleftharpoons E^{m+}(s) + mC^+(z)$
   $E^{m+}(s) + ne^- \rightleftharpoons E^{m-n}(s)$

where (z) indicates the zeolite matrix, (s) indicates the solution phase, and C^+ is the electrolyte cation. Similar processes occurring at clay-modified electrodes have been discussed by Kaviratna and Pinnavaia [60]. The major distinction between these mechanisms concerns the site of electron transfer to the intrazeolite electroactive moiety, E^{m+}. In mechanism 1 this is exclusively an intrazeolitic process, most likely proceeding via an electron-hopping mechanism or mediated electron transfer. In mechanism 2 the electroactive ion first diffuses and/or ion-exchanges out of the zeolite followed by its reduction at the electrically conductive portion of the electrode.

In both cases, if E^{m+} is a charge-balancing extraframework ion, a counterion (C^+) from the electrolyte solution enters the zeolite pores in order to maintain charge balance. The inclusion of cations in highly aluminous zeolites as non-charge-balancing ion pairs has been shown to be negligible [64,65] under normal ion-exchange conditions and typical electrolyte concentrations used in electrochemistry (i.e., < 0.1 M). Thus for zeolites Y and A (Si/Al $\sim$ 2.4 and 1.0, respectively) the inclusion of electroactive, non-charge-balancing ion pairs will not play an important role. Shaw et al. [27] showed that mechanism 2 was operative for the reduction of a variety of electroactive probes including heptyl viologen, hexammineruthenium (III), hexacyanoferrate (II), and Cu(II). The mechanism operating in the case of the reduction of methyl viologen dication was unclear. Furthermore, Gemborys and Shaw [24] also demonstrated that the currents observed during the reduction of methyl viologen in zeolite Y depended on the size of the electrolyte cation. In agreement with this work Baker and Senaratne [59] have shown that in the absence of a non-size-excluded electrolyte cation (i.e., C^+ above), the electrochemistry of intrazeolite cations is suppressed as the necessary ion exchange occurring in the first step of mechanism 2 is blocked. Suppression of the electroactivity of a CoY-modified electrode in acetonitrile containing TBA$^+$ is shown in Fig. 7.7. These authors have reported several examples of the influence of size exclusivity on the response of zeolite-modified electrodes. These will be discussed in Section 7.6.

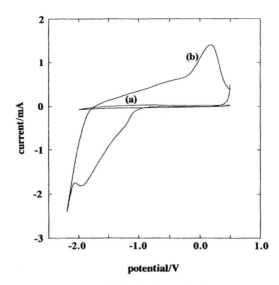

Figure 7.7. Cyclic voltammograms of Co(II)Y-modified electrodes in acetonitrile. Scan rate, 20 mV s^{-1}. Supporting electrolyte: (a) 0.1 M TBABF$_4$, (b) 0.1 M LiClO$_4$. The cobalt ion electroactivity is heavily suppressed in the absence of a non-size-excluded electrolyte cation. (Reprinted with permission from Ref. 68.)

The pursuit of systems showing electron transport via mechanism 1 has evoked considerable interest. Success in this venture will enable the electrochemist to harness the size and shape selectivity of the zeolite and the unique layered structure of clays. There are several reports in the open literature describing electron transfer to intrazeolite moieties [25,66,67]. In these cases careful interpretation is required to distinguish mechanisms 1 and 2 unequivocally. A convenient way of distinguishing between the two mechanisms is to study the effects of a charge- or size-excluded electrolyte moiety on the electrochemical reaction [68]. Some aspects of electron transfer to intrazeolite species will be discussed later in this chapter.

7.6 Analytical Applications

Zeolites in particular are well known for their molecular discriminating properties. Numerous examples of molecular sieving and shape-selective catalysis exist in the zeolite literature [6,69,70]. These are then the more "traditional" [4,41] applications that are often linked with zeolites. A promising yet less conventional use [41] is to employ the zeolite as a chemical sensor. In what follows we will first present recent work from our laboratory where the detection of trace quantities of solution-phase analytes using zeolite-modified electrodes has been demonstrated. We also address some of the more important issues of ion-exchange and solvation that markedly influence their electrochemistry (also see Section 7.8). We follow this by a short discussion of other electroanalytical applications of zeolite-modified electrodes, which have also been reviewed recently by Rolison [2]. Finally, we review the electroanalytical application of clay-modified electrodes.

The electrodes used in the following studies were prepared using the method described in Section 7.3.1 [26]. The first example involves the determinations of trace levels of cations in aqueous solution [59]. We first recall that the mechanism by which intrazeolite electroactive cations may be reduced involves ion-exchange of the intrazeolite cation with an electrolyte ion. As stated in Section 7.5 and shown in Fig. 7.7, the absence of a non-size-excluded electrolyte cation "suppresses" the electrochemical activity of the intrazeolite cation by hindering ion-exchange and/or diffusion. Increasing the concentration of a non-size-excluded electrolyte co-cation gradually lifts the suppressed condition. As we have shown [59], the currents observed in linear sweep voltammetry (stripping analysis) of such systems show a linear dependence on the concentration of the solution-phase non-size-excluded cation, thus displaying size selectivity of the type demonstrated by Barrer and James [71]. This is shown in Fig. 7.8 for a silver-ion-exchanged Y-modified electrode in tetrabutylammonium perchlorate. The response of the electrodes to non-size-excluded alkali-metal cations was linear in concentration over the range 1–10 ppm for Li^+, 1–35 ppm for Na^+, and 1–25 ppm for K^+. Thus the zeolite electrode can be used to detect cations amperometrically that cannot normally be detected in this manner because of the restraints imposed by the electrochemical stability of aqueous electrolyte systems. Indirect determinations of this type have received considerable recent interest [72–74] and may also be of use as detection systems in ion chromatography [74–76].

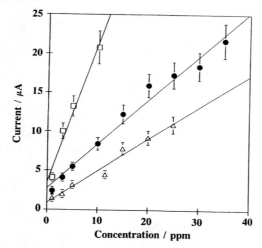

Figure 7.8. Peak currents observed in anodic stripping voltammetry for AgY electrodes in methanol containing 0.1 M TBAP plotted as a function of alkali-metal cation concentration. □ Li^+, ● Na^+, △ K^+. Scan rate, 10 mV s^{-1}. (Reprinted with permission from Ref. 59, © 1992, American Chemical Society.)

The second example from our laboratory involves the selective quantitative determination of trace amounts of water in organic solvents. The need for on-line water sensors, for example, to be used in process control, has been discussed recently [77]. Although this section is specifically devoted to electroanalytical applications, it is worth considering at this time the broader implications of these studies involving the response of zeolite-modified electrodes in nonaqueous solvents. Aqueous ion-exchange in zeolites has been studied extremely thoroughly [11,15,78] over the last 30 years and many isotherms for binary exchanges have been published. In contrast, there is a paucity of information on the nonaqueous ion-exchange properties of molecular sieve zeolites. A recent STN search of the *Chemical Abstracts* data base covering the time period from 1968 to August 1992 revealed 26 papers on this subject. Since the behavior of zeolite-modified electrodes is critically dependent on ion-exchange processes (see Section 7.5), the issues of ion-exchange and solvation effects in zeolites when exposed to nonaqueous electrolyte systems must be addressed. Many of the potential uses of zeolites in electrochemical environments will require an understanding of these properties. The sensor described later marks a step in this direction.

A suppression of the electrochemical activity of intrazeolite charge-balancing cations, similar to that described for aqueous systems (*vide supra*), is also observed in nonaqueous solvents [79]. As shown in Fig. 7.9, the faradaic currents observed at silver zeolite A-modified electrodes in DMF (Fig. 7.9a) were much smaller than those observed in aqueous solution (Fig. 7.9b). This was attributed to the diminished counter-diffusion rate of the exchanging species (Ag^+ and Li^+) resulting mainly from the (1) increased size of the cation–solvent complex, (2) hydrophilic nature of the zeolite, and (3) extent of solvation of the zeolite. Additions of aliquots of water to the dry solvent resulted in an increase in the observed faradaic currents due to the enhanced ion exchange involving hydrated species.

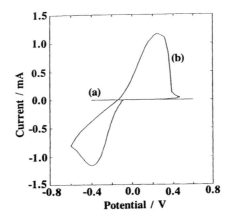

Figure 7.9. Cyclic voltammograms of AgA-modified electrodes in 0.1 M $LiClO_4$. (a) DMF, (b) water. Scan rate, 20 mV s^{-1}. (Reprinted with permission from Ref. 79.)

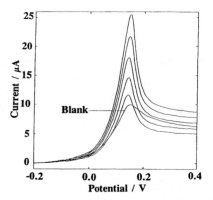

Figure 7.10. Anodic stripping voltammograms of Ag^+-exchanged zeolite A in dry DMF with 0.1 M $LiClO_4$. Water added to the blank in 1 ppm increments. Scan rate, 20 mV s^{-1}. (Reprinted with permission from Ref. 79.)

In Fig. 7.10, the anodic stripping voltammograms of a Ag^+-exchanged zeolite A-modified electrode in dry DMF containing 0.1 M $LiClO_4$ are shown. Each voltammogram corresponds to a successive increase in the electrolyte water concentration by 1 ppm. The calibration curve constructed from these data was linear ($r = 0.997$) up to 10 ppm. The smallest observed increment in water concentration was 100 ppb. The response of an electrode to the addition of 100 ppb of water is shown in Fig. 7.11. These data indicate that the detection limit is about 20 ppb, which is close to two orders of magnitude superior to the Karl Fischer method and other moisture sensors recently described [80,81]. In addition, the sensor incorporates size selectivity and hydrophilicity, a distinct advantage over widely used polymer-

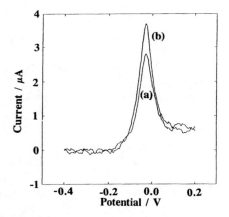

Figure 7.11. Anodic stripping voltammograms of Ag^+-exchanged zeolite A in dry DMF containing 0.1 M $LiClO_4$. (a) Blank, (b) 100 ppb water added. Scan rate, 20 mV s^{-1}. (Reprinted with permission from Ref. 79.)

based capacitive or resistive water sensors, which are in general not specific for water [80]. The electrochemistry observed for AgA zeolites is dominated by the aqueous phase even at trace water concentrations. Experiments performed with varying amounts of intrazeolite water (dehydration followed by controlled water addition) and different electrolytes were suggested as important areas for future study.

The presence of "spectator" species such as intrazeolite oxygen and water (*vide infra*) can substantially affect the analyses. For the water sensor described earlier, the effect of intrazeolite water must be addressed. In our nonaqueous experiments the zeolite was partially dehydrated by virtue of its route through the dry box antechamber. Partial dehydration of zeolites affects the ionic mobility of the extraframework cations. Stamires [82] has shown that there is a change of ionic conductivity (proportional to the diffusion coefficient [83]) of close to five orders of magnitude in going from the fully dehydrated to the fully hydrated form. Moreover, the change in ionic conductivity was observed to occur in stages. This was interpreted in terms of the preferential hydration of one type of extraframework cation. In zeolite A the first extraframework cations to become mobile were the four sodium ions located in the eight rings (at five water molecules per unit cell) followed by the remaining eight in the six-membered rings, which provided smaller contributions to the ionic conductivity.

Several other examples of the use of zeolite-modified electrodes in electroanalytical chemistry, including voltammetric determinations of silver [32], lead, and mercury [84,85], have appeared and have been reviewed [2]. An important point raised in the review, with which we concur, was that "although the presence of a zeolite at the electrochemical interphase can have a profound influence on the electrochemistry, the many subtle effects involved must be understood before successful analyses can be achieved." Thus far, only a few examples concerning the application of zeolite-modified electrodes in solution-phase electroanalyses have appeared. Nonetheless, the control that can be exercised on the zeolite in terms of pore size, framework and extraframework species, hydrophilicity, acidity, and location of electroactive centers (e.g., surface confined, small or large cavities) makes them serious candidates for study as selective electrochemical sensors. As aptly put by Lockhart [86], "A chemical Rip van Winkle awakening today from his extended sleep and examining the chemical scene, would surely be most struck by the chemist's ability to design molecules for a purpose." Zeolites, with their highly ordered structures, are indeed intriguing materials to further this viewpoint and show much promise for selective chemical sensing.

We now examine the electroanalytical applications of clay-modified electrodes. At this point we note that zeolite and clay membranes have been used in potentiometric analyses, as first described by Marshall [87,88]. This area has been reviewed recently [89] and will not be further described in this article.

The voltammetric determination of some organic nitro compounds was reported by Hernandez et al. [50,51]. Dinocap, a mixture of two isomers, 2-(1-methylheptyl)-4,6-dinitrophenyl crotonate and 4-(1-methylheptyl)-2,6-dinitrophenyl crotonate, was determined using sepiolite- and hectorite-modified electrodes. Using differential pulse voltammetry a linear response to the analyte was obtained resulting in ppb detection limits. Trace analysis at clay-modified electrodes has also been reported by Wang and Martinez [90]. Montmorillonite electrodes were shown to respond to Fe(III) in solution with a detection limit of 0.2 ppm. In this study the modified electrode was immersed in the test solution, under open circuit, and then subjected to differential pulse voltammetry. After the scan the electrode was transferred to a sodium carbonate solution in order to clean the electrode. Six successive voltammograms recorded at a concentration of 2 ppm are shown in Fig. 7.12, together with the corresponding scans for the clean electrode. The effect of the supporting electrolyte was also investigated. For Fe(III) analytes, KCl was found to give the best results. Ions that were found to provide no interference to the Fe(III) determinations at the 10 ppm level included Cr(III), Zn(II), Cu(II), Mn(II), Co(II), Ca(II), Cd(II), and Ag(I). Interference was, however, observed from aluminum, uranium, and gold ions.

Montmorillonite-modified electrodes have been used for ion-exchange voltammetry [91] in which the uptake of a test analyte, $Ru(NH_3)_6^{3+}$, was followed at trace concentrations using cyclic voltammetry. The data, collected in 0.01 M Na_2SO_4 containing 10^{-5} M $Ru(NH_3)_6^{3+}$, are shown in Fig. 7.13. Note that no cathodic or anodic peaks were observed at a bare Pt electrode (Fig. 7.13 upper trace), immediately pointing to a considerable enhancement in signal attainable for this analyte at the clay-modified electrode. The optimum performance of a clay-modified electrode for a particular analyte is a function of many variables. For example, the porosity of the clay film toward the analyte can be controlled by manipulation of the supporting electrolyte owing to the dependence of the clay interlayer spacing between the clay sheets on the type and concentration of the intercalated cation. The effect of the

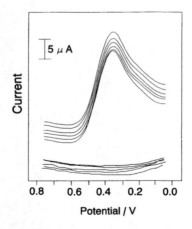

Figure 7.12. Differential pulse voltammograms of 2 mg L^{-1} iron (top), at a montmorillonite-modified electrode, together with the corresponding "clean" voltammograms (bottom). Pulse amplitude and scan rate were 50 mV and 10 mV s^{-1}. Ag/AgCl reference electrode. (Reprinted with permission from Ref. 90, © 1989, VCH Publishers.)

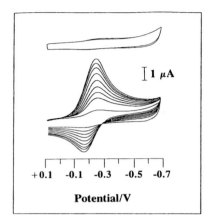

Figure 7.13. Cyclic voltammograms of 10^{-5} M $Ru(NH_3)_6^{3+}$ in 0.01 M Na_2SO_4. (Top) Bare Pt electrode. (Bottom) Clay-modified electrode. (Reprinted with permission from Ref. 91, © 1990, VCH Publishers.)

supporting electrolyte concentration on the clay-modified-electrode response is summarized in Fig. 7.14. Note that the response was at a maximum for low supporting-electrolyte concentrations, as expected from ion-exchange considerations. This study also considered the possibility of interlayer spacings affecting the response of the electrode (*vide supra*). Of note here is the distinct interlayer spacings that occur with sodium as opposed to potassium ions.

In KCl a single interlayer spacing predominates over the entire concentration

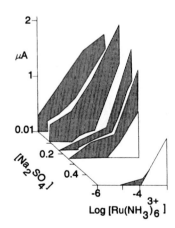

Figure 7.14. A three-dimensional plot of the maximum reduction peak for $Ru(NH_3)_6^{3+}$ at clay-modified and bare electrodes as a function of the electrolyte (Na_2SO_4) and analyte ($Ru(NH_3)_6^{3+}$) concentration. The lined portion of the diagram represents the difference in peak heights between the clay-modified and bare electrode. (Reprinted with permission from Ref. 91, © 1990, VCH Publishers.)

range, whereas with sodium ions three distinct spacings have been observed by x-ray diffraction. From 1 to 6 M NaCl the interlayer distance is 5.5 Å, from 0.3 to 1 M it is 9.5 Å, and below 0.3 M the layers expand continuously with decreasing concentration. The response of the clay electrode to the $Ru(NH_3)_6^{3+}$ analyte as a function of the supporting electrolyte concentration indeed reflected this behavior, as shown in Fig. 7.15. In this figure the ratio of the reduction peak currents for 4 mM $Ru(NH_3)_6^{3+}$ at the clay-modified electrode to those at the bare electrode are plotted as a function of the supporting-electrolyte concentration. For Na^+ there are three regions of interest. From 5 to 1.6 M the ratio is less than 1, reflecting exclusion of the analyte. From 1.6 to about 0.8 M Na^+ the ratio rises to close to 1, indicating that neither enhancement nor exclusion occurs, and for $[Na^+] \leq 0.8$ M there are progressively larger currents as the sodium ion concentration decreases. In KCl the results were rather different. At high concentrations there was exclusion of the analyte, but instead of two sharp increases in the current ratio there was a gradual increase over a broad concentration range. These results were thus consistent with the fact that there are no hydrational changes in K^+-exchanged montmorillonite in contrast to the sodium form. This study indicates that there is therefore potential for discriminatory sensing in the solution phase by using the molecular-sieving properties of clays. Prospects for sub picomole determinations using square wave or differential pulse techniques certainly exist.

7.7 Electrocatalysis

Zeolites and clays are perhaps best known as heterogeneous catalysts. The zeolites catalyze processes such as hydrocarbon cracking, methanol-to-gasoline conversion,

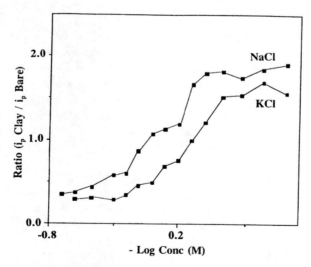

Figure 7.15. A plot of the maximum reduction current for 4 mM $Ru(NH_3)_6^{3+}$ at a clay-modified electrode as a function of the concentration of KCl or NaCl in solution (see text). (Reprinted with permission from Ref. 91, © 1990, VCH Publishers.)

and disproportionation and alkylation of aromatics (the latter two processes occurring in the presence of ZSM-5), becoming classic examples in the zeolite literature. As well as displaying high activity and low coking, ZSM-5 has often been cited as an example of a shape-selective catalyst [69,70]. Clays were used extensively as commercial catalysts before they were largely replaced by the more thermally stable and selective zeolite catalysts. Nonetheless, the ability to "pillar" clays by exchanging them with cations of well-defined size (see Section 7.3.2) has rekindled interest. As well as imparting some molecular-sieving properties to the clays, thermal stability is improved and size- and shape-selective catalysis becomes possible [92]. The use of clays and zeolites in electrocatalysis is an intriguing possibility. Clays and zeolites are known to bind or stabilize species with catalytic activities, and several examples of electrocatalysis at clay- and zeolite-modified electrodes have appeared in the literature [53,57]. These are described later. A further interesting aspect of the use of clays in electrocatalytic processes lies in their chiral selectivity. This aspect of clay electrochemistry was outlined in the 1990 review by Fitch [3]. Recently, Villemure and Bard [93] have reported the electrochemistry of $Ru(bpy)_3^{2+}$ enantiomers at montmorillonite-modified electrodes.

Electrocatalytic processes were first observed at clay-modified electrodes by Ghosh et al. [94]. In this study the voltammetric behavior of the zwitterionic species $Ru[bpy]_2[bpy(CO_2)_2]$ at a sodium hectorite/RuO_2–modified electrode was investigated. The data showed that the ruthenium complex imparted catalytic activity, as is illustrated in Fig. 7.16, which shows the cyclic voltammetry of $Ru[bpy]_2[bpy(CO_2)_2]$ at (1) a bare SnO_2 electrode, (2) a SnO_2/clay electrode, and (3) an SnO_2/clay/RuO_2 electrode. Curve 4 is the cyclic voltammogram of a SnO_2/clay/RuO_2 electrode in pure supporting electrolyte. All traces were obtained in 0.1 M Na_2SO_4. The shape of curve 3, when RuO_2 is present in the clay, is characteristic of a chemical reaction following the electron transfer step that regenerates the electroactive moiety (i.e., it is a catalytic EC' mechanism [95]). The chemical step was attributed to RuO_2-mediated oxidation of water by an electrogenerated Ru(III) complex. In this study the philosophy was to use the clay as a support material for the catalytic centers, which were incorporated into the clay films. The organization of catalytic centers and substrates at clay- and zeolite-modified electrodes has many distinct advantages. In addition to locally high concentrations of reactants being produced at the electrode, spatial ordering leading to electron trapping, and vectorial electron transport can also be achieved [52,25]. These aspects of zeolite- and clay-modified electrodes will be discussed later in this review.

The first example of the clay coating playing a primary rather than a supporting role in an electrocatalytic reaction was given by Oyama and Anson [96]. Here the electroreduction of H_2O_2 at montmorillonite-modified electrodes was achieved via electron transfer to Fe^{3+} impurities in the clay framework. These are found in octahedral sites replacing aluminum ions of the parent structure. Although these ions are immobile and are expected to display electrochemical silence, the presence of the mobile $Ru(NH_3)_6^{3+}$ and the electrogenerated $Ru(NH_3)_6^{2+}$ in the clay allows the mediated electron transfer to Fe(III). In trace quantities the latter is known to

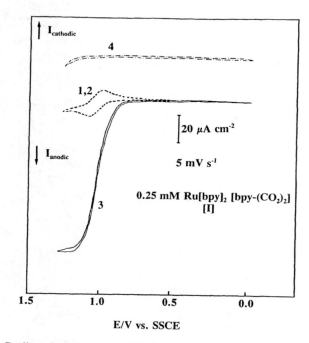

Figure 7.16. Cyclic voltammograms of $Ru[bpy]_2[bpy(CO_2)_2]$ at a (1) bare SnO_2 electrode (----), (2) sodium hectorite/PVA (or sodium hectorite/Pt/PVA)-coated electrode (....), and (3) sodium hectorite/RuO_2/PVA-coated SnO_2 electrode (_____). (4) Cyclic voltammogram of a sodium hectorite/RuO_2/PVA-coated SnO_2 electrode in pure supporting electrolyte (-··-). The supporting electrolyte was 0.1 M Na_2SO_4. Solution pH was 7 in all experiments and the electrode area was 1.2 cm². Scan rate, 5 mV s^{-1}. (Reproduced with permission from Ref. 94.)

enhance the rate of the reaction between $Ru(NH_3)_6^{2+}$ and H_2O_2 in homogeneous solution. Thus the clay plays an active rather than a supporting role in the electrocatalytic reaction. Voltammograms for the reduction of H_2O_2 at a rotating disk electrode are shown in Fig. 7.17. It is evident by comparison of Fig. 7.17E with A–D that rapid peroxide reduction occurs only when the clay is present at the electrode together with the ruthenium charge shuttle in solution. Further evidence was also presented in this paper for the clay-immobilized Fe^{3+} impurity to be the catalytic center, and the interested reader is directed to the original paper for further details [96].

As stated earlier, clays and zeolites are attractive substrates for organization and self-assembly of electroactive species close to an electrode surface. The first report on the use of clay-modified electrodes to organize an electrocatalytic reaction was given in 1988 by Rusling et al. [44,97]. Here the dehalogenation of 4,4'-dibromobiphenyl was achieved as a natural extension of previous work from this research group [98,99]. The idea was to use a surfactant (hexadecyltrimethylammonium bromide) to solubilize the nonpolar organic molecule in close proximity to

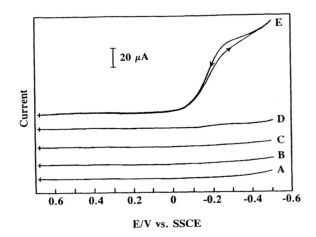

Figure 7.17. Current potential responses for the reduction of H_2O_2 at pyrolytic graphite electrodes rotated at 900 rpm. (A) Uncoated electrode, electrolyte composition 0.2 M CF_3COONa buffered to pH 2.9. (B) repeat of (A) after electrode was coated with sodium montmorillonite. (C) uncoated electrode, electrolyte composition 0.2 M CF_3COONa (pH 2.9) + 10^{-5} M $Ru(NH_3)_6^{3+}$. (D) repeat of (C) after addition of 0.2 M H_2O_2; (E) repeat of (D) after electrode was coated with sodium montmorillonite. Scan rate, 10 mV s^{-1}. (Reproduced with permission from Ref. 96.)

an electrocatalytic mediator, $Co(bpy)_3^{2+}$, which was incorporated into a bentonite clay surface. Catalytic reduction of the 4,4'-dibromobiphenyl proceeded via an electrochemically generated $Co(bpy)_2^-$ complex, suggesting enhanced catalytic activity for the clay-modified electrodes in comparison to uncoated electrodes. Bulk electrolysis was performed at potentials corresponding to the production of the $Co(bpy)_2^-$ species, and the product solutions were analyzed by HPLC. The sole products were 4-bromobiphenyl and biphenyl. This system was further studied in some detail using UV absorption, x-ray fluorescence, SEM-EDX, SIMS, and XPS, characterizing the reactants and surfactant adsorbed onto the clay-modified electrode [99].

An elegant study of the use of pillared clays in the assembly and operation of an electrocatalytic system was made by Rong et al. [52]. As mentioned earlier and in Section 7.2.3, pillaring of clays is a viable synthetic technique for modifying the adsorption properties of the clay. In this study montmorillonite was pillared with $(Al_{13}O_4(OH)_{28})^{3+}$ and was then bound to SnO_2 and Pt electrodes via a thin (two to four monolayer) coating of a polymerized silane, as described in Section 7.3.1. This provides an extremely interesting medium at the electrode surface. The polymer is able to bind anions such as $Fe(CN)_6^{4-}$ and $Mo(CN)_8^{4-}$, while large cations such as $Os(bpy)_3^{2+}$ and $Ru(bpy)_3^{2+}$ can bind to the clay surface. This spatial ordering caused charge-trapping phenomena, which is now described.

A general scheme of ion assembly for electrochemical charge trapping and light-driven vectorial electron transport is given in Fig. 7.18 [52]. The cationic silane

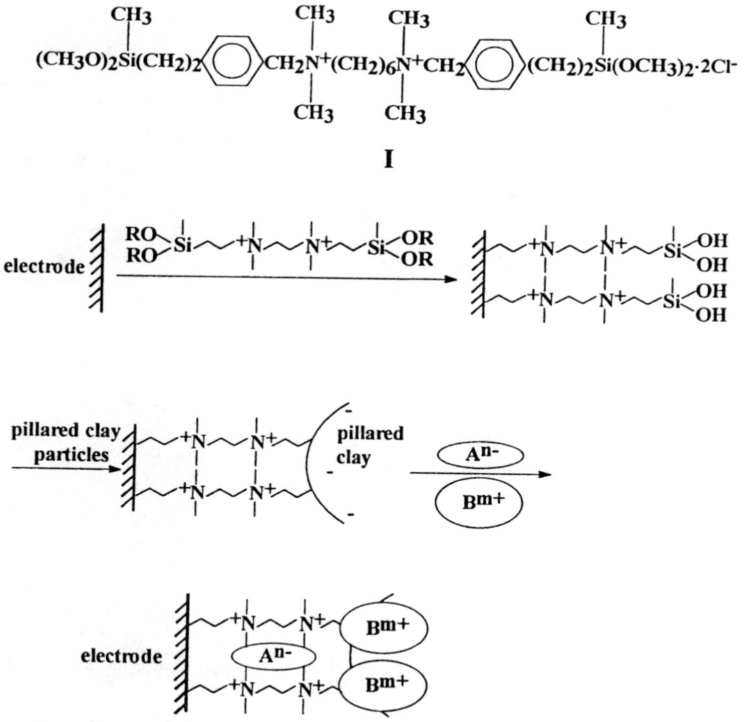

ELECTROCHEMISTRY WITH CLAYS AND ZEOLITES 363

Figure 7.18. Scheme for the production of clay-modified electrodes incorporating covalent tethering of the clay particles to the electrode. The whole assembly incorporates the clay and multiply charged anions (see text). Tethering is via the silane (top). (Reprinted with permission from Ref. 52, © 1990, American Chemical Society.)

molecule (I) is used to tether a thin layer of clay particles to the electrode surface. The positive charge of the silane provides binding sites for multiply charged anions, and the clay binds the electroactive cations. Of the latter, those that are larger than the pillar height (9 Å) solely adhere to the external surface of the clay. In this respect the clay electrode provides spatial ordering of anions and cations superior to that of zeolite-modified electrodes. Cyclic voltammograms of these layers ion-exchanged with $Fe(CN)_6^{4-}$ and $Mo(CN)_8^{4-}$ are shown in Fig. 7.19 that indicate quasi-reversible redox processes typical for surface-bound ions. However, when the clay-modified electrodes were further exchanged with large electroactive cations that bind only to the external surface of the clay [e.g., $Os(bpy)_3^{2+}$], some interesting effects were observed, as shown in Fig. 7.20. Although the $Os(bpy)_3^{2+}$ was clearly bound to the clay, as shown by the green coloration of the electrode, cyclic voltammetry initiated at cathodic potentials (Fig. 7.20A and B) showed only waves attributable to the $Fe(CN)_6^{4-/3-}$ redox couple with no apparent contribution from the cationic couple even though the potential of the electrode was swept well past its formal potential

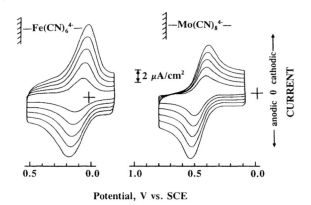

Figure 7.19. Cyclic voltammetry in 1 mM KH_2PO_4 of pillared-clay-modified electrodes ion-exchanged with (left) ferrocyanide and octacyanomolybdate. Scan rates, 20, 40, 60, 80, and 100 mV s^{-1}. (Reprinted with permission from Ref. 52, © 1990, American Chemical Society.)

(0.6 V v. SCE). However, holding the electrode at positive potentials (1.0 V) caused major changes in the voltammograms, evinced by a large enhancement in cathodic current. With very thin polymer films ($[\Gamma(Fe(CN)_6^{4-}] \leq 10^{-10}$ mol cm^{-2}), the onset of the enhanced cathodic current was observed near the formal potential of the $Os(bpy)_3^{3+/2+}$ couple.

This behavior was ascribed to a "charge-trapping" phenomenon brought about by the spatial ordering of the anions and cations at the electrode surface. Similar effects have been observed at electrodes modified with bilayer polymer films [100]. As the $Fe(CN)_6^{4-}$ is in this case held close to the electrode surface its redox reaction is facile [reaction (7.1)]; however, the redox processes of the $Os(bpy)_3^{3+}$ [reaction (7.2)] are slow because of the larger distance of the complex from the electrode surface.

$$Fe(CN)_6^{3-} + e \rightleftharpoons Fe(CN)_6^{4-} \tag{7.1}$$

$$Os(bpy)_3^{3+} + e \rightleftharpoons Os(bpy)_3^{2+} \tag{7.2}$$

The reduction of $Os(bpy)_3^{3+}$ is thus mediated by the $Fe(CN)_6^{4-/3-}$ couple via reaction (7.3):

$$Fe(CN)_6^{4-} + Os(bpy)_3^{3+} \rightleftharpoons Fe(CN)_6^{3-} + Os(bpy)_3^{2+}. \tag{7.3}$$

Although this reaction proceeds rapidly in the forward direction, it is extremely slow in the reverse direction, thus trapping $Os(bpy)_3^{2+}$. When potential is held at 1.0 V, both $Fe(CN)_6^{4-}$ and $Os(bpy)_3^{2+}$ are oxidized. However, during the cathodic scan only $Fe(CN)_6^{3-}$ is re-reduced. Although the electrochemical reduction of $Os(bpy)_3^{3+}$ is slow, it can be chemically reduced according to reaction (7.3), thereby regenerating $Fe(CN)_6^{3-}$. The cathodic peak currents of the $Fe(CN)_6^{4-/3-}$ couple, therefore, are much larger than the anodic peak currents. In this study spectroelectrochemical

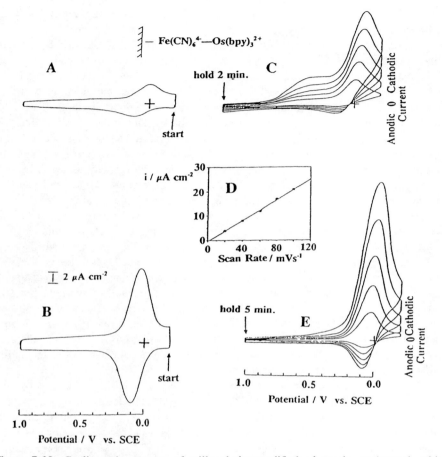

Figure 7.20. Cyclic voltammetry of pillared-clay-modified electrode exchanged with $Fe(CN)_6^{4-}$ and $Os(bpy)_3^{2+}$. A and B: scans initiated after prolonged hold at -0.2 V. C and E: scans initiated after hold at 1.0 V. In A and C, $\Gamma[Fe(CN)_6^{4-}] = 1.6 \times 10^{-10}$ mol cm^{-2}. In B and E, $\Gamma[Fe(CN)_6^{4-}] = 1.0 \times 10^{-10}$ mol cm^{-2}. Inset (D) shows the linear relationship between the scan rate and the cathodic peak current typical for a surface-confined species. (See cyclic voltammetry in E.) Scan rates are 40 mV s^{-1} (A), 100 mV s^{-1} (B) and 20, 40, 60, 80, and 100 mV s^{-1} (C and E). Potentials are with respect to SCE and the supporting electrolyte was 1 mM KH_2PO_4. (Adapted with permission from Ref. 52, © 1990, American Chemical Society.)

characterization of the charge-trapping phenomena was also performed, and the reader is directed to the original article for further detail. Charge-trapping phenomena occurring at zeolite-modified electrodes have also been demonstrated by Mallouk and co-workers [101], in studying the electrochemistry of metalloporphyrins and viologens at zeolite Y. It was shown that when a viologen ion was ion-exchanged into the bulk of the zeolite, and cobalt tetrakis(N-methyl-4-pyridyl) porphyrin was adsorbed on the outer surface, charge trapping phenomena were observed consistent with those observed at bilayers.

The electrocatalytic use of porphyrins anchored to the external surfaces of zeolites was first investigated by de Vismes et al. [102]. In this study $Mn^{III}TMpyP^{5+}$ [manganese meso-tetrakis(4-N-methylpyridyl)] was attached to the external surface of zeolite Y via ion-exchange. The oxidation of 2,6-di-tert-butylphenol was then studied using the zeolite-modified electrodes and the product distributions of 2,6,-di-tert-butyl-p-benzoquinone(Q) and diphenoquinone (DPQ) are given in Table 7.2. Although the selectivity of using the zeolite-modified electrode was no better than in the case of the homogeneously dispersed catalyst, the yields were substantially improved, suggesting that the catalyst was stabilized by the zeolite modifier.

In the latter example the electrochemical processes clearly occurred at the surface of the zeolite particles. The manganese porphyrin is too large to gain ingress into the zeolite pore system, and, as expected, the product distributions are unaffected by the presence of the zeolite. If one is to use the zeolite (or pillared clay) in any way as a participant in shape-selective electrocatalysis, then the electron transfer to the reactants must occur while they are resident in the channels and cages of the host. As we have seen in this section, there have been several successful approaches to achieving this goal. Another stemmed from the work of Rolison and co-workers [58,103] involving dispersion electrolysis. In this technique, electrochemistry occurs at feeder electrodes, between which the zeolite is held as a dispersion, by flowing an inert gas through the cell. Fleischmann et al. [104] have performed electrochemical reactions in poorly conducting and nonconducting media using this technique, which is of great interest to the zeolite or clay electrochemist.

The first example of the use of dispersion electrolyses with zeolites involved the use of a zeolite containing Pt(0) microstructures [58]. The latter can be produced in zeolite Y in various forms [105] by reduction of $Pt(NH_3)_4^{2+}$ in hydrogen at elevated temperatures. The majority of the platinum particles thus formed are constrained to the zeolite supercages and vary in diameter from 6 to 13Å. In addition some reduced platinum microparticles form on the external surfaces of the zeolite. As shown by XPS, these particles were 25 to 50 Å in diameter or smaller. Electrolysis was performed on water, and a comparison was made between the performance of Pt–zeolite dispersions and dispersions of platinum supported on alumina. The results showed that the zeolite dispersion was more effective than the alumina dispersion,

Table 7.2 Yield of Oxidation of 2,6 di-tertbutylphenol with Mn^{III} TMpy P^{5+} as a Catalyst.

Catalyst Phase for Porphyrin	Reducing Method	Reaction Time (hr)	Total Yield	Q (%)	DPQ (%)
Homogeneous solution	Bu$_4$NBH$_4$	70	98	35	65
Homogeneous solution	−0.3V vs. SCE	70	20	5	95
Heterogeneous suspension of porphyrin/zeolite	−0.3V vs. SCE	70	98	5	95

indicating the impact of a strong metal support interaction on the electrochemistry (see Rolison [1, Fig. 5]). However, under the low-field conditions used in these experiments (300 V cm^{-1}) there was no evidence of direct electrochemical participation of intrazeolite platinum. Electrolyses performed involving electroactive solutes capable of entering the zeolite cages and a size-excluded solute [i.e., bis(1,2-dibenzylcyclopentadienyl)iron(0)] established conclusively that the electrolysis occurred at the Pt particles on the external surface of the zeolite. This author nonetheless indicated that it should be possible, in principle, to develop a potential on metal particles (nanoelectrodes) sited in the zeolite interior using higher fields. In an extension of dispersion electrolysis to electrosynthesis Stemple and Rolison [106,107] used a zeolite-supported bimetallic catalyst (Pd^{II}Cu^{II}NaY) to oxidize propene to acetone at 0°C with a high selectivity. They also showed that when a nonspecific catalyst (NaY) is used, more than 10 products form, there being no reaction when the applied voltage is absent. If electrochemistry at intrazeolite centers is possible, the fundamental aspects including the electrochemical activity of metals in the quantum-size regime, investigations of metal support interactions and how they influence electrochemical processes, and the influence of the zeolite architecture on electrode reactions can be explored. Exciting prospects!

7.8 Ion Exchange in Clays and Zeolites

The ion-exchange process can play a major role in influencing the response of clay- and zeolite-modified electrodes. We saw in Section 7.5 that ingress of charge-balancing cations is of importance, and in Section 7.6 how the nature of the supporting electrolyte can influence the electrode response through ion-exchange processes. To turn this on its head we also recognize, therefore, that we can potentially arm ourselves with an excellent probe of the ion-exchange process itself. In situ studies of the uptake and release of electroactive species from clays and zeolites using relatively simple electrochemical instrumentation therefore become possible. There are several examples taken from the literature and our own work of both direct and indirect studies of ion-exchange using this methodology and these are now reviewed.

As outlined in Section 7.5, Shaw et al. [27] presented the possible mechanisms that could describe electrochemical processes occurring at zeolite-modified electrodes. In addition, a detailed study of the response of the electrodes to a variety of solution-phase species in the context of both their size and charge was made. A useful probe of the ion-exchange process proved to be the copper (II) ion pre-exchanged into zeolites A and Y. The latter contained 1.6 times as much copper as did the CuA zeolite, but demonstrated a peak current (for copper reduction) that was 3.3 times greater. The likely reason for this was thought to be the higher mobility of the copper ions in the Y zeolite, which could therefore exchange faster with the electrolyte cations. In addition to probing the ion-exchange between intrazeolite and electrolyte cations, this paper also showed how the competition between solution-phase (analyte) cations could be followed using the modified electrode. In Fig. 7.21

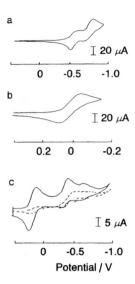

Figure 7.21. Cyclic voltammetry of NaY zeolite-modified electrodes in water. Supporting electrolyte 0.1 M KNO_3. The electrolyte solution contained the following electroactive species. (a) 1 mM MV^{2+}, (b) 1 mM $Ru(NH_3)_6^{3+}$, (c) 0.5 mM $Ru(NH_3)_6^{3+}$, and 2 mM MV^{2+}. Scans were recorded directly after immersion of electrode (dashed line) and 90 minutes later (solid line). Reference electrode Ag/AgCl. (Reprinted with permission from Ref. 27.)

the voltammograms observed for methyl viologen (MV^{2+}) and $Ru(NH_3)_6^{3+}$ at zeolite Y–modified electrodes 90 minutes after immersion of the electrodes in solution are shown. At first the ratio of the peak currents for the reduction of the two species matches that of their concentrations. After 90 minutes, however, the two peak currents are approximately the same, indicating a fourfold selectivity of $Ru(NH_3)_6^{3+}$ over MV^{2+}, even though the current response is in fact greater for MV^{2+} showing that the mobility (i.e., the rate of exchange) of MV^{2+} was greater. We see here an important distinction between the equilibrium selectivity of the zeolite for one ion over another as opposed to the rate of the early stages of ion-exchange.

A further example of the importance of the ion-exchange process in clay and zeolite-modified electrochemistry and how this can be probed concerns the electrochemistry of silver zeolite Y in alkyl ammonium electrolytes. Related data have also been reported concerning MV^{2+}–exchanged A and Y modified electrodes [24]. The effect of changing the electrolyte cation is shown in Fig. 7.22. The data show the maximum anodic currents observed as a function of the quaternary ammonium ion concentration (i.e., tetramethylammonium, tetraethylammonium, and tetrapropylammonium). In each case the background electrolyte was 0.1 M $TBABF_4$ in 50 vol. % water-methanol. As discussed in section 7.6 for alkali-metal cations [59], the currents increase linearly with concentration of the smaller electrolyte cation. The large change in slope between TEA^+ and TPA^+ indicates that the exchange of silver by TEA^+ is much faster than that by TPA^+. Although the factors pertaining to the

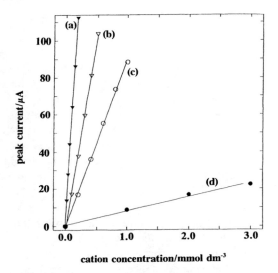

Figure 7.22. Anodic peak currents versus the concentration of (a) ammonium, (b) tetramethylammonium, (c) tetraethylammonium, and (d) tetrapropylammonium ions. In each case the background electrolyte was 0.1 M TBABF$_4$ in 50/50 water-methanol. (Reprinted with permission from Ref. 68.)

partial exclusion of cations by a zeolite are discussed in detail [68], the authors point out that the contributions to the differences in slopes between the alkyl ammonium cations are difficult to separate. Once again we alert the reader to the fact that the early stages of ion-exchange are probed in the electrochemical experiment and thus equilibrium ideas cannot be used satisfactorily.

The effect of zeolite co-cations on the response of zeolite-modified electrodes is also discussed in the preceding paper, specifically addressing the factors influencing the electroactivity of silver ions found in small and large cages. It is well known that, even in the hydrated form, cations occupy different sites in the zeolite framework [11], and in certain cases the electrochemistry has apparently shown sensitivity to this [26]. We have examined this phenomenon in more detail for silver-ion-exchanged zeolite Y, where it is possible through control of both the silver-ion loading and the zeolite co-cation to locate silver ions exclusively in the hexagonal prisms and sodalite cavities (see Fig. 7.2) [68]. The interconnected supercages of zeolite Y can usefully be termed the large-channel network, whereas the sodalite cages and hexagonal prisms can be thought of as comprising a small-channel network. Brown and Sherry [108,109] have proposed that the most likely pathway for ion exchange (and see mechanisms 1 and 2, Section 7.5) in zeolite Y of both large-channel and small-channel ("bound") cations is via the large-channel network. Small-channel cations must therefore initially exchange with more mobile cations in the large channels. That is, diffusion of cations through the small channels is very slow. This is not surprising, considering that cations must completely desolvate prior to their ingress into the hexagonal prisms.

The effect of this phenomenon is clearly reflected in the electrochemistry of

$Ag_{16}Cs_{40}Y$, which exclusively holds small-channel silver ions. Even in $CsNO_3$ electrolytes the electroactivity of the small-channel silvers is heavily suppressed compared to a $Ag_6Na_{50}Y$ sample containing both small- and large-channel silver ions [68], as shown in Fig. 7.23. Recent studies carried out using $AgNH_4Y$ and $AgNH_4X$ indicate that the small-channel Ag^+ ions cannot be electrochemically addressed when the electrolyte is $CsNO_3$ [110]. Results obtained for $AgNH_4X$ show that the exchange of small-channel Ag^+ ions with Na^+ is faster than that with either Li^+ or K^+. The rate of exchange, in this case, would be governed by the combined effects of dehydration energies and the ionic radii of the cations [111]. In contrast, the rate of ion-exchange of large-channel Ag^+ ions depends on the hydrated radii of the exchanging cations [27,59,68]. Thus the exchange of Ag^+ with K^+ is faster than that with either Li^+ or Na^+.

The known specificity of Ag^+ for small-channel sites particularly at low silver loadings [112,113] make these ideal systems for studying the importance of site occupancies and how these affect the electrochemical response. Temperature-dependent cyclic voltammetry and chronoamperometry have been particularly useful in this context [114], showing that there is a substantial change in the activation energies for diffusion through the zeolite pores as a function of silver loading, and this has been directly correlated with silver ion occupancy of either the small- or large-channel networks. The results once again underline the importance of solvation and ion-exchange phenomena when interpreting the electrochemical response of the electrodes.

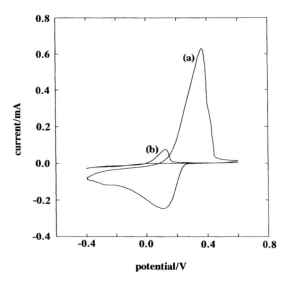

Figure 7.23. Cyclic voltammetry of (a) $Ag_6Na_{50}Y$ and (b) $Ag_{16}Cs_{40}Y$ electrodes in 0.1 M $CsNO_3$. The voltammograms clearly show the different behavior exhibited between samples containing either small- or large-channel silver ions (see text). Scan rate, 20 mV s^{-1}. (Reprinted with permission from Ref. 68.)

Recent studies using dispersion electrolysis [103] have shown that the application of potentials or fields to zeolite suspensions affects the rate of exchange of intrazeolite alkali metal cations with aqueous phase protons generated by the autoprotolysis of water. This is an intriguing observation in that the kinetics of the process depended on both the nature of the charge-balancing cation in the zeolite and the magnitude of the electric field. For zeolite Y, the kinetics did not follow a uniform trend, with either the ion size, the hydrated ionic radius, or the mobility of the ion. Nonetheless within the alkali-metal series, with the exception of Rb^+ the rate followed the order $Li^+ < Na^+ < K^+ < Cs^+$. Evidence was also presented to show that the rate of exchange depended on the structure and composition of the zeolite and the magnitude of the electric field used, in addition to the nature of the electrolyte cation. As well as being extremely significant in terms of electrosynthetic strategies that can be adopted using zeolites in electrochemical environments (see Section 7.7), measurements of this type will undoubtedly shed light on the key issues of solvation and ion-exchange, which will be important in understanding electrochemical phenomena involving zeolites in aqueous and nonaqueous media.

7.9 Molecular Wires

In Section 7.7 we discussed the possibility of electrochemically addressing intrazeolite centers by dispersion electrolysis. An alternative approach [1,115,116] is to synthetically place conducting polymers within the zeolite so that electrical communication with the outer environment is possible. In addition, the synthesis and characterization of conductive polymers in well-defined structurally ordered systems could assist in the successful development of models for the interpretation of the physical properties of conducting polymers. In the case of zeolites, because of the dimensions of the channels the polymer chains are limited to single chains or "molecular wires." In what follows we briefly outline the first attempts at the direct electrochemical growth of intramordenite polyaniline [117].

The cyclic voltammograms of anilinium-exchanged mordenite-modified electrodes recorded in aqueous tetrabutylammonium fluoride are shown in Fig. 7.24a. This sample was exchanged from its sodium form in anilinium hydrochloride, and from both NMR and AAS the exchange level was determined to be 53% of the total capacity. As the electrode was repeatedly scanned, there was a steady growth in both the anodic and cathodic branches of the voltammograms, and a noticeable color change of the electrode from white to blue-green. Each scan shown in the figure took about 2 minutes, the growth of the polymer proceeding slowly over about 2 hours. These observations logically point to the formation of some form of polyaniline. Indeed the formation of bulk polyaniline can be achieved electrochemically by repeated cyclic voltammetry of acidic solutions of aniline. Under the conditions used in this experiment, it is impossible for the intrazeolite anilinium to exit the zeolite via ion-exchange with the size-excluded TBA^+ ion indicating that the polymerization is intrazeolitic or is occurring through anilinium ions residing at exchange sites on the external surfaces of the zeolite particles. The latter is appar-

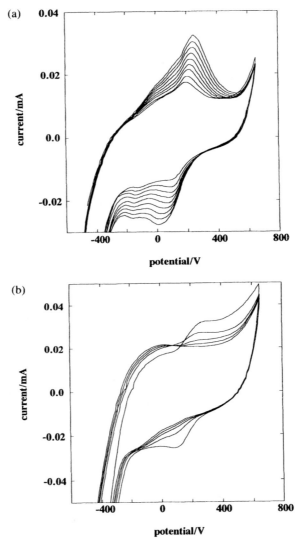

Figure 7.24. Cyclic voltammetry of mordenite structure types in water containing 0.1 M TBAF. The zeolites had previously been ion-exchanged with anilinium. (a) M-5 mordenite (b) AW 300 zeolite. In (a) the formation of polyaniline is apparent (see text). Scan rate, 20 mV s^{-1}. In (b) the current decreased as a function of time (see text).

ently inoperative since similar experiments using Linde AW 300 [11] zeolite (Fig. 7.25b), with a mordenite structure (pore size, 4Å), and EZ-500 zeolite with a Ferrierite structure (Engelhard Corporation, pore size, 4 Å) showed no apparent polymerization. These zeolites cannot exchange anilinium into their internal cage system because of pore size restrictions but exchange onto the external surface is indeed possible. Note that the current yields for the small pore samples decreased

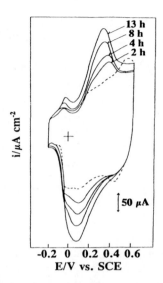

Figure 7.25. Cyclic voltammetry of montmorillonite-modified electrodes with intercalated aniline in 1 M HCl. Voltammograms taken at the initial stage of the potential sweeps (dashed line) and following preparation of the polyaniline-intercalated electrode (solid line) at times shown. Scan rate, 100 mV s^{-1}. Reference electrode, SCE. (Reprinted with permission from Ref. 120.)

with time as the anilinium was leached into solution. It is also of interest to note that polyaniline chemically synthesized in the mordenite zeolite using the procedures developed by Bein et al. [115,116] showed no electroactivity of the sort shown by either bulk polyaniline or the type just described. A possible cause is that the polymer chains prepared through intrazeolite chemical oxidations do not extend to the surface of the zeolite and are therefore electrochemically inaddressable. In fact, mordenite samples containing chemically synthesized intrazeolite polyaniline did at first show some small activity and polymer growth was evident. This was shown to be due to residual unreacted anilinium through FT-IR spectroscopy. Following removal of the anilinium by ion-exchange with sodium, the electroactivity was not apparent (i.e., no polymerization occurred) and the IR signature at 1495 cm^{-1} of anilinium [116] vanished.

The apparent formation of intrazeolite polyaniline was also indicated by the lack of sensitivity of the polymer to solution-phase anions. These are known to strongly affect the electrochemistry of bulk polyaniline [118,119]. Anions are excluded from the pore systems of zeolites under normal conditions. Note that the redox processes associated with intrazeolite polyaniline must be accompanied by ingress of solution-phase cations into the zeolite pore system in order to maintain charge balance. This apparently occurs via solution-phase non-size-excluded cationic impurities.

The electropolymerization of aniline intercalated in montmorillonite has also been reported by Inoue and Yoneyama [120], marking the first attempt to electrochemically polymerize clay-intercalated organic molecules. The clay was first loaded with aniline, which was then galvanostatically electrolyzed in 2 M HCl. The product was then characterized by cyclic voltammetry. During the electrolysis, the electrode changed its color from white to blue. This coloration could not be removed by washing in acetonitrile or ethanol, which led the authors to propose that intercalated polyaniline had been formed. Cyclic voltammetry of the as prepared polyaniline-intercalated clay electrode in HCl is shown in Fig. 7.25. Initially there was very little evidence of an electroactive product, but over a period of 13 hours a pair of redox peaks grew in the cyclic voltammograms. This was assigned to the electropolymerization of previously unreacted aniline.

7.10 Layered Double Hydroxides (Hydrotalcite Clays)

The majority of clay-modified electrodes studied have employed cation-exchangeable clays. However, layered double hydroxides (see Section 7.2.3) have also been used. The first such example was reported by Itaya et al. [121], who demonstrated that $Mo(CN)_8^{4-}$, $Fe(CN)_6^{3-}$, and $IrCl_6^{2-}$ were electrochemically active when incorporated into layered double hydroxide clays. Shaw and co-workers [122] have also studied the layered double hydroxide (LDH) surface with ferricyanide and phenol electrochemical probes. In this report, $Zn_x^{2+}Al_y^{3+}(Cl^-)_z(OH)_{2x+3y-z} \cdot mH_2O$ and its Mg analogue were employed. A representation of the possible modes of interaction of these probe molecules with the layered double hydroxide is shown in Fig. 7.26. A comparison of the electrochemistry observed for Prussian blue (PB) films with that of the LDH-modified electrode in ferricyanide containing electrolytes showed that a Prussian-blue-like film was formed upon repeated electrochemical cycling of the film. Although the films were not as sensitive to the size of the hydrated electrolyte ions as true PB films, the films formed on the LDH did show some sensitivity in both peak shape and currents as the electrolyte cation was varied through the sequence K^+, Na^+, and Li^+. The mode of interaction of the ferricyanide ion with the surface of the LDH was also investigated in this study. The conclusion was that the interaction was predominantly with zinc and aluminum ions of the clay. The LDH-modified electrode was also useful as a probe of the surface hydroxide activity, through studies of the electro-oxidation of phenol. It is interesting to note that phenol oxidation was catalyzed by the Mg containing clay but not by that containing zinc. These results were used to assign effective pH values of greater than 11.2 and 8.3, respectively, for the surfaces of these solids.

Pillaring of LDH clays has also been achieved, and a preliminary electrochemical study of LDHs pillared with metatungstate has been reported recently [123]. The incorporation of the metatungstate anion pillaring agent was followed electrochemically as this replaced the terephthalate dianion in the hydrotalcite $Zn_2Al(OH)_5(TA) \cdot xH_2O$ (TA = terephthalate). The evolution of the first reduction wave of the metatungstate anion $(H_2W_{12}O_{40})$ with time is plotted in Fig. 7.27. Curve 1 of this figure shows the behavior of a freshly polished glassy carbon electrode in a solution containing the metatungstate pillar. This shows that the current remains essentially constant for many hours. In contrast, the current increased over a period of several hours when the electrode was modified with the

ELECTROCHEMISTRY WITH CLAYS AND ZEOLITES

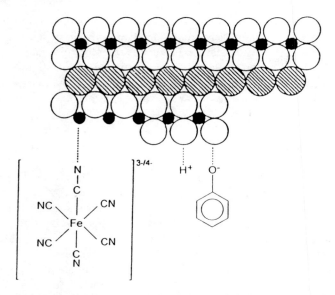

Figure 7.26. A schematic representation of a layered double hydroxide indicating the possible interactions of solution-phase species with the clay surface. Open circles = OH^-; filled circles = M^{2+}, M^{3+}; shaded circles = hydrated anions. The dotted lines represent possible interactions rather than specific modes of bonding. (Reprinted with permission from Ref. 122.)

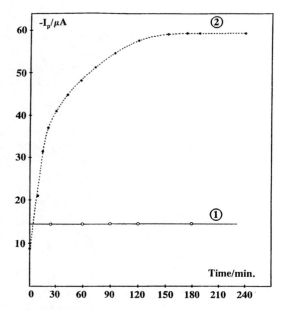

Figure 7.27. The evolution of the first reduction peak of $H_2W_{12}O_{40}^{6-}$ with time on a (1) bare and (2) LDH-modified electrode (see text). Scan rate 50 mV s^{-1}. (Reprinted with permission from Ref. 123.)

LDH. Note that shortly after the modified electrode was exposed to the pillaring agent, the observed current was actually smaller than that at the bare electrode. Nonetheless, the metatungstate anion penetrates the film quite rapidly and is concentrated into the clay film. When the response of the clay-modified electrode had stabilized, the modified electrode was removed to a solution containing only supporting electrolyte where the response remained constant for more than 20 hours.

7.11 Studies of Diffusion

Fitch [3] has collated the electrochemically derived diffusion coefficients for electroactive species in clays. These are much smaller than the corresponding values in solution, as expected for a species that interacts strongly with the clay surfaces. For zeolite-modified electrodes, there is little data of this type available. Examples, however, do exist for $Co(CpCH_3)_2^{+/0}$, and Ag^+ in zeolite Y [25,35]. Transport and diffusion issues in zeolites and clays have been discussed further by Rolison [1] and Fitch [3] and will not be addressed further here. Recent studies by Fitch and co-workers [124,125] on the diffusion of $Fe(CN)_6^{3-}$ through montmorillonite clays have shown that there is a correlation between the reduction currents and the interlayer spacing of the clay. It was suggested that anion sensing could be achieved as a result.

7.12 Conclusion

Electrochemical processes occurring in the presence of zeolites and clays show enormous potential in many areas of electrochemistry. This review has but scratched the surface of these modified electrodes and their fascinating chemistry. With the development of more reproducible coatings, either via direct synthetic growth or other means, coupled with in situ spectroscopic characterization (e.g., Raman, EPR) there will no doubt be much progress in the understanding of the electrochemistry exhibited by both clay- and zeolite-modified electrodes. In an applied sense there also seem to be good prospects for the development of novel electro-synthetic routes using dispersion electrolysis and for the design of highly specific solution-phase chemical sensors.

References

1. Rolison, D. R. *Chem. Rev.* **1990**, *90*, 867.
2. Rolison, D. R.; Nowak, R. J.; Welsh, T. T.; Murray, C. G. *Talanta* **1991**, *38*, 27.
3. Fitch, A. *Clays and Clay Miner* **1990**, *38*, 391.
4. Herron, A. In "Inclusion Compounds: Inorganic and Physical Aspects of Inclusion," Atwood, J. L.; Davies, J. E. D.; MacNicol, D. D., Eds. Oxford University Press, New York, 1991, Vol. 5, pp. 90–102.

5. Davis, M. E. *Chem. Ind.* **1992**, 137.
6. Dyer, A. "An Introduction to Zeolite Molecular Sieves." John Wiley, Chichester, 1988.
7. Kulesza, P. J.; Faszynska, M. *J. Electroanal. Chem.* **1988**, *252*, 461.
8. Jiang, M.; Zhou, X.; Zhao, Z. *J. Electroanal. Chem.* **1990**, *287*, 389.
9. Jacobs, P. A.; Martens, J. A. "Synthesis of High-Silica Aluminosilicate Zeolites." Studies in Surface Science and Catalysis. Vol. 33. Elsevier, Amsterdam, 1987.
10. Occelli, M. L.; Robson, H. E. "Zeolite Synthesis." ACS Symposium Series. Vol. 398. American Chemical Society, Washington, D.C., 1989.
11. Breck, D. W. "Zeolite Molecular Sieves." R. E. Krieger, Malabar, FL, 1984.
12. Iwamoto, M.; Yamaguchi, K. I.; Akutagawa, S.; Kagawa, S. *J. Phys. Chem.* **1984**, *88*, 4195.
13. Davis, M. E.; Saldarriaga, C.; Montes, C.; Garces, J.; Crowder, C. *Zeolites* **1988**, *8*, 362.
14. Estermann, M.; McCusker, L. B.; Baelocher, C.; Merrouche, A.; Kessler, H. *Nature* **1991**, *352*, 320.
15. Barrer, R. M. "Hydrothermal Chemistry of Zeolites." Academic Press, London, 1982.
16. Van Olphen, H. "An Introduction to Clay Colloid Chemistry for Clay Technologists, Geologists and Soil Scientists." John Wiley, New York, 1963.
17. Tanabe, K.; Misono, M.; Ono, Y.; Hattori, H. "New Solid Acids and Bases: Their catalytic properties." Studies in Surface Science and Catalysis. Vol. 51. Elsevier, Amsterdam, 1989.
18. Plee, D.; Borg, F.; Gatineau, L.; Fripiat, J. J. *J. Am. Chem. Soc.* **1985**, *107*, 2362.
19. Yamayanka, S.; Brindley, G. W. *Clays Clay Miner.* **1979**, *27*, 119.
20. Yamayanka, S.; Doi, T.; Hattori, M. *Mat. Res. Bull.* **1984**, *19*, 161.
21. Christiano, S. P.; Wang, J.; Pinnavaia, T. J. *Inorg. Chem.* **1985**, *24*, 1222.
22. Reichle, W. T. *Chemtech* **1986**, *16*, 58.
23. de Vismes, B.; Bedioui, F.; Devynck, J.; Bied-Charreton, C. J. *J. Electroanal. Chem.* **1985**, *187*, 197.
24. Gemborys, H. A.; Shaw, B. R. *J. Electroanal. Chem.* **1986**, *208*, 95.
25. Li, Z.; Mallouk, T. E. *J. Phys. Chem.* **1987**, *91*, 643.
26. Baker, M. D.; Zhang, J. *J. Phys. Chem.* **1990**, *94*, 8703.
27. Shaw, B. R.; Creasy, K. E.; Lanczycki, C. J.; Sargent, J. A.; Tirhado, M. J. *J. Electrochem. Soc.* **1988**, *135*, 869.
28. Johansson, G.; Risinger, L.; Faelth, L. *Anal. Chim. Acta* **1977**, *119*, 25.
29. Demertzis, M.; Evmiridis, N. P. *J. Chem. Soc. Faraday Trans. 1* **1986**, *82*, 3647.
30. Hernandez, P.; Alda, E.; Hernandez, L. *Fresenius' Z. Anal. Chem.* **1987**, *327*, 676.
31. Creasy, K. E.; Shaw, B. R., *Electrochim. Acta* **1988**, *33*, 551.
32. Wang, J.; Martinez, T. *Anal. Chim. Acta* **1988**, *207*, 95.
33. El Murr, N.; Kerkeni, R.; Sellami, A.; Ben Taarit, Y. *J. Electroanal. Chem.* **1988**, *246*, 461.
34. Shaw, B. R.; Creasy, K. E. *J. Electroanal. Chem.* **1988**, *243*, 209.
35. Periera-Ramos, J. P.; Messina, R.; Perichon, J. *J. Electroanal. Chem.* **1983**, *146*, 157.
36. Li, Z.; Lai, C.; Mallouk, T. E. *Inorg. Chem.* **1989**, *28*, 178.

37. Murray, C. G.; Nowak, R. J.; Rolison, D. R. *J. Electroanal. Chem.* **1984**, *164*, 205.
38. Calzaferri, G.; Hadener, K.; Li, J. *J. Chem. Soc., Chem. Comm.* **1991**, 653.
39. Davis, S. P.; Borgstedt, E. V. R.; Suib, S. L. *Chem. Mater.* **1990**, *2*, 712.
40. Jansen, J. C.; van Bekkum, H. In "Proceedings of the Ninth International Zeolite Conference, Montreal, 1992," von Ballmoos, R.; Higgins, J. B.; Treacy, M. M. J., Eds. Butterworth-Heinemann, Stoneham, MA, in press.
41. Ozin, G.; Kuperman, A.; Stein, A. *Angew. Chemie Int. Ed. Engl.* **1989**, *28*, 359.
42. Fitch, A.; Lavy-Feder, A.; Lee, S. A.; Kirsch, M. T. *J. Phys. Chem.* **1988**, *92*, 6665.
43. King, R. D.; Nocera, D. G.; Pinnavaia, T. J. *J. Electroanal. Chem.* **1987**, *236*, 43.
44. Rusling, J. F.; Shi, C.-N.; Suib, S. L. *J. Electroanal. Chem.* **1988**, *245*, 331.
45. Brahimi, B.; Labbe, P.; Reverdy, G. *J. Electroanal. Chem.* **1989**, *267*, 343.
46. Castro-Acuna, C. M.; Fan, F.-R. F.; Bard, A. J. *J. Electroanal. Chem.* **1987**, *234*, 347.
47. Rudzinski, W. E.; Figueroa, C.; Hoppe, C.; Kuromoto, T. Y.; Root, D. *J. Electroanal. Chem.* **1988**, *243*, 367.
48. Hu, N.; Rusling, J. F. *Anal. Chem.* **1991**, *63*, 2163.
49. Villemure, G.; Bard, A. J. *J. Electroanal. Chem.* **1990**, *282*, 107.
50. Hernandez, L.; Hernandez, P.; Lorenzo, E.; Ferrera, Z. S. *Analyst* **1988**, *113*, 621.
51. Hernandez, L.; Hernandez, P.; Lorenzo, E. In "Contemporary Electroanalytical Chemistry," Ivaska, A.; Lowenstein, A.; Sara, R., Eds. Plenum, New York, 1990.
52. Rong, D.; Kim, Y. I.; Mallouk, T. E. *Inorg. Chem.* **1990**, *29*, 1329.
53. Ghosh, P. K.; Bard, A. J. *J. Am. Chem. Soc.* **1983**, *105*, 5691.
54. Ege, D.; Ghosh, P. K.; White, J. R.; Equey, J. F.; Bard, A. J. *J. Am. Chem. Soc.* **1985**, *107*, 5644.
55. Rudzinski, W. E.; Bard, A. J. *J. Electroanal. Chem.* **1986**, *199*, 323.
56. Carter, M. T.; Bard, A. J. *J. Electroanal. Chem.* **1987**, *229*, 191.
57. Liu, H.-Y.; Anson, F. C. *J. Electroanal. Chem.* **1986**, *184*, 411.
58. Rolison, D. R.; Hayes, R. E.; Rudzinski, W. E. *J. Phys. Chem.* **1989**, *93*, 5524.
59. Baker, M. D.; Senaratne, C. *Anal. Chem.* **1992**, *64*, 697.
60. Kaviratna, P. De S.; Pinnavaia, T. J. *J. Electroanal. Chem.* **1992**, *332*, 135.
61. Edens, G. J.; Fitch, A.; Lavy-Feder, A. *J. Electroanal. Chem.* **1991**, *307*, 139.
62. Petridis, D.; Falaras, P.; Pinnavaia, T. J. *Inorg. Chem.* **1992**, *31*, 3530.
63. Xiang, Y.; Villemure, G. *Can. J. Chem.* **1992**, *70*, 1833.
64. Barrer, R. M.; Walker, A. J. *Trans. Faraday Soc.* **1964**, *60*, 171.
65. Barrer, R. M.; Meir, W. M. *J. Chem. Soc.* **1958**, 299.
66. Bedioui, F.; Boysson, E. D.; Devynk, J.; Balkus, K. J. *J. Chem. Soc. Faraday Trans.* **1991**, *87*, 383.
67. Bharathi, S.; Phani, K. L. N.; Joseph, J.; Pitchumani, D.; Jeyakumar, G.; Rao, P.; Rangarajan, S. K. *J. Electroanal. Chem.* **1992**, *334*, 145.
68. Baker, M. D.; Senaratne, C.; Zhang, J. *J. Chem. Soc. Faraday Trans.* **1992**, *88*, 3187.

69. Dessau, R. M. *J. Catal.* **1982**, *77*, 304.
70. Csicsery, S. M. *Chem. Br.* **1985**, *21*, 473.
71. Barrer, R. M.; James, S. D. *J. Phys. Chem.* **1960**, *64*, 421.
72. Amos, L.; Duggal, A.; Ragones, P.; Bocarsly, A. B.; Fitzgerald-Bocarsly, P. A. *Anal. Chem.* **1988**, *60*, 245.
73. Thomsen, K. N.; Baldwin, R. P. *Anal. Chem.* **1989**, *61*, 2594.
74. Tarter, J. G., Ed. "Ion Chromatography," Chromatographic Science Series, Vol. 37. Marcel Dekker, New York, 1987.
75. Small, H. "Ion Chromatography." Plenum, New York, 1989.
76. Dasgupta, P. K. *J. Chrom. Sci.* **1989**, *27*, 422.
77. Carr-Brion, K. "Moisture Sensors in Process Control." Elsevier, New York, 1986.
78. Sherry, H. S. In "Ion Exchange," Marinsky, J. A., Ed. Marcel Dekker, New York, 1969, Vol. 2, pp. 89–134.
79. Senaratne, C.; Baker, M. D. *J. Electroanal. Chem.* **1992**, *332*, 357.
80. Huang, H.; Dasgupta, P. K. *Anal. Chem.* **1990**, *62*, 1935.
81. Huang, H.; Dasgupta, P. K.; Ronchinsky, S. *Anal. Chem.* **1991**, *63*, 1570.
82. Stamires, D. N. *J. Chem. Phys.* **1962**, *36*, 3174.
83. Freeman, D. C.; Stamires, D. N. *J. Chem. Phys.* **1961**, *35*, 799.
84. Cassidy, J.; O'Donoghue, E. O.; Breen, W. *Analyst* **1989**, *114*, 1509.
85. Cassidy, J.; Breen, W.; O'Donoghue, E.; Lyons, M. E. G. *Electrochim. Acta* **1991**, *36*, 383.
86. Lockhart, J. C. In "Inclusion Compounds: Inorganic and Physical Aspects of Inclusion," Atwood, J. L.; Davies, J. E. D.; MacNicol, D. D., Eds. Oxford University Press, New York, 1991, Vol. 5, pp. 345–358.
87. Marshall, C. E. *J. Phys. Chem.* **1939**, *43*, 1155.
88. Marshall, C. E.; Bergman, W. E. *J. Am. Chem. Soc.* **1941**, *63*, 1911.
89. Coetze, C. J. In "Inorganic Ion Exchangers in Chemical Analysis," Qureshi, M.; Varshney, K. G., Eds. CRC Press, Boca Raton, Fl, 1991.
90. Wang, J.; Martinez, T. *Electroanalysis* **1989**, *1*, 167.
91. Weiglos, T.; Fitch, A. *Electroanalysis* **1990**, *2*, 449.
92. Pinnavaia, T. J. *Science* **1983**, *220*, 365.
93. Villemure, G.; Bard, A. J. *J. Electroanal. Chem.* **1990**, *283*, 403.
94. Ghosh, P. K.; Mau, A. W.-H.; Bard, A. J. *J. Electroanal. Chem.* **1984**, *169*, 315.
95. Bard, A. J.; Faulkner, L. R. "Electrochemical Methods: Fundamentals and Applications." John Wiley, New York, 1980.
96. Oyama, N.; Anson, F. C. *J. Electroanal. Chem.* **1986**, *199*, 467.
97. Rusling, J. F.; Shi, C.-N.; Gosser, D. K.; Shukla, S. S. *J. Electroanal. Chem.* **1988**, *240*, 201.
98. Kamau, G. N.; Rusling, J. F. *J. Electroanal. Chem.* **1988**, *240*, 217.
99. Shi, C.; Rusling, J. F.; Wang, Z.; Willis, W. S.; Winiecki, A. M.; Suib, S. L. *Langmuir* **1989**, *5*, 650.

100. Abruna, H. D.; Denisevich, P.; Umana, M.; Meyer, T. J.; Murray, R. W. *J. Am. Chem. Soc.* **1981**, *103*, 1 and references cited therein.

101. Li, Z.; Wang, C. M.; Persaud, L.; Mallouk, T. E. *J. Phys. Chem.* **1988**, *92*, 2592.

102. de Vismes, B. D.; Bedioui, F.; Devynck, J.; Bied-Charreton, C. B.; Peree-Fauvet, M. *Nouv J. Chim.* **1986**, *10*, 81.

103. Rolison, D. R.; Stemple, J. Z.; Curran, D. J. In "Proceedings of the Ninth International Zeolite Conference, Montreal, 1992," von Ballmoos, R.; Higgins, J. B.; Treacy, M. M. J., Eds. Butterworth-Heinemann, Stoneham, MA, in press.

104. Fleischmann, M.; Ghorogchian, J.; Rolison, D. R.; Pons, S. *J. Phys. Chem.* **1986**, *90*, 6392.

105. Gallezot, P.; Alarcon-Diaz, A.; Dalmon, J.-A.; Renouprez, A. J.; Imelik, B. *J. J. Catal.* **1975**, *39*, 334.

106. Stemple, J. Z.; Rolison, D. R. In "Extended Abstracts and Program" Ninth International Zeolite Conference, Montreal, July 5–10, 1992; Higgins, J. B.; von Ballmoos, R.; Treacy, M. M. J., Eds. Butterworth-Heinemann, Stoneham, MA, Paper No. RP47.

107. Rolison, D. R.; Stemple, J. Z. *J. Chem. Soc. Chem. Commun.* **1993**, 25.

108. Brown, L. M.; Sherry, H. S.; Krambeck, F. K. *J. Phys. Chem.* **1971**, *75*, 3846.

109. Brown, L. M.; Sherry, H. S. *J. Phys. Chem.* **1971**, *75*, 3855.

110. Senaratne, C.; Baker, M. D. Manuscript in preparation.

111. Barrer, R. M.; Davies, J. A.; Rees, L. V. C. *J. Inorg. Nucl. Chem.* **1969**, *31*, 2599.

112. Costenoble, M. L.; Maes, A. *J. Chem. Soc. Faraday Trans. 1* **1978**, *74*, 131.

113. Maes, A.; Cremers, A. *J. Chem. Soc. Faraday Trans. 1* **1978**, *74*, 136.

114. Baker, M. D.; Senaratne, C.; Zhang, J. *J. Phys. Chem.* (submitted)

115. Bein, T.; Enzel, P. *Angew. Chem., Int. Ed. Engl.* **1989**, *28*, 1692.

116. Bein, T.; Enzel, P. *Synth. Metals* **1988**, *29*, E163.

117. Baker, M. D.; Yu, X. Manuscript in preparation

118. Duic, L.; Mandic, Z. *J. Electroanal. Chem.* **1992**, *335*, 207.

119. Wang, B.; Tang, J.; Wang, F. *Synth. Metals* **1986**, *13*, 329.

120. Inoue, H.; Yoneyama, H. *J. Electroanal. Chem.* **1987**, *233*, 291.

121. Itaya, K.; Chang, H.-C.; Uchida, I. *Inorg. Chem.* **1987**, *26*, 624.

122. Shaw, B. R.; Deng, Y.; Strillacci, F. E.; Carrado, K. A.; Fessahaie, M. G. *J. Electrochem. Soc.* **1990**, *137*, 3136.

123. Keita, B.; Belhouari, A.; Nadjo, L. *J. Electroanal. Chem.* **1991**, *314*, 345.

124. Lee, A. A.; Fitch, A. *J. Phys. Chem.* **1990**, *94* 4998.

125. Fitch, A.; Du, J. *J. Electroanal. Chem.* **1991**, *319*, 409.

INDEX

β-alumina as Na$^+$ electrolyte, 66, 69
Adhesion of polymer films, 148
Adsorbed
  in contact with an electrolyte, 46
  on porous sintered oxide electrodes, 47
  radical pairs, 35
  redox systems, 33
Adsorption
  and desorption of reactant species, 25
  and electrocatalytic reaction sites, 25
  of organic compounds on oxide electrodes, 229, 230
  sites of electrode surfaces, 7
Alkali metal cations, 353, 368, 371, 374
Aluminosilicates, 340, 368–370
Ambipolar diffusion, 159
Anion exchanger, 166
Anionic clays, 344, 374
Anodic dissolution of UO$_2$, 313–20
Arrhenius exponential term, 78
Arrhenius plot of conductivity in a polyelectrolyte, 75
Association ions in polymers, 158
Atomic ionization potentials, 13

Batteries using conjugated polymers, 88
Beidellite, 344
Binders, 345, 346
Brucite, 344, 346

Cadmium
  iodide structure, 14
  phosphide, 49
  sulfide, 54

Carbon dioxide
  reduction, 2, 24, 36, 40
  reduction to methanol, 24
Carbon insertion materials, 117
Catalysis
  of oxygen reduction, 26
  of such multielectron transfer reactions, 42
Cation exchanger, 166
Ceramic oxide nuclear fuels
  composition, 299
  fabrication, 299
  fission products in, 299
  plutonium dioxide, 297, 299, 301, 305
  structural properties, 299, 301
  thorium dioxide, 297, 301
  uranium carbide, 299, 301, 305
  uranium nitride, 299, 301, 305
Chabazite, 340
Chalcogenide bonding, 10
Charge transfer, 2, 6, 9, 26, 44
  at the semiconductor/electrolyte interface, 36
  controlled by hopping processes, 49
  in the picosecond and subpicosecond range, 49
  parameters of preceeding reactants, 53
  via cluster centers, 30
Chemical
  bonding, 10, 25, 49
  coordination, 25
  derivatization, 32
Chemisorption, 3
  and electrocatalysis, 25
  of metal ions, 21
  of nitrogen, 42
Chlorine evolution reaction, 238, 244
Chlor-alkali cells, 207, 209

Cloverite, 343
Cluster, 14, 17, 20, 21, 25, 26, 27
 calculation in the pyrite system, 14
 chalcogenide ($Mo_2Re_4Se_8$), 30
 compounds (Chevrel phase), 15
 materials for oxygen reduciton, 25
 materials for (photo)electrocatalytic purposes, 28
$Co(bpy)_3^{3+}$, 362
Colloids
 of CdS and PbS, 48
 of semiconductors and metals, 49
Conduction band
 in clusters, 25
 of d-states, 14
 of transition metal, 14
Conductive polymers, 339, 345, 347, 371
Conductivity in polyelectolytes, 76
Corrosion, 21, 54
 and charge carrier transport, 49
 enhanced under illumination, 30
 mixed potential, 26
 stability of the catalyst interface, 26
Co_3O_4, 210, 217
$Cr(bpy)_3^{3+}$, 349
Cr_2O_3, 210
Crystal symmetry (octahedral and trigonal prismatic), 12
 field split transition metal, 15
Current
 generation mode, 42, 43
 voltage characteristics, 48
Cyclic voltammetry, 182, 188, 198

Dahms-Ruff model, 193
Decomposition potential, 21
 of neutral carbonyl metals, 48
Defects at the electrode surface, 19
Dehydration energy, 370
Density of states (DOS), 10
 distribution for the Mo_6S_8, 17
Deposition, 22
 of a layer with an amorphous structure, 27
 of noble metals, 24
Differential Electrochemical Mass Spectroscopy (DEMS), 41
Diffusion, 9, 343, 376
Diffusion in
 polyelectrolytes, 76
Diffusion in polymers, 87
 of electrons, 191
 of ions, 158
 of neutral species, 152
 of salts, 159

Dimensionally stable anodes (DSA^R), 207
Dispersion electrolysis, 366, 367, 371
Dissociation numbers in polyelectrolytes, 77
Distribution constant, 150
Distribution potential, 156
Dissociation in polymers, 158
Donnan potential, 169
Double layer capacity, 9

Efficiency of
 collection, 46, 55
Electric vehicles, 98
Electrical conductivity
 of UO_2, 301, 303, 309, 321–22, 326, 331
Electrocatalysis, 10, 24, 30, 359, 360, 362, 366
 factors of, 259–62
Electrochemical
 characterization of oxides, 231
 energy storage, 66
 impedance spectroscopy, 39, 33
 instability of FeS_2 electrode, 43
 interfacial behavior, 15
 mass transport, 47
 oxidation products, 21
 reactions, 9
 rectification, 184
 structure-sensitive, 9
Electrode
 employed in the sensitization, 9
 potential, 6, 30, 41
 stability (oxide), 262–71
 surface morphology, 7
Electrolysis, 52, 53
Electron hopping, 189, 191
Electron transfer
 at semiconductor/electrolyte interfaces, 3, 5
 catalysis via surface states, 5
 involving autocatalysis, 53
 mechanism, 348, 351, 352, 362, 364, 367, 369
 mechanism via chemical bonds, 55
 path through specialized proteins, 46
 reactions from the edges of valence, 36
Electron trapping, 360, 362–65
Electroneutrality coupling, 155, 190, 197
Electronic conductivity, 68, 194, 197
Electronic diffusion coefficient, 192
Electroorganic reactions, 255–58
Electropolymerization, 86
Electrostatic free energy, 156
Energy level diagram
 Fe_2O_3, 304
 GaP, 304
 NiO, 304
 SiC, 304

SnO$_2$, 304
TiO$_2$, 304
UO$_2$, 302, 309
ZnO, 304
Entropy
  fluxes, 54, 55
Extraframework cations, 342

Fe^{3+}, 360, 361
Fe(bpy)$^{2+}$, 348–51
Fe(CN)$_6^{4-}$, 350, 351, 352–65, 372, 373
Fermi level, 10, 15, 17, 26, 33
  function, 4
Film
  by MOCVD or sputtering techniques, 48
  of the p-type polymer, 47
Flatband potentials for UO$_2$, 304, 307, 310, 321, 323
Fractal
  dimension, 8
  geometry, 7
  objects, 10
  surface properties of solids, 9
Free energy of transfer, 150
Fuel cell, 26, 27, 53

Gallery sites, 349
Glass transition temperature in polyelectrolytes, 77
Gutman donor and acceptor numbers, 70

H$_2$W$_{12}$O$_{40}$, 374–76
Hectorite, 344, 360, 356
Helmholtz potential drop, 3, 33
Heptyl viologen, 351
Hexagonal prims, 369
Hydrated radius, 370
Hydrogen
  evolution reaction (HER), 7, 36, 250–53
  insertion reaction, 24
  peroxide formation, 26
Hydrotalate, 344, 346, 374

Immobilized redox sites, 189, 193
Impedance response of a Li$_x$C$_6$ electrode, 121
Impedance spectra
  UO$_2$, 318–20
Insertion compounds for lithium batteries, 111
Insertion electrodes for rocking chair batteries, 116
Intercalation
  of proton in MnO$_2$, 66
  reaction, 68, 69
Interface metal-polymer, 142, 166
Interface polymer-electrolyte, 141, 168

Interfacial
  clusters or cluster-metal associations, 26
  complexes, 55
  coordination chemical mechanism, 3, 55
  electrochemistry, 54
  kinetics, 33
  region between semiconductor and solution, 33
  states, 5, 6, 17, 33, 43
Interlayer spacing, 357–59
Ion chromoatgraphy, 353
Ion exchange, 339, 343, 348, 352, 367
Ion sensitive electrods, 141, 162
Ionic conductivity, 68, 159
Ionic radius, 370
Ionophores, 160
Ions in polymers, 154
Ir(C1)$_6^{2-}$, 374
IRO$_2$, 210, 211, 214, 215
Isopotential point, 349

Karl Fischer, 355
Kinetics
  interfacial behavior, 33–34
  of O$_2$ reduction, 28

LaCoO$_3$, 210
Langmuir-Blodgett films, 141
LaNiO$_3$, 210
Large channel network, 369
Layered
  compounds, 12, 19, 43
  double hydroxides, 344, 374–76
  lithium metal oxides, 125
Lead-acid battery, 68
Lifetime of a solid-state battery, 95, 100
Light
  absorbing sensitizing molecules, 46
  absorption, 50
  collection through non-imaging optics, 52
  collection through scattering processes, 53
  concentration via scattering phenomena, 50
  reflection from a semiconductor interface, 50
  scattering, 55
Lithium
  aluminum alloy, 111
  batteries, 65, 112
  cobalt oxide, 126
  insertion (interculation), 113
  manganese dioxide battery, 132
  nickel oxide, 128
  polymer-electrolytes, 72
  rocking chair batteries, 111, 115
Local relaxation, 169

Majority carrier, 45, 47
Manganese oxide, 132
Medium-effect coefficient, 150
Membrane, 141, 142, 365
Metal
  aggregates, 24
  d-states, 2, 10, 12, 14, 23, 24
    -centered photoelectrochemistry, 55
    -metal bonding, 14
    -metal interaction, 14
  sulfide semiconductors, 39
Metalloporphyrins, 365, 366
Methanol, 27, 40
  containing electrolyte, 26
Methyl viologen, 348, 351, 368
Microstructure, 50
Mixed conductors, 142
Mixed potential, 19
Mixed-valent state, 190, 195
$Mo(CN)_8^{4-}$, 362, 363, 372
Modified
  electrodes, 31
  p-Si electrodes, 36
  semiconductor photoanodes, 39
  semiconductor surface, 24
Molecular wires, 339, 371
Molten salts, 92
Montmorillonite, 343, 344, 347–49, 357, 359, 360, 362, 376
Mordenite, 342, 343, 371–73
Mott semiconductor, 15
Multielectron
  charge transfer reaction, 26
  transfer catalysis, 42, 55
  transfer processes, 42

Nafion, 39
Nanoelectrodes, 367
Nernst distribution, 150
Nernst-Donnan Potential, 155
Nernst-Planck equation, 159
Nickel-cadmium battery, 111
$NiCo_2O_4$, 210
Nitrogen fixation, 2, 36, 41
Nontronite, 344

Oligomere, 144
$Os(bpy)_3^{2+}$, 348, 349, 362–64
Overoxidation, 200
Oxidation
  potential of water, 22
  products of organic substances, 22

Oxide
  compounds, 22
  electrodes, 24, 46, 207
  electrode stability, 262–72
  physicochemical characterization of, 212
  semiconductor, 47
  single crystal electrode, 236, 237
  surface characterization of, 213
Oxide-solution interface, 219
Oxygen evolution reaction (OER), 36, 39, 244–50
Oxygen reduction, 2, 21, 25, 26, 253–55
  in methanol, 26
  performed with $Mo_4Ru_2Se_8$ electrodes, 26
  properties of $Mo_4Ru_2Se_8$, 26

Parsons-Zobel plot for an oxide electrode, 225
Partition coefficient, 150
Partitioning into polymers
  and electrochemistry, 166, 167, 173, 184
  equilibrium, 149, 168
  redox species, 173, 180
  selectivity, 151, 162, 172
  water, 150
$Pb_2Ir_2O_{7-y}$, 210
$Pb_2Ru_2O_7\text{-}y$, 210
PdO_4, 210
Perovskite, 254
Photoconversion process, 42
Photocurrent
  action spectra, 15, 30
  UO_2, 307, 309, 311–12, 326, 328
Photoelectrocatalysis, 33, 36, 39
  of hydrogen evolution, 33
Photoelectrochemical
  behavior of layer compounds, 45
  cells, 46
  characteristics, 43
  interfacial reactivity, 14
  investigation of Re6Se8C12, 29
  stability of dyes, 47
  studies on d-band transition metal, 6
  treatment of electrodes, 50
Photoelectrochemistry, 2, 7, 50, 55
  Using electrodes prepared from colloids, 49
Photoelectrodes
  for current generation, 54
  for energy conversion and catalysis, 2
Photoreaction
  based on strong interactions, 5
  based on weak interactions, 3
Photovoltaic
  conversion, 33
  electricity, 53

INDEX

Pillared clays, 344, 360, 362, 363, 366, 374, 376
Platinum, 366, 346, 347, 357, 362, 367
Point of zero charge of electrocatalytic oxides, 222, 223
Polyaniline, 371–74
Polyelectrolyte gel, 165
Polyethers
  phase diagram for $PCO-LiCF_3SO_3$, 76
  plasticizers in gel electroyltes, 84
  poly(iminoethylene), 72
  poly(oxyethylene), 72
  poly(oxypropylene), 72
  polyacteylene, 85
  polyaniline, 85
  polymers
    conjugated, 88, 95
    nonconjugaged, 89
  polymers as electrode materials, 85, 89
  polyparaphenylene, 85
  polypyrrole, 85
  polythiophene, 85
Polymeric
  materials for Lithium batteries, 65
Polymeric cations, 344
Polymerisation, 143
Polymer
  characterization, 143
  coated electrodes, 142, 143, 173, 177
  conducting, 141, 189
  conjugated, 189
  cross-linking, 145
  crystallinity, 145
  electrolyte, 142
  electrolytes, 66
  electron conducting, 143, 189
  films, 141
  fixed-charge, 165
  hydrophilic, 151
  hydrophobic, 151
  ionic, 165
  permeability, 148
  phase transision, 148
  redox, 189, 191
Polymers as solvating media, 71
Pore size, 340, 343, 372
Potentiostatic intermittent titration technique, 124
Power densities for polymer based cels, 90
Protective coatings, 141
Pyrite structure, 15, 42, 45
Pyrophillite, 343

Quantization phenomena, 49
Quantum yield, 43
Quantum efficiencies, 46, 47
  for sensitization, 46

Radical chemistry, 22
Radical photoelectrochemistry, 54
Randles circuit, 130
Rate constants, 28
Rate determining step, 33
Reaction rate, 7, 10, 33
Redox hopping, 192
Redox reactions
  on UO_2, 86, 321–31
Redox states in polymers, 142, 189, 191, 196
Reductive dissolution of oxide, 269
Rh_2O_3, 210
rocking-chair battery, 69, 111
$Ru(bpy)3^{2+}$, 348, 349, 360, 362
$Ru(NH_3)_6^{3+}$, 357–61
RuO_2, 210, 211, 214, 215

Saponite, 344
Sauconite, 344
Schottky, 6, 33, 36
  barrier, 33
Secondary building units, 342, 343
Self-assembled monolayers, 141
Self assembly, 360, 362
Self discharge of a battery, 95
Semi-crystalline polymers, 146
Semiconductor
  catalyst, 42
  compounds, 42, 44
  electrochemistry, 6
  layer, 48
  surface, 3, 5, 17, 22, 33, 34, 36, 42, 49
  suspensions, 41
Sensitization, 46–49, 53
  of cells based on TiO_2, 55
  process of dyes, 7
Sensors, 347, 352–56
Sepiolite, 356
Shape selectivity, 360
Si/Al ratio, 343, 350
Size exclusion, 353, 351, 352
Small channel network, 369
Smectite clays, 343, 344, 351
SnO_2, 210
Sodalite cage, 340, 342, 343, 369
Sodium aluminate, 69
Sodium-sulfur batteries, 65
Solar energy
  conversion, 7, 47, 50, 53
  converting electrodes, 52
  to hydrogen, 36

Solid state battery, 66
Solid-state ionizing solvents, 71
Solvating polymer architecture, 82
Solvation, 355, 356, 369
Space charge region, 19, 33
Spatial ordering, 360, 362–64
Spin-coating, 347
Spinels, 254
Stability
 against coagulation, 49
 and reactivity of interfaces, 54
 in presence of suitable redox agents, 45
 of electrode materials, 54
 of layer-type semiconducting surfaces, 14
 of sensitization cells, 55
Stability windows for lithium batteries, 93
Structure
 and morphology, 9
 of layered transiton metal chalcogenides, 12
 of pyrite materials, 15
 on pyrite sulfides, 14
Supercage, 342, 343, 366, 369
Supercapacitors, 233
Suppressed electroactivity, 354, 370, 352
Surface
 area of colloid, 49
 attached ruthenium complexes, 47
 derivatization, 21
 electrodes for sensitization, 49
 modifications of electrodes, 54
 morphology of oxides, 55
 states, 3, 5, 6, 7, 39, 45, 46
Surfactants, 360, 347

Ta_2O_5, 210
Talc, 343
Thermodynamic properties
 UO_2, 305–07
Thermodynamics, 52, 53
Thin film
 battery, 92, 93
 inorganic or organic, 21
TiO_2, 210, 211
TiS_2, 114
Toxicity hazard of lithium based solid state batteries, 103
Transition metal
 catalysts, 42
 centers, 43
 chalcogenides, 10, 12, 22, 45
 clusters, 27
 complex, 41
 compound, 27
 dichalcogenides, 14
 oxides, 24, 207
 states, 45
Transport in polymers
 diffusion-migration, 175, 197
 electrons, 189, 191
 ions, 158, 173
 neutral species, 149, 152
Transport numbers in polymers, 77, 79, 80

Uranium oxides
 potential-pH diagram, 305
 $UO_2(OH)_2$, 301, 312, 314
 $UO_{2.25}$ (U_4O_9), 301, 305, 307, 316, 327, 329
 $UO_{2.33}$ (U_3O_7), 298, 301, 305, 307, 310–12, 317, 325
 $UO_{2.5}$ (U_2O_5), 298, 301, 310–12
 $UO_{2.67}$ (U_3O_8), 298, 301, 305, 307, 310–12
 UO_{2+x}, 298, 301, 310, 316, 327, 329, 331
 $UO_2 \cdot 2H_2O$, 298, 305, 306, 307, 310, 312
 UO_3, 301, 312, 314
Uranium dioxide
 anodic dissolution, 313–20
 complexation equilibria, 308, 312, 317
 corrosion, 307–25
 defect clusters, 300–301
 dissolution currents, 314–17, 320
 donor-acceptor states, 325–27
 electrical conductivity (resistivity), 301, 303, 309, 321–22, 326, 331
 electrical properties, 301
 electrochemical reactivity, 331–32
 energy level diagram, 302, 309
 Fermi level, 302, 304
 fission product dopants, 303
 flatband potential, 304, 307, 310, 321, 323
 fluorite structure, 299, 301, 311
 geological disposal, 298, 321
 high field ion conduction, 311
 hole injection into, 307, 309
 hydrothermal extraction, 298, 321, 322
 impedance spectra, 318–20
 interstitial sites, 299, 301, 303, 311
 niobium dopants, 303
 n-type, 326
 number density of charge carriers, 303
 O^{2-} incorporation, 311
 OH^- adsorption, 311
 orthorhombic structure, 311, 312
 photocurrents, 307, 309, 311–12, 328
 polaron hopping, 301, 303
 p-type, 303, 307, 311, 325
 rare earth dopants, 303, 326–27
 recrystallization, 311, 312
 redox chemistry, 297, 298

redox reactions on, 321–31
SIMFUEL, 303, 326, 327
solubility, 297, 298, 305, 306, 312
soluble species, 305–7
surface states, 310, 311, 321
thermodynamic properties, 305–7
vacuum energy level, 302
voltammetric response, 307, 310, 314, 323–24
work function, 302, 304
XPS analysis, 310–12

van der Waals gap, 44
van der Waals surfaces, 45
Vogel-Tamman-Fulcher equation, 73, 76

Water oxidation, 2, 21
Williams-Landel-Ferry equation, 73, 76
Work function of metals, 33
Work of transfer, 150

XPS, 15

Zeolite A, 342, 344, 345, 348, 356, 367, 368
Zeolite co-cations, 369
Zeolite films, 340, 347
Zeolite Y, 340, 342, 343, 347, 365–71, 376
Zeotypes, 340, 343
Zinc-carbon dry cell, 111
ZrO_2, 210